Groups – Korea '94

Groups – Korea '94

Proceedings of the International Conference,
held at Pusan National University, Pusan, Korea,
August 18–25, 1994

Editors

A. C. Kim
D. L. Johnson

Walter de Gruyter · Berlin · New York 1995

Editors

A. C. Kim
Department of Mathematics
Pusan National University
Pusan 609-735
Korea

D. L. Johnson
Department of Mathematics
Nottingham University
Nottingham NG7 2RD
England

1991 Mathematics Subject Classification: 20-06

Library of Congress Cataloging-in-Publication-Data

Groups−Korea '94 , proceedings of the international conference held at Pusan National University, Pusan, Korea, August 18−25, 1994 / editors, A. C. Kim, D. L. Johnson.
  p.   cm.
  ISBN 3-11-014793-9 (alk. paper)
  1. Group theory−Congresses. I. Kim, A. C. (Ann Chi),
1938−   . II. Johnson, D. L.
QA174.677   1995
512′.2−dc20   95-34312
                                               CIP

Die Deutsche Bibliothek − Cataloging-in-Publication-Data

Groups − Korea '94 : proceedings of the international con-
ference, held at Pusan National University, Pusan, Korea,
August 18 − 25, 1994 / ed. A. C. Kim ; D. L. Johnson. −
Berlin ; New York : de Gruyter, 1995
  ISBN 3-11-014793-9
NE: Kim, An-ji; Pusan-Taehakkyo

Preface

The Third International Conference on the Theory of Groups, Groups – Korea 1994, was held at Pusan National University, Pusan, 18–25 August 1994. There were 31 invited one-hour lectures and 23 contributed half-hour seminar-type talks. "Groups – Korea 1994" was financially supported by

the Australian Academy of Science (AAS),
the British Council in Seoul (BC),
the Deutsche Forschungsgemeinschaft (DFG),
the Deutscher Akademischer Austauschdienst (DAAD),
the Commission on Development and Exchanges (CDE, IMU),
the International Science Foundation (ISF, USA),
the Korea Science and Engineering Foundation (KOSEF),
the Korean Mathematical Society (KMS), and
Pusan National University (PNU).

We record here our sincere thanks to these institutions, and their kind officers: Professor Pierre Bérard (CDE), Miss Bonnie Bauld (AAS), Dr Jost-Gert Glombitza (DFG), Mr Eleanor Gorman (ISF), Dr Patrick Hart (BC), Mr Moo Nam Lee (KOSEF), Dr Ulrich Lins (DAAD), and Professor Moo Ha Woo (KMS). We are also grateful to the President of PNU, Dr Hyuk Pyo Chang, for encouraging the organizers to sustain their passion for the meeting.

A. C. Kim
D. L. Johnson

Table of Contents

viii

On the Rank of a Finite Product of Two p-Groups

*Bernhard Amberg and Lev S. Kazarin**

1. Introduction

If the finite p-group $G = AB$ is the product of two subgroups A and B whose Prüfer ranks are bounded by r, then the Prüfer rank of G is bounded by a polynomial function of r; see [5] and [1]. Although no bound is given there explicitly, the analysis of the proof of this theorem leads to polynomial bounds of relatively high degree (see [1]). In the following we shall give better bounds. A natural way to do this is to obtain first a bound for the normal rank of $G = AB$ which will immediately give a bound for the Prüfer rank of G by Lemma 2.6 below. Our bound for the normal rank of $G = AB$ depending on the Prüfer ranks of A and B is close to being linear. Even if this result may not be best possible, it will be useful in the study of the structure of finite products of groups with low rank.

Recall that a group X has **Prüfer rank** $r = r(X)$ if every finitely generated subgroup of X can be generated by r elements and r is the least such integer. The **normal rank** $r_n(X)$ of X is the maximum of the minimal number of generators of each normal subgroup of X.

Our main result is the following.

Theorem 1.1. *Let the finite p-group $G = AB$ be the product of two of its subgroups A and B. Let $r_0 = \min\{r(A), r(B)\}$ and $r_1 = r(A) + r(B)$. Then the normal rank $r_n(G)$ satisfies the following inequality:*

$$r_n(G) \leq r_0(\lceil \log_p r_n(G) \rceil + 1 + \lceil \log_p r_0 \rceil \lceil \log_2 2r_0 \rceil + \delta_{2p}) + r_1.$$

The inequality in this theorem may look unusual, but it shows that for any $\epsilon > 0$ and for sufficiently large r_0 we have the following almost linear bound

$$r_n(G)^{1-\epsilon} \leq r_0(3 + \lceil \log_p r_0 \rceil \lceil \log_2 2r_0 \rceil) + r_1.$$

If the two subgroups A and B are abelian, Theorem 1.1 can be improved as follows.

*The authors like to thank the Departments of Mathematics of the Universities of Mainz and Yaroslavl for their excellent hospitality during the preparation of this paper

Theorem 1.2. *Let the finite p-group $G = AB$ be the product of two abelian subgroups A and B. Let $r_0 = \min\{r(A), r(B)\}$ and $r_1 = r(A) + r(B)$. Then the normal rank $r_n(G)$ of G satisfies the inequality*

$$r_n(G) \leq r_0 \lceil \log_p r_n(G) \rceil + r_1.$$

If the finite p-group $G = AB$ is the product of two cyclic subgroups A and B, then it follows from Theorem 1.2 that $r_n = r_n(G) \leq \lceil \log_p r_n \rceil + 2$. This implies $r_n \leq 3$ and even $r_n \leq 2$ for $p > 3$. Note however that there exists a finite 2-group of normal rank 3 which is a product of two of its cyclic subgroups. (see [2], Aufgabe 28, p. 341).

The notation is standard and can be found in [2] and [1]. If X is a finite p-group, we note in particular

$$\Omega_i(X) = \text{subgroup generated by all elements } g \text{ in } X \text{ such that } g^{p^i} = 1.$$

$$\mho_i(X) = \text{subgroup generated by all } g^{p^i} \text{ with } g \in X.$$

The **exponent** $\exp(X)$ of X is the largest order of its elements. If a minimal generating system of X consists of m elements then we write $d(X) = m$. If α is a real number, then $\lceil \alpha \rceil = m$ is the smallest integer such that $\alpha \leq m$. δ_{ij} denotes the Kronecker symbol.

2. Preliminaries

The first lemma is well-known.

Lemma 2.1. *If G is a finite p-group with nilpotency class c, then the derived length of G does not exceed $\lceil \log_2 c \rceil + 1$.*

Proof. See [4], 5.1.12.

Lemma 2.2 (Alperin). *Let G be a finite p-group, n be an integer such that $p^n \neq 2$ and let A be a maximal element in the set of all abelian normal subgroups with exponent $\leq p^n$. If $x \in C_G(A)$ and $x^{p^n} = 1$, then $x \in A$.*

Proof. See [2], p. 341.

Lemma 2.3. *Let G be a regular finite p-group with Prüfer (or normal) rank $r > 1$ and exponent p^v. Then*

$$|G| \leq p^{vr(\lceil \log_2(r-1) \rceil + 2)}.$$

Furthermore, if $p = 2$ then $|G| \leq 2^{vr}$.

Proof. It is obvious that for each i the group $H = \mho_i(G)/\mho_{i+1}(G)$ has exponent less or equal to p. If $A = C_H(A)$ is an abelian self-centralizing normal subgroup of H with rank less or equal to r, then H/A is isomorphic to a subgroup of $\text{Aut}(A) \subseteq GL(r, p)$. If $p = 2$, then G is abelian (see [2], p. 327) and $H = A$. By Theorem 16.3 of [2], p. 382, H/A has nilpotency class at most $r - 1$. By Lemma 2.1 the derived length of H/A does not exceed $\lceil \log_2(r - 1) \rceil + 1$. Therefore the derived length of H does not exceed $\lceil \log_2(r - 1) \rceil + 2$, since A is abelian. By [2], Theorem on p. 327, we have that $\mho_{\nu+1}(G) = 1$. The lemma is proved.

Lemma 2.4 (Thompson). *Let G be a finite p-group where $p \neq 2$ is a prime. If every abelian normal subgroup of G can be generated by s elements, then a minimal generating system of G contains at most $s(s + 1)/2$ elements.*

Proof. See [2], p. 343.

Lemma 2.5. *Let G be a finite p-group. Suppose that if $p > 2$ then every element of order p of G lies in its center, and if $p = 2$ then every element of order ≤ 4 of G lies in its center. Then the following holds:*
(i) $d(G) \leq d(Z(G)) = d(\Omega_1(G))$,
(ii) $|G| \leq p^{\nu r}$ *where r is the Prüfer (or normal) rank of G and $p^\nu = \exp(G)$.*

Proof. We use an idea of Blackburn to prove both statements simultaneously (see [2], p. 342).

Suppose that $A = \Omega_1(G)$ if $p > 2$ and $A = \Omega_2(G)$ if $p = 2$. Let B be an element with maximal order among the normal subgroups X of G containing A with elementary abelian factor group X/A. If $b \in B$ and $g \in G$, then $b^g = bc$ for some $c \in B$. As $b^p \in A \leq Z(G)$, we have $b^p = (b^p)^g = (b^g)^p = (bc)^p = b^p c^p [c, b]^{\binom{p}{2}}$. It follows that if $p > 2$ then $\binom{p}{2} \equiv 0 \pmod{p}$ and $[c, b]^{\binom{p}{2}} = 1$. In this case we have also $c^p = 1$ and $c \in A$. If $p = 2$ then $c^2[c, b] = 1$. We prove now that $c^4 = 1$. As

$$x^2 y = yx^2 = xyx[y, x] = yx^2[y, x]^2$$

for each pair $x, y \in B$, we have $B' \subseteq \Omega_1(A)$. Hence it follows from $c^2[c, b] = 1$ that $c^4 = 1$ and so $c \in Z(G)$. Therefore $[B, G] \subseteq \Omega_1(A)$ in each case and $B/\Omega_1(A)$ is an abelian group of exponent p for $p > 2$ and a group of exponent 4 for $p = 2$. Furthermore, this group is not contained in any larger abelian normal subgroup of exponent p (for $p > 2$) or of exponent 4 (for $p = 2$) of the group $G/\Omega_1(A)$. By Lemma 2.2 we have

$$B/\Omega_1(A) = \Omega_1(G/\Omega_1(A)) \subseteq Z(G/\Omega_1(A)) \text{ if } p > 2 \text{ and}$$

$$B/\Omega_1(A) = \Omega_2(G/\Omega_1(A)) \subseteq Z(G/\Omega_1(A)) \text{ if } p = 2.$$

If $p > 2$ then by Theorem 12.2 of [2], p. 342, we have

$$d(B) \leq d(\Omega_1(B)) = d(\Omega_1(A)) = d(\Omega_1(G)).$$

As $\exp(G/\Omega_1(G)) \le p^{\nu-1}$ it is easy to prove by induction that $|G| \le p^{\nu r}$.

For the remainder of the proof we may suppose now that $p = 2$. If G is abelian, it is easy to see that both statements of Lemma 2.5 hold. Assume now that Lemma 2.5 holds for all groups whose order is less than the order of G. If $B \ne G$ then $d(B) \le d(\Omega_1(B)) = d(Z(G))$ and $\exp(G/\Omega_1(B)) = \exp(G/\Omega_1(G)) \le 2^{\nu-1}$ by induction. Since $B/\Omega_1(B) = B/\Omega_1(A)$ plays the same role as $A = \Omega_2(G)$ in G for $G/\Omega_1(A)$, then $|G/\Omega_1(G))| \le 2^{(\nu-1)r}$ where $r = d(\Omega_1(G)) \ge d(G/\Omega_1(G))$. This proves (ii). If $\Omega_1(B) = \Omega_1(A) \subseteq \Phi(G)$, then $d(G/\Omega_1(G)) = d(G)$ and so (i) is also proved.

Now suppose that $\Omega_1(B) = \Omega_1(A) = \Omega_1(G) \not\subseteq \Phi(G)$. Then there exists a maximal subgroup M of G such that $M\Omega_1(A) = G$ with $\Omega_1(A) \not\subseteq M$. Obviously we have $M \cap \Omega_1(A) = \{x | x \in M, x^2 = 1\} = \Omega_1(Z(M))$. By induction

$$d(M) \le d(Z(M)) = d(\Omega_1(Z(M)) = d(M \cap Z) \le d(Z(G)) - 1.$$

Therefore $d(G) \le d(M) + 1 \le d(Z(G))$, and we are done.

Suppose now that $B = G$. As $(xy)^4 = x^4 y^4$ for each pair of elements x, y of G, then the map $g \to g^4$ is a homomorphism from G into $\Omega_1(G) = \Omega_1(A)$. It is easy to see that the kernel of this homomorphism is $A = \Omega_2(G)$. By [2], p. 272, we have that $\Omega_1(G) = \Phi(G)$ and if $\Omega_1(G) = A$ then $d(G) = d(G/A) \le d(\Omega_1(A)) = d(Z(G))$. Hence $\Omega_1(G) \ne A$ and in particular there exists an element of $\Omega_1(A)$ which does not have a root of degree 4 in G.

The set of all elements of $\Omega_1(A)$ having a root of degree 4 in G is a subgroup D of G which has a complement C in $\Omega_1(A)$ such that $A = A_1 \times A_2$ where $\Omega_1(A_1) = D$ and $\Omega_1(A_2) = C = A_2$. If $g^4 \in A_2$ for some $g \in G$ then $g^4 \in C$ and so $g^4 = 1$. Hence $\Omega_2(G/A_2) = A/A_2 \subseteq Z(G/A_2)$. By induction we have

$$d(G/A_2) \le d(\Omega_1(Z(G/A_2))) = d(A_1).$$

It follows that $d(G) \le d(A_1) + d(A_2) = d(Z(G))$. This proves Lemma 2.5.

Lemma 2.6. *If every abelian normal subgroup of the finite p-group G can be generated by at most s elements, then the Prüfer rank of G is at most $1/2(s + s^2)$ for $p > 2$ and at most $3/2(s^2 + s)$ for $p = 2$.*

Proof. For $p > 2$ this is a theorem of Thompson (see [2], Satz 12.3, p. 343). If $p = 2$ the proof follows from a slight modification of this theorem. One has only to use Lemma 2.2 for $p^n = 4$ and the arguments of Thompson.

Next we will obtain a bound for the order of a finite p-group in terms of its rank and its exponent.

Lemma 2.7. *Let G be a finite p-group with Prüfer rank r and exponent p^ν. Then the following inequalities hold.*
(i)   *If $p > 2$ then $|G| \le p^{\nu r + r(\lceil \log_p r \rceil)(\lceil \log_2 2r \rceil)}$.*
(ii) *If $p = 2$ then $|G| \le 2^{\nu r + r + r(\lceil \log_2 r \rceil)(\lceil \log_2 2r \rceil)}$.*

Proof. Let A be the largest abelian normal subgroup of G of exponent p for $p > 2$ and of exponent ≤ 4 for $p = 2$. Suppose first that $\exp(A) = p$. Then clearly the factor group $H = G/C_G(A)$ can be embedded into $\text{Aut}(A) \leq GL(r, p)$. Since G is a p-group it is isomorphic to a subgroup of a Sylow p-subgroup of $GL(r, p)$. By Theorem 16.3 of [2], p.382, the group H has nilpotency class less or equal to $r - 1$ and its derived length is less or equal to $\lceil \log_2(r - 1) \rceil + 1$. Since every p-element $a \in GL(r, p)$ has a normal Jordan form with Jordan matrices of size less or equal to r then $a - 1$ is a nilpotent element and $(a - 1)^r = 0$. If $\lceil \log_p r \rceil = m$, then $(a - 1)^{p^m} = 0$ which implies $a^{p^m} = 1$. Hence the exponent of H does not exceed p^m. Each factor $H^{(i)}/H^{(i+1)}$ of the commutator series of H has order not exceeding p^{mr} and so $|H| \leq p^{mr(\lceil \log_2(r-1) \rceil + 1)}$. Since $C_G(A)$ contains each element of order p in its center by Lemma 2.2, then by Lemma 2.5(i) we have $|C_G(A)| \leq p^{vr}$ and so the first assertion is proved.

Suppose now that $\exp(A) = 4$. Then $H = G/C_G(A)$ is isomorphic to a subgroup of a group of invertible (r, r)-matrices with entries in $\mathbb{Z}_4$. Let U be the subgroup of this group consisting of all matrices (a_{ij}) such that $a_{ij} \equiv \delta_{ij} \bmod 2$. It is easy to see that the inverse image V of U in G is normal in G and the group $U = V/C_G(A)$ is abelian. Since $u^2 = 1$ for each matrix $u = (a_{ij})$ in U we have $|U| \leq 2^r$. Now $G/V \subseteq GL(r, 2)$ and we may use the previous arguments. Hence

$$|G| \leq |G/V||U||C_G(A)| \leq 2^{r(\lceil \log_2 r \rceil \lceil \log_2 2r \rceil) + r + vr}.$$

Corollary 2.8. *Let G be a finite p-group with Prüfer rank r and exponent p^v. Then*

$$|G| \leq p^{vr(2 + \lceil \log_2 r \rceil) + r\delta_{2p}}.$$

This result corresponds to Lemma 2.3 for regular p-groups. Note that there is a similar formula in [5], but its proof is not correct.

3. Some Special Cases

The proofs of our theorems will be reduced to the following special situation of a triply factorized group.

Lemma 3.1. *Let the finite p-group $G = AN = BN = AB$ be the product of two subgroups A and B and an elementary abelian normal subgroup N of G such that $A \cap N = B \cap N = 1$. If the Prüfer rank of one of the subgroups A and B is bounded by r, then*

$$d(N) \leq r(\lceil \log_p d(N) \rceil + 1 + \lceil \log_p r \rceil \lceil \log_2 2r \rceil + \delta_{2p}).$$

Proof. Obviously we have $|G| = |A||B||A \cap B|^{-1} = |A||N| = |B||N|$. Hence $|A| = |B| = |N||A \cap B|$. Let $|N| = p^n = m$ and let the elements of N be $c_i = a_i b_i$

with $a_i \in A$ and $b_i \in B$ where $1 \le i \le m$. We show first that $\{a_i | 1 \le i \le m\}$ (respectively $\{b_i | 1 \le i \le m\}$) is a full system of representatives of A (respectively of B) with respect to the subgroup $H = A \cap B$. Assume that on the contrary for some $1 \le i \ne j \le m$ we have $a_i H = a_j H$ with $i \ne j$. Then $c_i^{-1} c_j = b_i^{-1} a_i^{-1} a_j b_j \in N \cap B = 1$, so that $c_i = c_j$, a contradiction. Similarly, $b_i H \ne b_j H$ if $i \ne j$. As $m = |A : H| = |B : H|$ the assertion about the representatives is proved.

It is easy to see that for each choice of a system of representatives $a_1, a_2, \ldots, a_m$ of A for the subgroup H there is a system of representatives $b_1, b_2, \ldots, b_m$ of B such that $N = \{a_i b_i | 1 \le i \le m\}$. The subgroup $D = C_A(N)$ is normal in A, so that DH is a subgroup of A. We may choose a system of representatives $a_1, a_2, \ldots, a_m$ in A such that $\bigcup_{i=1}^{k} a_i H = DH$ for some $k \le m$.

Let $c_i = a_i b_i$ and $c_j = a_j b_j$ be elements in N where $1 \le i, j \le k$. It is easy to see that $[a_i, c] = [a_j, c] = 1$ for each $c \in N$. Hence $[a_i, b_i] = [a_j, b_j] = 1$ for $1 \le i, j \le k$. As $c^p = 1$ for each $c \in N$ and $c_i c_j = c_j c_i$ for each pair i, j then we have $a_i^p b_i^p = 1$ and so $a_i b_i a_j b_j = a_j a_i b_i b_j = a_j b_j a_i b_i = a_i a_j b_j b_i$ for each $1 \le i, j \le k$. Therefore $a_i^p \subset H$, $[a_j, a_i] - [b_j^{-1}, b_i^{-1}] \in H$ for $1 \le i, j \le k$. By [2], p. 272, it follows that $R = \Phi(D) \subseteq H$. Obviously R is normal in G and $G/R \simeq (AR/R)(BR/R) \simeq (A/R)(NR/R) \simeq (B/R)(NR/R)$, where G/R satisfies the conditions of Lemma 3.1. Without loss of generality we may suppose now that $D = C_A(N)$ is an elementary abelian group of rank at most r, and $\bar{A} = A/D$ is isomorphic to a subgroup of $\text{Aut}(N) \simeq GL(n, p)$. Consider now N as a natural $\bar{A}$-module over $F = GF(p)$. Let $\exp(A) = p^\nu$. Then $p^{\nu-1} \le \exp(\bar{A}) \le p^\nu$. Therefore the minimal polynomial of each $\bar{a} \in \bar{A}$ divides $x^{p^\nu} - 1 = (x - 1)^{p^\nu}$. In this case $u = \bar{a} - 1$ is a nilpotent element. It is not difficult to see that $u^d = 0$ for some $d \le \dim_F N = n$. Hence if an integer α satisfies the inequality $p^{\alpha-1} < n \le p^\alpha$ then we have $(1 + u)^{p^\alpha} = \bar{a}^{p^\alpha} = 1$. Thus $-1 + \nu \le \alpha$. Since $\alpha = \lceil \log_p n \rceil$ by Lemma 2.7 we have the inequality

$$
\begin{aligned}
n &= \log_p |N| = \log_p |A : H| \le \log_p |A| \\
&\le \nu r + r \lceil \log_p r \rceil \lceil \log_2 2r \rceil + r \delta_{2p} \\
&\le \lceil \log_p n \rceil r + r \lceil \log_p r \rceil \lceil \log_2 2r \rceil + r \delta_{2p} + r \\
&= (\lceil \log_p n \rceil + 1 + \lceil \log_p r \rceil \lceil \log_2 2r \rceil + \delta_{2p}) r.
\end{aligned}
$$

The lemma is proved.

Lemma 3.2. *Let the finite p-group $G = AN = BN = AB$ be the product of two subgroups A and B and an elementary abelian normal subgroup N of G. Let the Prüfer rank of A be bounded by r and each element of order p of A lie in its center for $p > 2$ and each element of order ≤ 4 of A lie in its center for $p = 2$. Then $d(N) \le r(\lceil \log_p d(N) \rceil + 1)$.*

For the proof one only has to replace Lemma 2.7 by Lemma 2.5 in the proof of Lemma 3.1.

The following lemma is obvious.

Lemma 3.3. *Let $G = AB$ be the product of two normal p-subgroups A and B. If the Prüfer rank of A is r_1 and the Prüfer rank of B is r_2 then the Prüfer rank of G does not exceed $r_1 + r_2$.*

4. Proof of the Main Results

4.1. Proof of Theorem 1.1

Assume that Theorem 1.1 is false, and let the finite p-group $G = AB$ be a counterexample with minimal order. Let $r_0 = \min\{r(A), r(B)\}$ and $r_1 = r(A) + r(B)$. Let N be a normal subgroup of G with maximal rank. The subgroup $\bar{N} = N/\Phi(N)$ of the factor group $\bar{G} = G/\Phi(N)$ has the same rank as N. Clearly $r(\bar{A}) \leq r(A)$ and $r(\bar{B}) \leq r(B)$ where $\bar{A} = A\Phi(N)/\Phi(N)$ and $\bar{B} = B\Phi(N)/\Phi(N)$. Hence $\min\{r(\bar{A}), r(\bar{B})\} = \bar{r}_0 \leq r_0$ and $\bar{r}_1 = r(\bar{A}) + r(\bar{B}) \leq r_1$. If $\Phi(N) \neq 1$ then $r(\bar{N}) = r(N)$ satisfies the inequality

$$r_n(G) = r(N) \leq \bar{r}_0(\lceil \log_p r(N) \rceil + 1 + \lceil \log_p \bar{r}_0 \rceil \lceil \log_2 2\bar{r}_0 \rceil + \delta_{2p}) + \bar{r}_1.$$

Since $\bar{r}_0 \leq r_0$ and $\bar{r}_1 \leq r_1$ then $r(N) = r_n(G)$ satisfies the required inequality in Theorem 1.1.

Hence we may assume that $\Phi(N) = 1$ and so N is an elementary abelian normal subgroup of G. Now suppose that $AN = H \neq G$ or $BN = H \neq G$. It is clear that $H = (A \cap H)(B \cap H), r(A \cap H) + r(B \cap H) \leq r_1$ and $\min\{r(A \cap H), r(B \cap H)\} \leq r_0$. Thus $r_n(G) = r(N)$ satisfies the conclusion of Theorem 1.1, a contradiction. Therefore we may assume that $AN = BN = AB = G$. Since N is abelian, the subgroups $A \cap N$ and $B \cap N$ are normal in G, so that also $C = (A \cap N)(B \cap N)$ is normal in G. By Lemma 3.3 $r((A \cap N)(B \cap N)) \leq r(A) + r(B) = r_1$. If $\bar{G} = G/C$, then $\bar{G} = \bar{A}\bar{B} = \bar{A}\bar{N} = \bar{B}\bar{N}$ where $\bar{A} = AC/C, \bar{B} = BC/C, \bar{N} = NC/C$ and $\bar{A} \cap \bar{N} = 1 = \bar{B} \cap \bar{N}$. In particular $\bar{A} \simeq \bar{B} \simeq \bar{G}/\bar{N}$. By Lemma 3.1 we have

$$d(\bar{N}) \leq r_0(\lceil \log_p d(\bar{N}) \rceil + 1 + \lceil \log_p r_0 \rceil \lceil \log_2 2r_0 \rceil + \delta_{2p}).$$

Furthermore, we have

$$r_n(G) = r(N) = d(N) \leq d(\bar{N}) + r(C) \leq d(\bar{N}) + r_1.$$

Since $\lceil \log_p d(\bar{N}) \rceil \leq \lceil \log_p d(N) \rceil = \lceil \log_p r_n(G) \rceil$ the theorem follows.

4.2. Proof of Theorem 1.2

Assume that Theorem 1.2 is false, and let the finite p-group $G = AB$ be a minimal counterexample where the two subgroups A and B are abelian. Let $r_0 =$

$\min\{r(A), r(B)\}$ and $r_1 = r(A) + r(B)$. As in the proof of Theorem 1.1 it is easy to reduce the proof to the case $G = AN = BN = AB$ where N is an elementary abelian normal subgroup of G with maximal rank. Since $C(N) = N(C(N) \cap A)$ is also abelian then $C(N) \cap A \leq N$ and $C(N) = N$. Hence G/N is isomorphic to a subgroup of $\mathrm{Aut}(N) = GL(n, p)$ where $n = r(N) = d(N)$. Moreover, the subgroups $A \cap N$, $B \cap N$ and $A \cap B$ are central in G so that $D = A \cap B = A \cap N \cap (B \cap N)$. From $|G| = |A||B|/|D| = |A||N|/|A \cap N| = |B||N|/|B \cap N|$ it follows that $|G/N|^2 = (|A|/|A \cap N|)(|B|/|B \cap N|) = |G||D|/(|A \cap N||B \cap N|)$. If $Z = (A \cap N)(B \cap N)$, then $|Z||G/N| = |N|$. Hence if $|Z| = p^x$ then the Jordan form of each element of the group $G/N = A/(A \cap N)$ has at least x Jordan matrices. Now the maximal size of a Jordan matrix is less than $n - x + 2$. It follows from the proof of Lemma 3.1 that $\log_p(\exp(G/N))$ does not exceed $\lceil \log_p(n - x + 1) \rceil$. By Lemma 2.5 this implies $\log_p(|G/N|) \leq r_0 \lceil \log_p(n - x + 1) \rceil$ and $n - x \leq r_0 \lceil \log_p(n - x + 1) \rceil$. We have $n \leq r_0 \lceil \log_p(n-x+1) \rceil + x$ with $x \leq r_1$. Now the function $f_i = r_0 \lceil \log_p(n-x+1) \rceil + x$ is increasing in the interval $0 \leq x \leq r_1$. Thus $n \leq \max(f_i) = r_0 \lceil \log_p(n-r_1+1) \rceil + r_1$. This proves Theorem 1.2.

References

[1] Amberg, B., Franciosi, S., and de Giovanni, F., Products of groups, Clarendon Press, Oxford (1992).

[2] Huppert, B., Endliche Gruppen I, Springer, Berlin (1967).

[3] Kazarin, L., On groups with factorization, Soviet Math. Dokl. 23 (1981), 19–22.

[4] Robinson, D. J. S., A course in the theory of groups, Springer, New York (1982).

[5] Zaitsev, D. I., Factorizations of polycyclic groups, Mat. Zametki 29 (1981), 481–490.

Locally Soluble Products of Two Minimax Subgroups

Bernhard Amberg and Yaroslav P. Sysak*

1. Introduction

Lennox and Roseblade in [4] and Zaitsev in [13] have shown that a soluble group $G = AB$, which is the product of two polycyclic subgroups A and B, is likewise polycyclic. Moreover, Wilson in [11] and independently Sysak in [9] proved that a soluble product of two minimax subgroups is likewise a minimax group. These authors obtained similar theorems for the finiteness conditions "finite Prüfer rank" and "finite abelian section rank" (see [9] and [12]).

The question arises whether these results can be extended to locally soluble products of two subgroups (see [2], Question 10). Obviously by the theorem of Lennox, Roseblade and Zaitsev also locally soluble products of two polycyclic groups are polycyclic. But even locally finite-soluble products of two subgroups with finite abelian section rank need not have finite abelian section rank (see [8], Theorem 1, p. 4).

In this note we consider locally soluble products of minimax groups. Recall that a group G is a *minimax group* if it has a finite series whose factors satisfy the minimum or the maximum condition for subgroups.

Theorem 1.1. *If the locally soluble group $G = AB$ is the product of two minimax subgroups A and B, then G is a soluble minimax group*

The proof of Theorem 1.1 will be reduced to the case when G is hyperabelian by the following result. Recall that a group G if *residually of bounded finite Prüfer rank* if there exist normal subgroups N_i of G with $\bigcap N_i = 1$ and a positive integer k such that the Prüfer ranks r_i of the factor groups G/N_i satisfy $r_i \leq k$ for every i in the index set I. Here a group is said to have *finite Prüfer rank* r if all its finitely generated subgroups can be generated by r elements and r is the least positive integer with this property.

---

*The second author likes to thank the Department of Mathematics of the University of Mainz, Germany, for its excellent hospitality during the preparation of this paper in 1993. He would also like to thank the International Science Foundation for the possibility to attend the Conference "Groups - Korea 1994"

Theorem 1.2. *If the locally soluble group G is residually of bounded finite Prüfer rank, then G is hyperabelian.*

The proof of Theorem 1.2 depends on the following proposition about the endomorphism ring of an abelian group of finite Prüfer rank, which is of independent interest.

Proposition 1.3. *Let M be an abelian group of finite Prüfer rank r. Then the endomorphism ring* End *M satisfies the standard polynomial of degree* 2r

The results of this note have earlier been published as Preprint No. 2 (November 1993) of the Preprint-Reihe des Fachbereichs Mathematik der Johannes Gutenberg-Universität Mainz. The notation is standard and can be found in [2], [5], [7] and [6]. In particular the Prüfer rank of the group G will be denoted by $r(G)$.

2. Proof of Proposition 1.3

Recall that the standard polynomial of degree n is the polynomial

$$S_n(x_1, \ldots, x_n) = \sum_{\pi \in \mathrm{Sym}(n)} (\mathrm{sgn}\, \pi) x_{\pi 1} \cdots x_{\pi n}.$$

The ring R satisfies the standard polynomial of degree n if $S_n(r_1, \ldots, r_n) = 0$ for all elements $r_1, r_2, \ldots, r_n$ of R. It is easy to see that the property that a ring satisfies the standard polynomial for some degree n is inherited by subrings, factor rings and cartesian products. The theorem of Amitsur and Levitzki says that the ring $M_n(R)$ of $n \times n$-matrices with coefficients in the commutative ring R satisfies the standard polynomial of degree $2n$ (see [6], Theorem 1.4.1).

Proof of Proposition 1.3. Assume first that the abelian group of finite Prüfer rank M is radicable. If M is a p-group or torsion-free, then End M is isomorphic to the ring of matrices $M_r(K)$ over the field K of p-adic numbers or of rational numbers, respectively. Hence End M satisfies the standard polynomial of degree $2r$ by the theorem of Amitsur and Levitzki. Clearly if M is periodic, then End M also satisfies this polynomial identity. Therefore we may suppose that the maximal periodic subgroup T of M satisfies $1 \subset T \subset M$.

The endomorphism rings End T and End M/T satisfy the standard polynomials of degree m and n respectively, where $m = 2r(T)$ and $n = 2r(M/T)$. We will show that End M satisfies the standard polynomial of degree $n + m$. The restriction of an endomorphism α of M onto T is an endomorphism of T, the subring Hom(M, T) is an ideal of End M and the factor ring End $M/$Hom(M, T) is isomorphic with End M/T. Therefore if $\alpha_1, \ldots, \alpha_m, \alpha_{m+1}, \ldots, \alpha_{m+n}$ are arbitrary endomorphisms of M and t is an element in T, it follows that $S_m(\alpha_1, \ldots, \alpha_m)(t) = 0$ and $S_n(\alpha_{m+1}, \ldots, \alpha_{m+n})$ belongs to Hom(M, T). Hence for every element a in M we have

$$S_m(\alpha_1, \ldots, \alpha_m)(S_n(\alpha_{m+1}, \ldots, \alpha_{m+n})(a)) = 0.$$

This implies $S_{m+n}(\alpha_1, \ldots, \alpha_m, \alpha_{m+1}, \ldots, \alpha_{m+n}) = 0$, since

$$S_{m+n}(x_1, \ldots, x_{m+n}) = \sum_{\pi \in S} (\operatorname{sgn} \pi) S_m(x_{\pi 1}, \ldots, x_{\pi m}) S_n(x_{\pi(m+1)}, \ldots, x_{\pi(m+n)}),$$

where S is the set of all permutations of the symmetric group $\operatorname{Sym}(m + n)$ such that for every subset of $N = \{1, \ldots, m + n\}$ with m elements $i_1, \ldots, i_m$ there is a permutation π in S with $\pi 1 = i_1, \ldots, \pi m = i_m$ and $\pi(m + i)$ is the minimal number in the complement set of $\{\pi 1, \ldots, \pi m, \pi(m + 1), \ldots, \pi(m + i - 1)\}$ in N. Since $r(M) = r(T) + r(M/T)$, it follows that the ring $\operatorname{End} M$ satisfies the standard polynomial of degree $m + n = 2r(T) + 2r(M/T) = 2(r(T) + r(M/T)) = 2r$. This concludes the proof of Proposition 1.3 for radicable groups.

Suppose now that M is arbitrary and let $\bar{M}$ be the radicable hull of M. Then $r(\bar{M}) = r(M)$. By a theorem of Dlab (see [3], Satz 5)

$$\operatorname{End} M \simeq \operatorname{End}_M(\bar{M}, M) / \operatorname{End}_0(\bar{M}, M),$$

where $\operatorname{End}_M(\bar{M}, M)$ is the subring of all endomorphisms of $\bar{M}$ which map M into itself and $\operatorname{End}_0(\bar{M}, M)$ is the ideal of $\operatorname{End}_M(\bar{M}, M)$ consisting of all endomorphisms of $\bar{M}$ which map M onto 0. We have shown above that the ring $\operatorname{End} \bar{M}$ satisfies the standard polynomial of degree $2r$. In particular the subring $\operatorname{End}_M(\bar{M}, M)$ and so also its factor ring $\operatorname{End}_M(\bar{M}, M) / \operatorname{End}_0(\bar{M}, M) \simeq \operatorname{End} M$ satisfy the standard polynomial of degree $2r$. This proves Proposition 1.3.

3. Proof of Theorem 1.2

For the proof of Theorem 1.2 we need the following lemmas.

Lemma 3.1. *Let the group G be the cartesian product of hyperabelian groups of bounded Prüfer rank. Then there exists a normal subgroup M of G which is nilpotent of class at most 2 such that the factor group G/M is embedded in the multiplicative group of a ring with the standard polynomial identity.*

Proof. Let H be a hyperabelian group and let N be a maximal normal subgroup of G with class at most 2. Then $Z(N) = C_H(N)$ (see for example [7], Chapter 2, proof of Proposition 3). The intersection $C_H(Z(N)) \cap C_H(N/Z(N))$ is a nilpotent normal subgroup of G with class at most 2 by a lemma of Kaluznin (see [7], Chapter 1, Proposition 10). Since $N = C_H(N/Z(N)) \cap C_H(Z(N))$, the factor group H/N is a subgroup of the direct product of the groups $H/C_H(N/Z(N))$ and $H/C_H(Z(N))$. Now these groups are embedded in $\operatorname{End} N/Z(N)$ and $\operatorname{End} Z(N)$, respectively. If $r(H) \leq r$, then the rank of the abelian groups $N/Z(N)$ and $Z(N)$ is likewise at most r. By Proposition 1.3 the rings $\operatorname{End} N/Z(N)$ and $\operatorname{End} Z(N)$ and therefore also their direct product satisfy the standard polynomial of degree $2r$. Thus the factor group H/N is embedded in the multiplicative group of a ring which satisfies the standard polynomial of degree $2r$.

Now let $G = \text{Cr}_{i \in J} H_i$ be the cartesian product of hyperabelian groups H_i with $r(H_i) \leq r$, and let M_i be a maximal nilpotent normal subgroup of H_i with class at most 2. Then $M = \text{Cr}_{i \in J} M_i$ is a nilpotent normal subgroup of G of class at most 2 such that the factor group G/M has the desired property.

Lemma 3.2. *A locally soluble subgroup of the multiplicative group of a ring with polynomial identity is hyperabelian.*

Proof. Let R be a ring with polynomial identity. By Proposition 1.6.25 of [6] the nil radical $N = N(R)$ of R contains a non-zero nilpotent ideal of R or $N = 0$. If I is a nilpotent ideal of R, then $1 + I$ is a nilpotent normal subgroup of the multiplicative group R^* of R (see [7], Chapter 1, Proposition 9). Hence $1 + N$ is a normal subgroup of R^* which has an ascending invariant series of R^* with abelian factors. By Theorem 1.6.27 of [6] the factor ring R/N is embedded in the ring of matrices $M_r(\mathbb{Z}[x])$ for some degree r, as it has trivial nil radical. Since the factor group $R^*/(1+N)$ is a linear group of degree r over the noetherian commutative ring $\mathbb{Z}[x]$, every locally soluble subgroup of $R^*/(1 + N)$ is soluble (see [10], 13.12). It follows that every locally soluble subgroup of R^* is hyperabelian.

Proof of Theorem 1.2. Let the locally soluble group G be residually of bounded finite Prüfer rank. Every locally soluble group with finite Prüfer rank is hyperabelian (see [5], Vol. 2, p. 179). Therefore the group G is isomorphic with a subgroup H of a cartesian product C of hyperabelian groups with bounded finite Prüfer rank. It follows from Lemma 3.1 that C contains a nilpotent normal subgroup M such that C/M is embedded in the multiplicative group of a ring with a polynomial identity. By Lemma 3.2 the locally soluble subgroup $HM/M \simeq H/(H \cap M)$ of C/M is hyperabelian. Since $H \cap M$ is nilpotent, also H and its isomorphic copy G are hyperabelian. This proves Theorem 1.2.

4. Proof of Theorem 1.1

A group G satisfies the **weak minimum condition for subgroups** if every descending chain of subgroups S_i has only finitely many infinite indices $|S_{i+1} : S_i|$. The weak minimum condition for normal subgroups is defined accordingly.

Lemma 4.1 (Amberg [1], Theorem 2.5). *If the group $G = AB$ is the product of two subgroups A and B with weak minimum condition for subgroups, then G satisfies the weak minimum condition for normal subgroups.*

Proof. Let U and V be normal subgroups of G such that $U \subset V$ and the indices $|AV : AU|$ and $|(A \cap V) : (A \cap U)|$ are finite. Then the following indices are finite:

$$|V : U(A \cap V)| = |V : (V \cap AU)| = |AV : AU|$$

and

$$|U(A \cap V) : U| = |(A \cap V) : (A \cap U)|.$$

Therefore the following index is also finite as a product of two finite indices:

$$|V : U| = |V : U(A \cap V)||U(A \cap V) : U| = |AV : AU||(A \cap V) : (A \cap U)|.$$

It is now easy to derive the lemma from this fact.

Proof of Theorem 1.1. It suffices to show that the locally soluble group G is hyperabelian, since then G is a minimax group by [9], Corollary A. Since every epimorphic image of G is likewise a locally soluble product of two minimax subgroups we only need to show that the group $G \neq 1$ has a non-trivial abelian normal subgroup. By Lemma 4.1 the group G satisfies the weak minimum condition for normal subgroups. Therefore there exists a normal subgroup N of G such that either

(1) N is a minimal normal subgroup of G, or

(2) for all normal subgroups M of G contained i N the factor group N/M is finite and the intersection of all these normal sub groups M is trivial.

In case (1) the minimal normal subgroup N of the locally soluble group G is abelian (see [5], Vol. 1, Corollary 1 to Theorem 5.27). Therefore we may suppose that N satisfies condition (2). The factorizer of N has the triple factorization

$$X = X(N) = NA_1 = NB_1 = A_1B_1$$

where $A_1 = A \cap BN$ and $B_1 = B \cap AN$ are minimax groups. Let M be a normal subgroup of G such that $M \subseteq N$ and N/M is finite. Then

$$\bar{X} = X/M = \bar{N}\bar{A}_1 = \bar{N}\bar{B}_1 = \bar{A}_1\bar{B}_1$$

where $\bar{N} = N/M$, $\bar{A}_1 = A_1M/M$ and $\bar{B}_1 = B_1M/M$. Here $\bar{N}$ is a finite normal subgroup of $\bar{X}$. Since $\bar{A}_1$ and $\bar{B}_1$ are soluble minimax groups and in particular have finite Prüfer ranks, also $\bar{X}$ has finite Prüfer rank bounded by a function of the Prüfer ranks of A and B; see [2], Theorem 4.3.5. Therefore every such factor group X/M has bounded Prüfer rank. Hence X is residually of bounded finite Prüfer rank. By Theorem 1.2 the group X is hyperabelian and so by the Theorem of Sysak and Wilson it is a soluble minimax group; see [9], Corollary A. This implies that N contains a non-trivial abelian normal subgroup of G. Theorem 1.1 is proved.

References

[1] Amberg, B., Factorizations of infinite groups, Habilitationsschrift, Univ. Mainz (1973).

[2] Amberg, B., Franciosi, S., and de Giovanni, F., Products of groups, Clarendon Press, Oxford (1992).

[3] Dlab, V., Die Endomorphismenringe abelscher Gruppen und die Darstellung von Ringen durch Matrizenringe, Czech. Math. J. 7 (1957), 485–519.

[4] Lennox, J. C., and Roseblade, J. E., Soluble products of polycyclic groups, Math. Z. 170 (1980), 153–154.

[5] Robinson, D. J. S., Finiteness conditions and generalized soluble groups, Vol. 1 and 2, Springer, Berlin (1972).

[6] Rowen, L. H., Polynomial identities in ring theory, Academic Press, New York (1980).

[7] Segal, D., Polycyclic groups, Cambridge University Press, New York (1980).

[8] Sysak, Y. P., Products of periodic groups, Preprint 82.53, Akad. Nauk Ukrain. Inst. Mat. Kiev (1982).

[9] Sysak, Y. P., Radical modules over groups of finite rank, Preprint 89.18, Akad. Nauk Ukrain. Inst. Mat. Kiev (1989).

[10] Wehrfritz, B. A. F., Infinite linear groups, Springer, Berlin (1973).

[11] Wilson, J. S., Soluble products of minimax groups and nearly surjective derivations, J. Pure Appl. Algebra 53 (1988), 297–318.

[12] Wilson, J. S., Soluble groups which are products of groups of finite rank, J. London Math. Soc. (2) 40 (1989), 405–419.

[13] Zaitsev, D. I., Factorizations of polycyclic groups, Mat. Zametki 29 (1981), 481–490.

A Group-Theoretic Reduction of
J. H. C. Whitehead's Asphericity Question

W. A. Bogley and M. N. Dyer

Abstract. J. H. C. Whitehead asked in 1941 whether subcomplexes of aspherical two-complexes are aspherical. The question remains unanswered as of this writing. In this note we use a theorem of J. Howie to show that Whitehead's question can be reduced to two problems in combinatorial group theory. Some partial results are surveyed.

1991 Mathematics Subject Classification: Primary 57M20; Secondary 20F19, 20F22

1. Introduction

This article is concerned with group-theoretic aspects of the following topological question, which was posed by J. H. C. Whitehead in 1941 [W41]: "Is any subcomplex of an aspherical, 2-dimensional complex itself aspherical?" A 2-dimensional complex is a CW complex in which each cell has dimension at most two; in short, what we will call a *two-complex*. A connected space is *aspherical* if its universal covering is contractible. For a connected two-complex X, this is equivalent to saying that the second homotopy group $\pi_2 X$ is trivial.

A survey of the extensive work that has been done on Whitehead's question appears in [B93]. The purpose of this article is to publicize the fact that Whitehead's question can be reduced to a pair of problems in combinatorial group theory. It is hoped that the group-theoretic formulations that are presented here will stimulate further work on the problem.

Interest in Whitehead's question can be motivated by the fact that the complement of any tame knot in the three-sphere has the homotopy type of a two-complex that can be embedded in a finite contractible two-complex. A positive solution to Whitehead's question therefore holds the promise of a (new) proof of the asphericity of knot complements. A footnote included in the midst of Whitehead's original question [W41, Footnote 30] suggests that this prospect may have been uppermost in Whitehead's mind at the time.

Our group-theoretic reduction of Whitehead's question is based on a topological reduction of the problem that appears in the following theorem due to J. Howie.

Theorem 1 ([H83]). *If the answer to Whitehead's question is NO, then there exists a connected two-complex L such that either*

1. *L is finite and contractible and $L - e$ is not aspherical for some open two-cell e of L, or*
2. *L is the union of an infinite ascending chain of finite connected nonaspherical subcomplexes $K_0 \subset K_1 \subset \cdots$, where each inclusion $K_{i-1} \subset K_i$ is nullhomotopic.*

$\square$

The situation in 1.1 will be referred to as the ***finite case***; 1.2 will be called the ***infinite case***. Of course, there is a converse to Howie's theorem in the sense that if there is a two-complex L with the properties described in either the finite or the infinite case, then the answer to Whitehead's question is NO. In addition, it has been shown by E. Luft [L94] that if there is a two-complex L of the sort described in the finite case, then there is also an example of the sort described in the infinite case. Thus, Whitehead's question actually reduces to the infinite case. This does not detract from the finite case however, which is still very interesting.

We will show that each of the two cases in Theorem 1 can be reduced to a problem in combinatorial group theory. The finite case leads to a problem (Theorem 3) concerning intersections of normal subgroups in finitely generated free groups. A partial result (Theorem 4) essentially solves the problem ***modulo the central series***, and leads to a question about residual nilpotence of certain groups. In the infinite case, we reduce Whitehead's question to one that concerns the existence of groups admitting certain ascending chains of normal subgroups. The particulars are given in Theorem 2.

Following this introductory section, the infinite case is discussed in Section 2. Section 3 treats the finite case. All spaces in this paper will be connected two-complexes. Basepoints for homotopy groups will be suppressed from the notation, but will always be taken to be a fixed zero-cell. If A and B are subgroups of a group G, then $[A, B]$ denotes the subgroup of G that is generated by all commutators $[a, b]$ ($a \in A$, $b \in B$), where $[a, b] = aba^{-1}b^{-1}$. If A and B are normal in G, then so is $[A, B]$, and in this case we also have $[A, B] \subseteq A \cap B$. The lower central series is defined inductively by $G_1 = G$ and $G_{n+1} = [G, G_n]$. All homology groups will be computed with integer coefficients.

2. The Infinite Case

The possibility of constructing an example as in the infinite case has been considered by M. Dyer [D92]. Suppose that a connected two-complex L is given as a union $K_0 \subset K_1 \subset \cdots \subset \bigcup_i K_i = L$ as in 1.2. Replacing each K_i by $K_i \cup L^{(1)}$, where $L^{(1)}$ denotes the one-skeleton of L, we have that for each $i \geq 1$, K_i is obtained from K_{i-1} by attaching two-cells (so that the inclusion-induced homomorphism $\pi_1 K_{i-1} \to \pi_1 K_i$ is surjective) and the inclusion-induced map $\pi_2 K_{i-1} \to \pi_2 K_i$ is trivial.

For an inclusion of two-complexes, the triviality of the induced map on second homotopy modules can be formulated in terms of the subgroup structure of the fundamental group of the subcomplex. Following [BD81], let X be a connected two-complex and let $N \leq \pi_1 X$. The two-complex X is **N-Cockcroft** if the lifted Hurewicz map $\pi_2 X \to H_2 X_N$ is trivial, where $X_N \to X$ is the covering corresponding to N. This property derives its name from its earliest consideration by W. H. Cockcroft in his work on Whitehead's question [C51]. Note that if X is N-Cockcroft and $N' \leq \pi_1 X$ contains some $\pi_1 X$-conjugate of N, then X is N'-Cockcroft. Also, X is Cockcroft $\Leftrightarrow$ X is $\pi_1 X$-Cockcroft, while X is aspherical $\Leftrightarrow$ X is $\{1\}$-Cockcroft. Our interest in the Cockcroft properties comes from the following elementary observation.

Lemma 1. *Suppose that X is a subcomplex of a connected two-complex Y. The inclusion-induced map $\pi_2 X \to \pi_2 Y$ is trivial if and only if X is $\ker i_\#$-Cockcroft, where $i_\# : \pi_1 X \to \pi_1 Y$ is the inclusion-induced homomorphism of fundamental groups.*

Proof. Let $p : \tilde{Y} \to Y$ be the universal covering and let $\bar{X}$ be a connected component of $p^{-1}(X)$; the restriction of p then determines the covering $\bar{X} \to X$ corresponding to $\ker i_\# \leq \pi_1 X$. Since $\pi_2 Y \to H_2 \tilde{Y}$ and $H_2 \bar{X} \to H_2 \tilde{Y}$ are both injective, it readily follows that $\pi_2 X \overset{0}{\to} \pi_2 Y \Leftrightarrow \pi_2 X \overset{0}{\to} H_2 \bar{X}$.   □

Quite a lot of work has been done on Cockcroft properties in recent years. Of particular group-theoretic interest is the fact, due independently to J. Harlander [H94] and to N. Gilbert and J. Howie [GH94], that for any two-complex X, there is a minimal subgroup H of $\pi_1 X$ such that X is H-Cockcroft. Such minimal subgroups are referred to as Cockcroft **thresholds** for X. Informally, it is appropriate to say that if X has a "small" Cockcroft threshold, then X is "nearly" aspherical.

If X is any topological space, it is obvious that a spherical map $S^2 \to X$ can be rendered nullhomotopic by attaching a three-cell to X: One simply uses the spherical map to attach the three-cell! Somewhat less obvious is the fact that essential spherical maps into two-complexes can be rendered nullhomotopic simply by adding **two-cells**. The following sort of example is fairly well known. Let X be the real projective plane, modeled on the presentation $(a : a^2)$ for the cyclic group of order two. Thus, X is constructed by attaching a disc to a circle S_a^1 by a two-fold wrap of the boundary circle of the disc onto S_a^1. One has that $H_2 X = 0$ and that $\pi_2 X$ is infinite cyclic, since X is covered by the two-sphere. Let Y be the two-complex modeled on the presentation $(a : a^2, a)$ for the trivial group. Thus, Y is obtained from X by attaching another disc to X, this time using a homeomorphism of the boundary circle in the disc with S_a^1. Now Y is simply connected (in fact Y has the homotopy type of the two-sphere), and so the Hurewicz homomorphism $\pi_2 Y \to H_2 Y$ is an isomorphism (of infinite cyclic groups). It follows that the inclusion-induced map $\pi_2 X \to \pi_2 Y$ factors through $H_2 X = 0$, and so this map is trivial. (Thus, X is $\pi_1 X$-Cockcroft.) However, this process can not be repeated in any fashion, for if Y is a subcomplex of any two-complex Z, then $\pi_2 Y \cong H_2 Y \neq 0$ embeds in $H_2 Z$, and so $\pi_2 Y \to \pi_2 Z$ is nontrivial. In other words, Y is not $\pi_1 Y$-Cockcroft.

Coupled with this example, Lemma 1 reveals the main difficulty in attempting to construct a two-complex $L = \bigcup_{i \geq 0} K_i$ of the sort described in the infinite case. Having constructed K_{i-1}, one must add (two-)cells in such a way that the resulting adjunction space K_i has a suitable Cockcroft property. We examine the requirements from a group-theoretic perspective.

Suppose that X is a connected two-complex and that

$$Y = X \cup \bigcup_{\alpha \in \mathcal{A}} c_\alpha^2 \text{ and } Z = Y \cup \bigcup_{\beta \in \mathcal{B}} d_\beta^2$$

are obtained from X by attaching two-cells. Set $G = \pi_1 X$. For each $\alpha \in \mathcal{A}$, let $a_\alpha \in G$ denote the (based) homotopy class of an attaching map for the two-cell c_α^2. The element a_α is well-defined up to conjugacy in G. In the same way, let $b_\beta \in G$ be the based homotopy class for an attaching map of d_β^2. We set

$$A = \ker(\pi_1 X \to \pi_1 Y) \text{ and } B = \ker(\pi_1 X \to \pi_1 Z)$$

so that $A \leq B \leq G$ where A and B are normal subgroups of G. Note that A is normally generated in G by $\{a_\alpha : \alpha \in \mathcal{A}\}$ and B is normally generated in G by $\{a_\alpha : \alpha \in \mathcal{A}\} \cup \{b_\beta : \beta \in \mathcal{B}\}$. We have that $\pi_1 Y = G/A$ and $\pi_1 Z = G/B$. The abelianized group $H_1 A = A/[A, A]$ is a (left) $\mathbb{Z}G/A$-module under conjugation in G:

$$g \cdot a[A, A] = gag^{-1}[A, A]$$

for all $g \in G$ and for all $a \in A$. This module is $\mathbb{Z}G/A$-generated by $\{a_\alpha[A, A] : \alpha \in \mathcal{A}\}$. Killing the action of the subgroup B/A of G/A, the group

$$A/[A, B] \cong \mathbb{Z} \otimes_{B/A} H_1 A$$

is a $\mathbb{Z}G/B$-module with generators $\{a_\alpha[A, B] : \alpha \in \mathcal{A}\}$.

Lemma 2. *If $\pi_2 X \overset{0}{\to} \pi_2 Y$, then $\pi_2 Y \overset{0}{\to} \pi_2 Z$ if and only if both of the following conditions are satisfied.*
1. *$A/[A, B]$ is a free $\mathbb{Z}G/B$-module with indexed basis $\{a_\alpha[A, B] : \alpha \in \mathcal{A}\}$.*
2. *$H_2 B \to H_2 B/A$ is injective.*

Proof. Let $\widetilde{X}$, $\widetilde{Y}$, and $\widetilde{Z}$ denote the universal covering complexes for X, Y, and Z, respectively. As shown in the following diagram of inclusions and covering projections, let $\overline{X}$ and $\widehat{X}$ denote the preimages of X in $\widetilde{Y}$ and $\widetilde{Z}$, respectively, and let $\overline{Y}$ denote the preimage of Y in $\widetilde{Z}$. All of these spaces are connected. The covering complexes $\overline{X}$ and $\widehat{X}$ of X are those corresponding to the subgroups A and B of $G = \pi_1 X$, respectively. The covering complex $\overline{Y}$ of Y is that corresponding to the subgroup B/A of $G/A = \pi_1 Y$. Assuming the $\pi_2 X \to \pi_2 Y$ is trivial, it follows that $\pi_2 X \to H_2 \overline{X}$ is trivial by Lemma 1. This implies that $\pi_2 X \to H_2\widehat{X}$ is the zero map, and it follows that the natural surjection $H_2\widehat{X} \to H_2\pi_1\widehat{X} = H_2 B$ is an isomorphism.

$$
\begin{array}{ccccc}
\widetilde{X} & & & & \\
\downarrow & & & & \\
\overline{X} & \subseteq & \widetilde{Y} & & \\
\downarrow & & \downarrow & & \\
\widehat{X} & \subseteq & \overline{Y} & \subseteq & \widetilde{Z} \\
\downarrow & & \downarrow & & \downarrow \\
X & \subseteq & Y & \subseteq & Z
\end{array}
$$

By Lemma 1, the map $\pi_2 Y \to \pi_2 Z$ is trivial if and only if $\pi_2 Y \to H_2\overline{Y}$ is the zero map. Using the fact that $H_2\widehat{X} \to H_2\overline{Y}$ is injective and $H_2\widehat{X} \to H_2 B$ is an isomorphism, a chase in the following commutative diagram (which has exact rows and columns) shows that $\pi_2 Y \to \pi_2 Z$ is trivial if and only if $H_2 B \to H_2 B/A$ is injective and the composite $(\pi_2 Y \to H_2\overline{Y} \to H_2(\overline{Y}, \widehat{X}))$ is the zero map.

$$
\begin{array}{ccccc}
\pi_2 Y & \longrightarrow & H_2(\widetilde{Y}, \overline{X}) & \longrightarrow & H_1\overline{X} \\
\downarrow & & \downarrow & & \downarrow \\
H_2\widehat{X} & \longrightarrow & H_2\overline{Y} & \longrightarrow & H_2(\overline{Y}, \widehat{X}) & \longrightarrow & H_1\widehat{X} \\
\downarrow & & \downarrow & & & & \\
H_2 B & \longrightarrow & H_2 B/A & & & &
\end{array}
$$

It remains to show that the latter condition is equivalent to the condition 1 of the lemma. To see this, note that the boundary map $H_2(\widetilde{Y}, \overline{X}) \to H_1\overline{X}$ can be identified with the $\mathbb{Z}G/A$-module epimorphism

$$\bigoplus_{\alpha \in \mathcal{A}}(\mathbb{Z}G/A)\, t_\alpha \to A/[A, A] \to 0$$

that carries the basis element t_α to $a_\alpha[A, A]$. This follows by excision since Y is obtained from X by attaching one two-cell for each element $\alpha \in \mathcal{A}$ and the covering $\widetilde{Y} \to Y$ has automorphism group G/A. In addition, the map $H_2(\widetilde{Y}, \overline{X}) \to H_2(\overline{Y}, \widehat{X})$ can be identified with the map

$$1 \otimes - : \bigoplus_{\alpha \in \mathcal{A}}(\mathbb{Z}G/A)\, t_\alpha \to \mathbb{Z} \otimes_{B/A} \bigoplus_{\alpha \in \mathcal{A}}(\mathbb{Z}G/A)\, t_\alpha \cong \bigoplus_{\alpha \in \mathcal{A}}(\mathbb{Z}G/B)\, t_\alpha.$$

This is because the covering $\widetilde{Y} \to \overline{Y}$ has automorphism group B/A.

With these identifications, consider the effect of applying the right exact functor $\mathbb{Z} \otimes_{B/A} -$ to the exact sequence

$$\pi_2 Y \to H_2(\tilde{Y}, \overline{X}) \to H_1 \overline{X} \to 0.$$

We find that the composite $(\pi_2 Y \to H_2 \overline{Y} \to H_2(\overline{Y}, \widehat{X}))$ is the zero map if and only if the map

$$\mathbb{Z} \otimes_{B/A} H_2(\tilde{Y}, \overline{X}) \to \mathbb{Z} \otimes_{B/A} H_1 \bar{X} \cong A/[A, B]$$

is an isomorphism, as in the condition 1 of Lemma 2. This completes the proof of the lemma.   □

Theorem 2. *If there is a two-complex L as described in Theorem 1.2, then there exists a finite connected nonaspherical two-complex K and an infinite ascending chain $\{1\} = N_0 < N_1 < N_2 < \cdots < \pi_1 K$ of normal subgroups of $\pi_1 K$ such that the following two properties hold.*
1. *K is N_1-Cockcroft.*
2. *There are subsets $\mathbf{r}_i \subseteq \pi_1 K$ $(i \geq 1)$ such that $\{r_i N_{i-1} : r_i \in \mathbf{r}_i\}$ normally generates N_i/N_{i-1} in $\pi_1 K/N_{i-1}$ and such that the following two conditions are satisfied for each positive integer i.*
    (a) *$N_i/N_{i-1}[N_i, N_{i+1}]$ is a free $\mathbb{Z}\pi_1 K/N_{i+1}$-module with indexed basis $\{r_i N_{i-1}[N_{i+1}, N_{i-1}] : r_i \in \mathbf{r}_i\}$.*
    (b) *$H_2(N_{i+1}/N_{i-1}) \to H_2(N_{i+1}/N_i)$ is injective.*
    Conversely, if such a two-complex K exists, then the answer to Whitehead's question is NO.

Proof. Suppose that we are given $K_0 \subseteq K_1 \subseteq \cdots \subseteq \bigcup_{i \geq 0} K_i = L$ as in Theorem 1.2. Replace each of the subcomplexes K_i by the union $K_i \cup L^1$, where L^1 denotes the one-skeleton of L. Let $K = K_0$; this two-complex is not aspherical. Each K_i is obtained from K_{i-1} by attaching two-cells and the inclusion-induced map $\pi_2 K_{i-1} \to \pi_2 K_i$ is trivial. For each positive integer i, let N_i be the kernel of the inclusion-induced epimorphism $\pi_1 K \to \pi_1 K_i$ and let $\mathbf{r}_i$ be a subset of $\pi_1 K$ consisting of one based homotopy class of an attaching map for each two-cell of $K_i - K_{i-1}$. Then $\{1\} = N_0 < N_1 < N_2 < \cdots < \pi_1 K$ is an ascending chain of normal subgroups of $\pi_1 K$ and $\{r_i N_{i-1} : r_i \in \mathbf{r}_i\}$ normally generates N_i/N_{i-1} in $\pi_1 K_{i-1} = \pi_1 K/N_{i-1}$.

Now $K = K_0$ is N_1-Cockcroft by Lemma 1, since $\pi_2 K_0 \to \pi_2 K_1$ is the zero map. Fixing a positive integer i, consider the triple

$$K_{i-1} \subseteq K_i \subseteq K_{i+1}$$

and set $G = \pi_1 K_{i-1} = \pi_1 K/N_{i-1}$, $A = N_i/N_{i-1}$, and $B = N_{i+1}/N_{i-1}$. Note that $A/[A, B] = N_i/N_{i-1}[N_i, N_{i+1}]$ is a module over the integral group ring of $G/B = \pi_1 K/N_{i+1}$ and that $B/A = N_{i+1}/N_i$. Since both of these inclusions induce the trivial map in second homotopy, Lemma 2 implies that the conditions 2(a) and

2(b) of Theorem 2 are satisfied.

For the converse, suppose that we are given K, N_i and $\mathbf{r}_i$ that satisfy the conditions 1 and 2 in Theorem 2. The two-complex K is not aspherical. We will show that the answer to Whitehead's question is NO by embedding K in an aspherical two-complex L. Let $K_0 = K$. For each $i \geq 1$ let K_i be obtained from K_{i-1} by attaching two-cells along based loops representing the elements $r_i \in \mathbf{r}_i \subseteq \pi_1 K_0$. By Lemma 1, condition 1 of Theorem 2 implies that the inclusion-induced map $\pi_2 K_0 \to \pi_2 K_1$ is trivial. Arguing inductively, given that $\pi_2 K_{i-1} \to \pi_2 K_i$ is the zero map, the conditions 2(a) and 2(b) of Theorem 2 imply that $\pi_2 K_i \to \pi_2 K_{i+1}$ is trivial by Lemma 2. We set $L = \bigcup_{i \geq 0} K_i$, where L is given the weak topology with respect to the closed subspaces K_i. The nonaspherical two-complex K is thus a subcomplex of the two-complex L, and L is aspherical by compact supports. For each spherical map $S^2 \to L$ has its image in a finite subcomplex of L, and hence in one of the subcomplexes K_i. This spherical map is then nullhomotopic in K_{i+1}, and hence in L. This shows that the answer to Whitehead's question is NO, and so completes the proof of the theorem. $\square$

We remark that in light of the result of E. Luft that was mentioned in the Introduction [L94], the answer to Whitehead's question is NO if and only if there is a two-complex K of the sort described in Theorem 2. If one is trying to construct such an example, then one seeks a group that contains an infinite ascending chain of normal subgroups with certain properties. Theorem 2 indicates that one need only look among finitely presented groups; this is because the two-complex K is finite. However, if one could find a two-complex K, not necessarily finite but which otherwise satisfies the conditions of the theorem, then the proof of the theorem shows how to embed K in an aspherical two-complex.

Before moving to the finite case, we may as well admit that the statement of Theorem 2 is not purely group-theoretic. This can be remedied artificially by defining a group G to be N-Cockcroft (where N is a subgroup of G) if there is a two-complex with fundamental group isomorphic to G and which is N-Cockcroft. In search of a nonaspherical subcomplex of an aspherical two-complex via the infinite route, we would then be asking for a (finitely presented) group G with an ascending chain $\{1\} = N_0 < N_1 < N_2 < \cdots$ of normal subgroups satisfying the conditions 2 and 2 of Theorem 2, and where G is N_1-Cockcroft but is not $\{1\}$-Cockcroft. Less formally, one seeks a group G that is not "aspherical" (i.e. is not $\{1\}$-Cockcroft), but which is "very nearly aspherical" in the sense that it contains a Cockcroft threshold that is small enough to sit underneath an ascending chain of a certain restricted type.

3. The Finite Case

If L is a finite contractible two-complex and $K = L - e$ is obtained by removing an open two-cell e from L, then the fundamental group $G = \pi_1 K$ is a group of deficiency one that has weight one. In other words, G has a finite presentation with one fewer relator than generators, and G is normally generated in itself by a single element. Moreover, it is not difficult to show that all such groups arise in this way. It can be shown that the finite case of Whitehead's question is equivalent to the question of whether finitely presented groups having deficiency and weight one must have cohomological dimension at most two. See [BDS83] for further discussion.

Another algebraic approach to the finite case of Whitehead's question leads to the study of groups of the form $F/[R, S]$ where R and S are normal subgroups of a finitely generated free group F. In order to state this reduction of the problem, we need some terminology.

Let F be a finitely generated free group with basis $\mathbf{x}$ and let $\mathbf{r}$, $\mathbf{s}$, and $\mathbf{t}$ be finite subsets of F. Let R, S, and T be their normal closures in F, respectively. Let K_r, K_s, and K_t be the two-complexes modeled on the group presentations $(\mathbf{x} : \mathbf{r})$, $(\mathbf{x} : \mathbf{s})$, and $(\mathbf{x} : \mathbf{t})$, respectively. Let $K = K_r \cup K_s$ and $L = K \cup K_t$. These are all finite connected two-complexes. Now L is simply connected if and only if $F = RST$. Further, L is contractible if and only if $F = RST$ and $|\mathbf{r}| + |\mathbf{s}| + |\mathbf{t}| - \text{rank } F$. In this case we say that $F = RST$ is an *efficient factorization* of F. The following result is proved in [B91] and in [B93].

Theorem 3 ([B91]). *The following two statements are logically equivalent.*
1. *Connected subcomplexes of finite contractible two-complexes are aspherical.*
2. *If R and S are distinct factors from an efficient factorization of a finitely generated free group, then $R \cap S \subseteq [R, S]$.*

□

We shall not reprove this result here, but it is worth mentioning the main ingredient. If $\mathbf{x}$, $\mathbf{r}$, and $\mathbf{s}$ are arbitrarily given as above and if corresponding two-complexes are constructed in the manner indicated, then by [GR81, Theorem 1] there is an exact sequence

$$\pi_2 K_r \oplus \pi_2 K_s \to \pi_2 K \to \frac{R \cap S}{[R, S]} \to 0.$$

At a fundamental level, this is the result that explains our interest in the group $Q = F/[R, S]$. Note that the subgroup $\Theta = (R \cap S)/[R, S]$ is naturally a module over the integral group ring of F/RS, via conjugation in F. This module has been studied in [B84] and in [HK91], to name two sources.

Returning to Whitehead's question, in the finite case there is a partial result on the group-theoretic problem.

Theorem 4 ([B91]). *If R and S are distinct factors from an efficient factorization of a finitely generated free group F, then*

$$R \cap S \subseteq \bigcap_{n \geq 1} [R, S] F_n.$$

It follows easily that the quotient $(R \cap S)/[R, S]$ embeds naturally in $Q_\omega = \bigcap_{n \geq 1} Q_n$. The proof of Theorem 4 essentially amounts to a determination of the structure of the Lie algebra that is built out of the lower central series of Q [B91, Theorem 2]. In particular it is shown that Q_n/Q_{n+1} is finitely generated and free abelian for all $n \geq 1$. Little seems to be known about Q_ω however. With Theorem 4, one might be led to ask for conditions under which the group Q is residually nilpotent (i.e. $Q_\omega = 1$).

There are many test cases to consider in the finite case. The model of any finite balanced presentation for the trivial group is a finite contractible two-complex. A large and interesting class of examples arises as follows. Let Γ be a finite tree with vertices $V\Gamma$ and (geometric) edges $E\Gamma$. Assume that each edge of Γ is oriented and is labeled by a vertex of Γ. (Thus Γ is a *labeled oriented tree* or LOT.) Associated to Γ is a group presentation

$$\mathcal{P}(\Gamma) = (V\Gamma : i(e)\lambda(e)t(e)^{-1}\lambda(e)^{-1}(e \in E\Gamma)).$$

Here, $i(e)$, $\lambda(e)$ and $t(e)$ denote the initial vertex, label and terminal vertex of the edge $e \in E\Gamma$, respectively. Let $K(\Gamma)$ denote the two-complex modeled on $\mathcal{P}(\Gamma)$.

It is not difficult to show that upon adding a single relation of the form $v = 1$, ($v \in V\Gamma$), there results a finite balanced presentation for the trivial group. If we denote the cellular model of the presentation $\mathcal{P}(\Gamma)$ by $K(\Gamma)$ (so that $K(\Gamma)$ is an *LOT complex*), then $K(\Gamma)$ is a connected subcomplex of a finite contractible two-complex. If one wishes to prove that there are no examples of the sort described in the finite case, one must therefore prove that LOT complexes are aspherical. J. Howie has proved some partial results in this area [H85]. Notably, if an LOT Γ has diameter less than four, then $K(\Gamma)$ is aspherical. A crucial element in Howie's proof is the fact that a tree of diameter three has at most two nonextremal vertices. The structure of larger trees can be far more varied. The complexes $K(\Gamma)$ therefore provide a wide open playing field.

This area includes the connection to knots that was mentioned in the introduction, for the complement of any tame knot in the three-sphere has the homotopy type of an LOT complex. See [H83, H85] for references and further discussion. It is an open question whether each proper subcomplex of a finite contractible two-complex has the homotopy type of a subcomplex of an LOT complex.

We close with the following problem, which is seen to contain a large and interesting portion of Whitehead's question. Let Γ be a labeled oriented tree. Let F be the free group on the set of vertices of Γ and let $\mathbf{r} \cup \mathbf{s}$ be a nontrivial partition of the set of edges of Γ. Let R (resp. S) be the normal closure of the set of all element of the form $i(e)\lambda(e)t(e)^{-1}\lambda(e)^{-1}$, where $e \in \mathbf{r}$ (resp. $e \in \mathbf{s}$). Is $R \cap S = [R, S]$? Equivalently, does $Q = F/[R, S]$ embed in $F/R \times F/S$? If not, then the answer to Whitehead's question is NO.

References

[B91]   W. A. Bogley, An embedding for π_2 of a subcomplex of a finite contractible two-complex, Glasgow Math. J. 33 (1991), 365–371.

[B93]   W. A. Bogley, On J. H. C. Whitehead's asphericity question, in: Two-dimensional Homotopy and Combinatorial Group Theory (C. Hog-Angeloni, W. Metzler, and A. J. Sieradski, eds.), London Math. Soc. Lecture Note Ser. 197, Cambridge University Press, 1993, 309–334.

[BD81]   J. Brandenburg and M. N. Dyer, On J. H. C. Whitehead's aspherical question I, Comment. Math. Helv. 56 (1981), 431–446.

[BDS83] J. Brandenburg, M. N. Dyer and R. Strebel, On J. H. C. Whitehead's aspherical question II, in: Low Dimensional Topology (S. Lomonaco, ed.), Contemp. Math. 20 (1983), 65–78.

[B84]   R. Brown, Coproducts of crossed *P*-modules: Applications to second homotopy groups and to the homology of groups, Topology 23 (1984), 337–345.

[C51]   W. H. Cockcroft, Note on a theorem by J. H. C. Whitehead, Quart. J. Math. Oxford Ser. (2) 2 (1951), 159–160.

[D92]   M. N. Dyer, Cockcroft 2-complexes, preprint, University of Oregon, 1992.

[GH94]   N. D. Gilbert and J. Howie, Threshold subgroups for Cockcroft 2-complexes, Comm. Algebra, to appear.

[GR81]   M. A. Gutiérrez and J. G. Ratcliffe, On the second homotopy group, Quart. J. Math. Oxford Ser. (2) 32 (1981), 45–55.

[H94]   J. Harlander, Minimal Cockcroft subgroups, Glasgow Math. J. 36 (1994), 87–90.

[HK91]   B. Hartley and Yu. V. Kuz'min, On the quotient of a free group by the commutator of two normal subgroups, J. Pure Appl. Algebra 74 (1991), 247–256.

[H83]   J. Howie, Some remarks on a problem of J. H. C. Whitehead, Topology 22 (1983), 475–485.

[H85]   J. Howie, On the asphericity of ribbon disc complements, Trans. Amer. Math. Soc. 289 (1985), 281–302.

[L94]   E. Luft, On 2-dimensional aspherical complexes and a problem of J. H. C. Whitehead, preprint, University of British Columbia, 1994.

[W41]   J. H. C. Whitehead, On adding relations to homotopy groups, Ann. of Math. 42 (1941), 409–428; Note on a previous paper, Ann. of Math. 47 (1946), 806–809.

Efficiency and Direct Products of Groups

Melanie J. Brookes, Colin M. Campbell
and Edmund F. Robertson

Abstract. We extend techniques introduced in [3] to obtain effici nt presentations for certain direct products and give some general results.

1991 Mathematics Subject Classification: 20F05

1. Introduction

Let $\mathcal{P}$ be the finite presentation $\langle X \mid R \rangle$. The **deficiency** of $\mathcal{P}$ is $|R| - |X|$ and, if $\mathcal{P}$ defines a finite group, the deficiency of $\mathcal{P}$ is non-negative. The **deficiency of a group** G, def G, is the minimum of the deficiencies of all finite presentations of G. It is well known that def $G \geq \mathrm{rk}(M(G))$ where $M(G)$ is the Schur multiplier of G. A group G is said to be **efficient** if def $G = \mathrm{rk}(M(G))$.

The efficiency of finite groups has been studied over many years; see for example [5], [11]. In particular the efficiency of direct products of groups, stimulated by questions asked by Wiegold in [11], has been studied by several authors; see for example [5], [7], and [8]. Recently a new approach to finding efficient presentations for certain direct products was suggested by Izumi Miyamoto and is used in [3] to show that, for p a prime, $PSL(2, p) \times SL(2, p)$ and $PSL(2, p) \times PSL(2, p) \times PSL(2, p)$ are efficient.

In this paper we extend Miyamoto's method and prove a more general theorem. We then apply the theorem to obtain other classes of efficient direct products of groups. The key idea in Miyamoto's method is contained in the following easily proved result:

Lemma 1.1 (Lemma 2.1 of [3]). *Let G be a group with $a, b \in G$ satisfying $a^\epsilon = (a^m b^\delta)^n$ where $\epsilon, \delta = \pm 1$, m and n integers. Then $\langle a, b \rangle$ is a cyclic subgroup of G and $a^{\epsilon - mn} = b^{\delta n}$.*

2. Some Direct Products of Groups Having Trivial Multiplier with Those Having Cyclic Multiplier

Our first result on efficient direct products generalises the efficiency of the group $PSL(2, p) \times SL(2, p)$.

Theorem 2.1. *Let G_1, G_2 be finite perfect groups with trivial centres, G_1 having multiplier C_2 and G_2 having trivial multiplier. Let G_1, G_2 have presentations of the form*

$$\langle a, b \mid a^{\alpha_1} = b^{\alpha_2} = (ab)^{\alpha_3} = w(a, b) = 1 \rangle$$
$$\langle x, y \mid x^{\beta_1} = y^{\beta_2} = (xy)^{\beta_3} = v(x, y) = 1 \rangle$$

with the α_i, β_i satisfying the congruences $\beta_3 \equiv \pm 1 \pmod{\alpha_1}$, $\beta_2 \equiv \pm 1 \pmod{\alpha_3}$, $\alpha_2 \equiv \pm 1 \pmod{\beta_1}$, $\alpha_3 \equiv \pm 1 \pmod{\beta_1}$. Let the group G, given by the presentation below, be perfect:

$$\langle a, b, x, y \mid (xy)^{\pm 1}((xy)^{(\beta_3 \mp 1)/\alpha_1} a)^{\alpha_1} = y^{\pm 1}(y^{(\beta_2 \mp 1)/\alpha_3} ab)^{\alpha_3} =$$
$$(ab)^{\pm 1}((ab)^{(\alpha_3 \mp 1)/\beta_1} x)^{\beta_1} = b^{\pm 1}(b^{(\alpha_2 \mp 1)/\beta_1} x)^{\beta_1} = a^{-\alpha_1} w(a, b)v(x, y) = 1 \rangle.$$

(The four congruences decide the choice of $\pm$ and $\mp$ in each of the four relations so that all of the powers within the relations are integer powers. In each relation one is chosen to be $+$ and the other $-$.)

Let G_1 be such that in the group presented by

$$\langle a, b \mid a^{\alpha_1} = s, \ b^{\alpha_2} = (ab)^{\alpha_3} = u, \ w(a, b) = t; \ s, u, t \text{ central involutions} \rangle$$

we have $s = t$ and let G_2 be such that

$$\langle x, y \mid x^{\beta_1} = y^{\beta_2} = u, \ (xy)^{\beta_3} = s, \ v(x, y) = 1; \ u, s \text{ central involutions} \rangle$$

is perfect. Then $G \cong G_1 \times G_2$ and so this direct product is efficient.

Proof. By Lemma 1.1, the following relations hold in G:

$$[ab, y] = [ab, x] = [b, x] = [a, xy] = 1,$$
$$b^{\alpha_2} = (ab)^{\alpha_3} = x^{-\beta_1} = y^{-\beta_2}, \ a^{\alpha_1} = (xy)^{-\beta_3}.$$

Let $H = \langle a, b \rangle$, $K = \langle x, y \rangle$. From the relations $[ab, x] = 1$ and $[b, x] = 1$ we have $[a, x] = 1$. Similarly we have $[a, y] = 1$ and $[b, y] = 1$. Hence $[H, K] = 1$. Let $D = \langle x^{\beta_1}, (xy)^{\beta_3}, v(x, y) \rangle$. Clearly $D \leq H \cap K \leq Z(G)$ and, since $G/D \cong G_1 \times G_2$ which has trivial centre, we must have that $D = H \cap K = Z(G)$. G is perfect with D central and $G/D \cong G_1 \times G_2$ so, by Lemma 4.1 of [12], D must be an epimorphic image of $M(G_1 \times G_2)$. Therefore D is either trivial or cyclic of order two. We also have that $G/H \cong K/D \cong G_2$ and $G/K \cong H/D \cong G_1$. Now, in H the following relations hold:

$$a^{\alpha_1} = s, \ b^{\alpha_2} = (ab)^{\alpha_3} = u, \ w(a, b) = t,$$

with s, u, t, central involutions. By hypothesis we then have $s = t$ and so, by the fifth relation of G, $v(x, y) = 1$. Now consider K given by

$$\langle x, y \mid x^{\beta_1} = y^{\beta_2} = u, \ (xy)^{\beta_3} = s, \ v(x, y) = 1; \ u, \ s \text{ central involutions} \rangle.$$

By hypothesis, this group is perfect and, since K is a central extension of G_2 by D, it is a stem extension of G_2 by D. We know that G_2 has trivial multiplier and so D must be trivial. Hence $G \cong G_1 \times G_2$ as required.   □

Corollary 2.2. *The direct product, $G_1 \times G_2$, in Theorem* 2.1 *can be presented by*

$$\langle c, d \mid d^{\mp(\alpha_3 \mp 1)-1} c^{\mp \alpha_1} ((d^{\mp(\alpha_3 \mp 1)-1} c^{\mp \alpha_1})^{(\beta_2 \mp 1)/\alpha_3} d^{\mp \beta_1})^{\alpha_1} =$$
$$c^{\mp(\beta_3 \mp 1)-1} d^{\mp \beta_1} ((c^{\mp(\beta_3 \mp 1)-1} d^{\mp \beta_1})^{\alpha_2 \mp 1)/\beta_1} d^{\pm(\alpha_3 \mp 1)+1})^{\beta_1} =$$
$$(c^{\pm(\beta_3 \mp 1)+1})^{-\alpha_1} w(c^{\pm(\beta_3 \mp 1)+1}, c^{\mp(\beta_3 \mp 1)-1} d^{\mp \beta_1}) v(d^{\pm(\alpha_3 \mp 1)+1}, d^{\mp(\alpha_3 \mp 1)-1} c^{\mp \alpha_1})$$
$$= 1 \rangle$$

and so has an efficient presentation on two generators.

Proof. Put $c = (xy)^{(\beta_3 \mp 1)/\alpha_1} a$, $d = (ab)^{(\alpha_3 \mp 1)/\beta_1} x$ to get $a = c^{\pm(\beta_3 \mp 1)+1}$, $b = c^{\mp(\beta_3 \mp 1)-1} d^{\mp \beta_1}$, $x = d^{\pm(\alpha_3 \mp 1)+1}$, $y = d^{\mp(\alpha_3 \mp 1)-1} c^{\mp \alpha_1}$ and substitute into the presentation for G.   □

Remarks. (i) This first remark concerns the conditions imposed on the $\alpha_i, \beta_j, i, j \in \{1, 2, 3\}$ in Theorem 2.1. In any of the four congruences, we can interchange the α_i and β_j without affecting the proof. Also, let $\theta \in S_3$. Then, letting θ act on the i or j does not affect the mechanics of the proof. As an example, suppose that $\beta_2 \equiv \pm 1 \pmod{\alpha_1}, \alpha_2 \equiv \pm 1 \pmod{\beta_1}, \alpha_3 \equiv \pm 1 \pmod{\beta_1}, \alpha_2 \equiv \pm 1 \pmod{\beta_3}$. We can modify the theorem as follows. The group G we consider is given by the presentation below:

$$\langle a, b, x, y \mid y^{\pm 1} (y^{(\beta_2 \mp 1)/\alpha_1} a)^{\alpha_1} = b^{\pm 1} (b^{(\alpha_2 \mp 1)/\beta_1} x)^{\beta_1} =$$
$$(ab)^{\pm 1} ((ab)^{(\alpha_3 \mp 1)/\beta_1} x)^{\beta_1} = b^{\pm 1} (b^{(\alpha_2 \mp 1)/\beta_3} xy)^{\beta_3} = a^{-\alpha_1} w(a, b) v(x, y) = 1 \rangle.$$

The only other modification is the condition we require of G_2. We want the group presented by

$$\langle x, y \mid x^{\beta_1} = (xy)^{\beta_3} = u, \ y^{\beta_2} = s, \ v(x, y) = 1; \ s, \ u \text{ central involutions} \rangle$$

to be perfect. Some of the examples given later will make use of this type of modification of the theorem.

(ii) We can extend the idea of the theorem and look at the direct product of two perfect groups with trivial centre, one with cyclic multiplier C_n, and the other group having trivial multiplier. We need only modify slightly the statement of Theorem 2.1.

Again let G_1 be the group with cyclic multiplier and presentation

$$\langle a, b \mid a^{\alpha_1} = b^{\alpha_2} = (ab)^{\alpha_3} = w(a, b) = 1 \rangle.$$

Let G_2 have trivial multiplier and presentation

$$\langle x, y \mid x^{\beta_1} = y^{\beta_2} = (xy)^{\beta_3} = v(x, y) = 1 \rangle$$

with the α_i, β_i satisfying the congruences given in Theorem 2.1. Let G_1 be such that in the group presented by

$$\langle a, b \mid a^{\alpha_1} = s, \ b^{\alpha_2} = (ab)^{\alpha_3} = u, \ w(a, b) = t; \ s, \ u, \ t \ \text{central of order dividing } n \rangle$$

we have $s^l = t^m$ for some $1 \leq l, \ m \leq n$. Let G, given by the presentation below, be perfect:

$$\langle a, b, x, y \mid (xy)^{\pm 1}((xy)^{(\beta_3 \mp 1)/\alpha_1} a)^{\alpha_1} = y^{\pm 1}(y^{(\beta_2 \mp 1)/\alpha_3} ab)^{\alpha_3} =$$
$$(ab)^{\pm 1}((ab)^{(\alpha_3 \mp 1)/\beta_1} x)^{\beta_1} = b^{\pm 1}(b^{(\alpha_2 \mp 1)/\beta_1} x)^{\beta_1} = a^{-\alpha_1 l} w(a, b)^m v(x, y) = 1 \rangle.$$

If the group presented by

$$\langle x, y \mid x^{\beta_1} = y^{\beta_2} = u, \ (xy)^{\beta_3} = s, \ v(x, y) = 1; \ u, \ s \ \text{central of order dividing } n \rangle$$

is perfect, then $G \cong G_1 \times G_2$ and so this direct product is efficient.

The next result uses the presentation for $PSL(2, p)$, $p \geq 5$, given in [10],

$$\langle a, b \mid a^2 = b^p = (ab)^3 = (ab^4 ab^{(p+1)/2})^2 = 1 \rangle.$$

Lemma 2.3. *Let L be the group given by the presentation:*

$$\langle a, b \mid a^2 = s, \ (ab^4 ab^{(p+1)/2})^2 = t, \ b^p = (ab)^3 = u; \ s, \ t, \ u \ \text{central involutions} \rangle,$$

for prime $p \geq 5$. Then $s = t$.

Proof. It is straightforward to check, using matrix methods, that $L/L' \cong C_2$, b having order two in the abelianisation of L. Therefore b has even order in L and u is nontrivial. Now $L/\langle u \rangle$ is a perfect group and is a stem extension of $PSL(2, p)$ with presentation:

$$\langle a, b \mid a^2 = s, \ (ab^4 ab^{(p+1)/2})^2 = t, \ b^p = (ab)^3 = 1; \ s, \ t \ \text{central involutions} \rangle.$$

So, $L/\langle u \rangle$ is either $PSL(2, p)$ or $SL(2, p)$. In fact this group is $SL(2, p)$ since the following generators for $SL(2, p)$ satisfy these relations:

$$a = \begin{pmatrix} 0 & -1 \\ 1 & 0 \end{pmatrix}, \quad b = \begin{pmatrix} 1 & 0 \\ 1 & 1 \end{pmatrix}.$$

We can check that $s = t$ in $L/\langle u \rangle$ using these matrix generators. Alternatively, we can consider the cases $s = 1, t \neq 1$ and $s \neq 1, t = 1$. In the case $s = 1$, since $SL(2, p)$ has only one involution, the central element, a is central and we have a contradiction since

$SL(2, p)$ is not abelian. In the case $t = 1$ the central element would be $ab^4ab^{(p+1)/2}$, and so, factorising by this element would give $PSL(2, p)$ with presentation:

$$\langle a, b \mid a^2 = 1, \ b^p = (ab)^3, \ ab^4ab^{(p+1)/2} = 1 \rangle.$$

The group with this presentation is generated by b and ab so b^p is central and, since $PSL(2, p)$ has trivial centre, we must have $b^p = 1$. Now we have $a^{-1}b^4a = b^{-(p+1)/2}$. Choose λ such that $4\lambda \equiv 1 \pmod{p}$. Then we have $a^{-1}ba = b^{-\lambda(p+1)/2}$ implying that $\langle b \rangle$ is a normal subgroup so again we have a contradiction.

Now we have $s = t$ in $L/\langle u \rangle$. Therefore $s \equiv t \pmod{u}$ in L. So, in L we have either $s = t$ or $t = su$. In the latter case, L would be given by the presentation:

$$\langle a, b \mid a^2 = s, \ (ab^4ab^{(p+1)/2})^2 = su, \ b^p = (ab)^3 = u; \ s, \ u \text{ central involutions} \rangle.$$

Now if we factor out by s we obtain

$$\langle a, b \mid a^2 = 1, \ (ab^4ab^{(p+1)/2})^2 = b^p = (ab)^3 = u; \ u \text{ central involution} \rangle,$$

which is perfect. Since u is non-trivial this group cannot be $PSL(2, p)$. Using reasoning used earlier in the proof, this group cannot $SL(2, p)$ either, since a would be the central involution forcing the group to be abelian. So, we have a contradiction when $t = su$ and we must have $s = t$ in L as required. $\square$

Corollary 2.4. *The following direct products are efficient:* $PSL(2, p) \times SL(2, 8)$, $PSL(2, p) \times SL(2, 16)$, $PSL(2, p) \times SL(2, 32)$, $PSL(2, p) \times SL(2, 64)$, $PSL(2, p) \times PSL(3, 3)$, $PSL(2, p) \times PSU(3, 3)$, $PSL(2, 25) \times SL(2, 8)$, $PSL(2, 25) \times SL(2, 16)$, $PSL(2, 25) \times PSL(3, 3)$, $PSL(2, 25) \times PSU(3, 3)$, $PSL(2, 27) \times SL(2, 8)$, $PSL(2, 27) \times SL(2, 16)$, $PSL(2, 27) \times PSL(3, 3)$, $PSL(2, 27) \times PSU(3, 3)$, *for prime* $p \geq 5$.

Proof. All we need to do is show that $PSL(2, p)$, $PSL(2, 25)$, $PSL(2, 27)$ have presentations satisfying the properties required of G_1 in Theorem 2.1, that $SL(2, 8)$, $SL(2, 16)$, $SL(2, 32)$, $SL(2, 64)$, $PSL(3, 3)$, $PSU(3, 3)$ have presentations satisfying the properties required of G_2, and that the corresponding groups G are perfect. We can use the following presentations:

$$PSL(2, p) = \langle a, b \mid a^2 = b^p = (ab)^3 = (ab^4ab^{(p+1)/2})^2 = 1 \rangle$$
$$PSL(2, 25) = \langle a, b \mid a^2 = b^{13} = (ab)^3 = (ab^3ab^{-4})^2 = 1 \rangle$$
$$PSL(2, 27) = \langle a, b \mid a^2 = b^{13} = (ab)^3 = (ab^3ab^{-3})^2 = 1 \rangle$$
$$SL(2, 8) = \langle x, y \mid x^2 = y^7 = (xy)^3 = (xy)^3(xy^{-3}xy^2xy^{-3})^2 = 1 \rangle$$
$$SL(2, 16) = \langle x, y \mid x^2 = y^{15} = (xy)^3 = xy^3x^{-1}y^{-5}xy^3xy^{10} = 1 \rangle$$
$$SL(2, 32) = \langle x, y \mid x^2 = y^{31} = (xy)^3 = xy^{28}xy^7xy^{-3}x^{-1}y^7 = 1 \rangle$$
$$SL(2, 64) = \langle x, y \mid x^2 = y^{65} = (xy)^3 =$$
$$xy^{-1}xyxy^{-4}xy^5xy^{-5}xy^5xy^{-4}xyxy^{-1}x^{-1}y^4 = 1 \rangle$$

$$PSL(3,3) = \langle x, y \mid x^2 = y^{13} = (xy)^3 =$$
$$(y^4xy^{-1}x)^2y^2xy^{-3}xy^{-3}xy^2xy^{-1}x = 1\rangle$$
$$PSU(3,3) = \langle x, y \mid x^3 = y^7 = (xy)^4 = xyx^{-1}y^{-2}xy^4xy^2 = 1\rangle.$$

It is easy enough to check that the relevant groups are perfect using the usual matrix methods. It is also straightforward to check the $s = t$ property required of G_1 for individual groups by using GAP, [9]. Lemma 2.3 shows that this property holds for $PSL(2, p)$.   □

Remarks. (iii) The presentations given above come from the following sources with slight modifications. The presentation for $PSL(2, p)$ comes from [10], that for $SL(2, 8)$ from [1], those for $SL(2, 16)$, $SL(2, 32)$, $SL(2, 64)$ from [2], and those for $PSL(3, 3)$ and $PSU(3, 3)$ from [6]. Presentations for $PSL(2, 25)$, $PSL(2, 27)$ were derived from those given in [4]. The presentations had to be modified, firstly to get them in the required form and, secondly, to accommodate the perfectness properties required of G and K.

(iv) When constructing the direct product $PSL(2, 27) \times PSL(3, 3)$, we need to modify the presentation for $PSL(2, 27)$ to

$$PSL(2, 27) = \langle a, b \mid a^2 = b^{13} = (ab)^3 = ab^3ab^{-3}a^{-1}b^3ab^{-3} = 1\rangle.$$

(v) When constructing the direct products involving $PSU(3, 3)$, $SL(2, 32)$ and $SL(2, 64)$ we must use the modification of Theorem 2.1 given in Remark (i). In the case of $PSU(3, 3)$ this is required to satisfy the α_i, β_j congruences and with $SL(2, 32)$ and $SL(2, 64)$ it is in order to make G and the central extension of G_2 perfect.

3. Direct Products of Some Groups with Multiplier Cyclic of Order Two

Now we look at the case where both groups in the direct product have cyclic multipliers.

Theorem 3.1. *Let G_1, G_2 be finite perfect groups with trivial centres, both G_1 and G_2 having multiplier C_2. Let G_1, G_2 have presentations of the form*

$$\langle a, b \mid a^{\alpha_1} = b^{\alpha_2} = (ab)^{\alpha_3} = w(a, b) = 1\rangle$$
$$\langle x, y \mid x^{\beta_1} = y^{\beta_2} = (xy)^{\beta_3} = v(x, y) = 1\rangle$$

with the α_i, β_i satisfying the congruences $\beta_3 \equiv \pm 1 \pmod{\alpha_1}$, $\beta_2 \equiv \pm 1 \pmod{\alpha_3}$, $\alpha_2 \equiv \pm 1 \pmod{\beta_1}$, $\alpha_3 \equiv \pm 1 \pmod{\beta_1}$ as before. Let the group G, given by the presentation below, be perfect:

$$\langle a, b, x, y \mid a^{\alpha_1} = (xy)^{\pm 1}((xy)^{(\beta_3 \mp 1)/\alpha_1}a)^{\alpha_1} = y^{\pm 1}(y^{(\beta_2 \mp 1)/\alpha_3}ab)^{\alpha_3} =$$

$$(ab)^{\pm 1}((ab)^{(\alpha_3 \mp 1)/\beta_1} x)^{\beta_1} = b^{\pm 1}(b^{(\alpha_2 \mp 1)/\beta_1} x)^{\beta_1} = w(a, b)v(x, y) = 1 \rangle.$$

Let G_1 be such that in the group presented by

$$\langle a, b \mid a^{\alpha_1} = 1, \ b^{\alpha_2} = (ab)^{\alpha_3} = u, \ w(a, b) = t; \ u, \ t \ central \ involutions \rangle$$

we have $t = 1$ and let G_2 be such that in the group presented by

$$\langle x, y \mid x^{\beta_1} = y^{\beta_2} = u, \ (xy)^{\beta_3} = v(x, y) = 1; \ u \ central \ involution \rangle$$

we have $u = 1$. Then $G \cong G_1 \times G_2$ and so $G \cong G_1 \times G_2$ is efficient.

Proof. Again putting $H = \langle a, b \rangle, K = \langle x, y \rangle$ we get from Lemma 1.1 that $[H, K] = 1$ and that the following relations hold in G:

$$b^{\alpha_2} = (ab)^{\alpha_3} = x^{-\beta_1} = y^{-\beta_2}.$$

Let $D = \langle b^{\alpha_2}, w(a, b) \rangle$ so that, again, $D = H \cap K = Z(G)$. G is perfect and so by [12] we have that D is an epimorphic image of $M(G_1 \times G_2)$. Hence D is either trivial, C_2, or $C_2 \times C_2$. We have that $G/H \cong K/D \cong G_2$ and $G/K \cong H/D \cong G_1$. Now, in H the following relations hold

$$a^{\alpha_1} = 1, \ b^{\alpha_2} = (ab)^{\alpha_3} = u, \ w(a, b) = t,$$

with u, t central involutions. By hypothesis we have $t = 1$ and so, by the sixth relation of G, $v(x, y) = 1$. Now K is given by

$$\langle x, y \mid x^{\beta_1} = y^{\beta_2} = u, \ (xy)^{\beta_3} = 1, \ v(x, y) = 1; \ u \ central \ involution \rangle$$

and so, by hypothesis we have $u = 1$. Hence D is trivial and $G \cong G_1 \times G_2$. □

Corollary 3.2. *The direct product $G_1 \times G_2$ can be presented by*

$$\langle c, d \mid (c^{\pm(\beta_2 \mp 1)+1} d^{(\alpha_2 \mp 1)/\beta_1})^{\alpha_1} =$$
$$(d^{-(\alpha_2 \mp 1)/\beta_1 + 1} c^{\mp \alpha_3})^{\pm 1}((d^{-(\alpha_2 \mp 1)/\beta_1 + 1} c^{\mp \alpha_3})^{(\beta_3 \mp 1)/\alpha_1} c^{\pm(\beta_2 \mp 1)+1} d^{(\alpha_2 \mp 1)/\beta_1})^{\alpha_1} =$$
$$d^{\mp(\beta_2 \mp 1)\pm 1}((c^{\pm(\beta_2 \mp 1)+1})^{(\alpha_3 \mp 1)/\beta_1} d^{-(\alpha_2 \mp 1)/\beta_1 + 1})^{\beta_1} =$$
$$w(c^{\pm(\beta_2 \mp 1)+1} d^{(\alpha_2 \mp 1)/\beta_1}, d^{\mp \beta_1})v(d^{-(\alpha_2 \mp 1)/\beta_1 + 1}, c^{\mp \alpha_3}) = 1 \rangle$$

and so has an efficient presentation on two generators.

Proof. Put $c = y^{(\beta_2 \mp 1)/\alpha_3} ab, \ d = b^{(\alpha_2 \mp 1)/\beta_1} x$ so that $a = c^{\pm(\beta_2 \mp 1)+1} d^{(\alpha_2 \mp 1)/\beta_1}$, $b = d^{\mp \beta_1}, \ x = d^{-(\alpha_2 \mp 1)/\beta_1 + 1}, \ y = c^{\mp \alpha_3}$ and substitute into the four generator presentation given in the theorem. □

Corollary 3.3 (to Theorem 3.1). *The following direct products have efficient presentations: $PSL(2, 25) \times PSL(2, 25)$, $PSL(2, 25) \times PSL(2, 27)$, $PSL(2, 27) \times PSL(2, 27)$, $PSL(2, p) \times PSL(2, 25)$, $PSL(2, p) \times PSL(2, 27)$, for prime $p \geq 5$.*

Proof. The proof is just a matter of checking that the conditions required of G_1 and G_2 in Theorem 3.1 are satisfied by the groups in the direct products above. The presentations for $PSL(2, 25)$ and $PSL(2, 27)$, given earlier, work in these cases also. The fact that $PSL(2, p)$ satisfies the properties required of G_1 is illustrated in Lemma 2.3. The perfectness of the corresponding groups G can easily be checked using matrix methods.   □

4. Efficient Presentations for Some Direct Cubes

Lemma 4.1 (Lemma 3.1 of [3]). *Let G be a group and a, b, c* $\in$ *G satisfy the relations*

$$a(ab^{-1})^2 = 1, \ c^y = (c^k ab^{-1})^6$$

where $y = \pm 1$, k *an integer. Then* $\langle a, b, c \rangle$ *is cyclic and the relations* $b^2 = a^3 = c^{6k-y}$ *hold in G.*

Theorem 4.2. *Let G_1, G_2, G_3 be finite perfect groups, each with trivial centre and multiplier C_2. Let G_1, G_2, G_3 have presentations*

$$\langle a, b \mid a^2 = b^l = (ab)^3 = w_1(a, b) = 1 \rangle$$
$$\langle x, y \mid x^2 = y^m = (xy)^3 = w_2(x, y) = 1 \rangle$$
$$\langle u, v \mid u^2 = v^n = (uv)^3 = w_3(u, v) = 1 \rangle$$

respectively where $l, m, n \equiv \pm 1$ (mod 6). *Let G_1, G_2, G_3 be such that in each of the group presentations below we have $s = t$:*

$$\langle a, b \mid a^2 = b^l = (ab)^3 = s, \ w_1(a, b) = t; \ s, \ t \ central \ involutions \rangle$$
$$\langle x, y \mid x^2 = y^m = (xy)^3 = s, \ w_2(x, y) = t; \ s, \ t \ central \ involutions \rangle$$
$$\langle u, v \mid u^2 = v^n = (uv)^3 = s, \ w_3(u, v) = t; \ s, \ t \ central \ involutions \rangle.$$

Let G be the group given by

$$\langle a, b, x, y, u, v \mid xy(xya^{-1})^2 = v^{\pm 1}(v^{(n\mp 1)/6}xya^{-1})^6 =$$
$$uv(uvx^{-1})^2 = b^{\pm 1}(b^{(l\mp 1)/6}uvx^{-1})^6 = ab(abu^{-1})^2 = y^{\pm 1}(y^{(m\mp 1)/6}abu^{-1})^6 =$$
$$xy(xyu^{-1})^2 = b^{\pm 1}(b^{(l\mp 1)/6}xyu^{-1})^6 = w_1(a, b)^{-1}w_2(x, y)w_3(u, v) = 1 \rangle.$$

Then, if G is perfect we have $G \cong G_1 \times G_2 \times G_3$ and so this direct product is efficient.

Proof. The proof is similar to the proof in [3] where it is proved that $PSL(2, p)^3$, $p \geq 5$ is efficient. Let $H = \langle a, b \rangle$, $K = \langle x, y \rangle$ and $L = \langle u, v \rangle$. By Lemma 4.1 we have $[H, K] = [H, L] = [K, L] = 1$ and the following relations holding in G:

$$v^n = a^2 = (xy)^3 = b^l = x^2 = (uv)^3 = y^m = u^2 = (ab)^3.$$

Let $D = \langle a^2, w_1(a, b), w_2(x, y) \rangle$. Now $G/D \cong G_1 \times G_2 \times G_3$ and so $D = Z(G)$. G is perfect and so, by Lemma 4.1 of [12], D is a homomorphic image of $C_2 \times C_2 \times C_2$.

By hypothesis we have $a^2 = w_1(a, b)$, $x^2 = w_2(x, y)$, $u^2 = w_3(u, v)$ but we also have $a^2 = x^2 = u^2$ and so, by the ninth relation of G, we have $a^2 = w_1(a, b) = x^2 = w_2(x, y) = u^2 = w_3(u, v) = 1$. Hence $D = 1$ and $G \cong G_1 \times G_2 \times G_3$ as required.   □

Corollary 4.3. $PSL(2, 25)^3$ and $PSL(2, 27)^3$ are efficient.

Proof. It is easy to check that, using the presentations given earlier, these groups have the properties required by Theorem 4.2.   □

Remark. (vi) We can also write down an efficient presentation for the direct product of any combination of three of $PSL(2, p)$, $PSL(2, 25)$, $PSL(2, 27)$.

References

[1] C. M. Campbell, M. D. E. Conder and E. F. Robertson, Defining relations for Hurwitz groups, Glasgow Math. J. 36 (1994), 363–370.

[2] C. M. Campbell, T. Kawamata, I. Miyamoto, E. F. Robertson and P. D. Williams, Deficiency zero presentations for certain perfect groups, Proc. Roy. Soc. Edinburgh Ser. A 103 (1986), 63–71.

[3] C. M. Campbell, I. Miyamoto, E. F. Robertson and P. D. Williams, The efficiency of $PSL(2, p)^3$ and other direct products of groups, to appear.

[4] C. M. Campbell, E. F. Robertson and P. D. Williams, On presentations of $PSL(2, p^n)$, J. Austral. Math. Soc. Ser. A 48 (1990), 333–346.

[5] C. M. Campbell, E. F. Robertson and P. D. Williams, Efficient presentations of the groups, $PSL(2, p) \times PSL(2, p)$, p prime, J. London Math. Soc. (2) 41 (1990), 69–77.

[6] J. J. Cannon, J. McKay and K. C. Young, The non-abelian simple groups G, $|G| < 10^5$, presentations, Comm. Algebra 7 (1979), 1397–1406.

[7] P. E. Kenne, Presentations for some direct products of groups, Bull. Austral. Math. Soc. 28 (1983), 131–133.

[8] P. E. Kenne, Minimal group presentations: a computational approach, Ph.D. Thesis, Australian National University, 1991.

[9] M. Schönert et al, GAP 3.3, Lehrstuhl D für Mathematik, RWTH Aachen, 1993.

[10] J. G. Sunday, Presentations of the groups $SL(2, m)$ and $PSL(2, m)$, Canad. J. Math 24 (1972), 1129–1131.

[11] J. W. Wamsley, Minimal presentations for finite groups, Bull. London Math. Soc. 5 (1973), 129–144.

[12] J. Wiegold, The Schur multiplier, Groups-St. Andrews 1981 (C. M. Campbell and E. F. Robertson eds.), London Math. Soc. Lecture Note Ser. 71, Cambridge University Press, Cambridge, 1982, 137–154.

Zeta Function of Finitely Generated Nilpotent Groups

Martin Dörfer and Gerhard Rosenberger

1. Introduction

This paper deals with the Euler product decomposition of the zeta function of a finitely generated nilpotent group. Although the result is well known, see e.g [2], [4] or [5], it is only proved in the case of a finitely generated torsion-free nilpotent group using the theory of profinite completions. Here we give an elementary and straightforward proof which also works if torsion elements occur. In a second section, some examples are mentioned to make use of the theorem above.

Our interest in zeta functions of finitely generated nilpotent groups comes from the following example.

If $G = \mathbb{Z}^2$ then

$$\zeta_G(s) = \zeta(s)\zeta(s-1) = \sum_{n=1}^{\infty} \sigma_1(n)n^{-s}$$

where

$$\sigma_1(n) = \sum_{\substack{d|n \\ d \geq 1}} d,$$

and $\zeta_G(s)$ converges for $\mathrm{Re}(s) > 2$.

$\zeta_G(s)$ satisfies the functional equation

$$R(2 - s) = -R(s)$$

where

$$R(s) = (2\pi)^{-s}\Gamma(s)\zeta_G(s)$$

and hence

$$f(\tau) = -\frac{1}{24} + \sum_{n \geq 1} \sigma_1(n)e^{2\pi i n \tau}$$

defines a modular integral of weight 2 with rational period function

$$q(\tau) = -\frac{1}{4\pi i \tau}.$$

We would like to find other finitely generated nilpotent groups which define automorphic integrals in an analogous manner via suitable functional equations for the corresponding zeta functions and to understand the connections between those finitely generated nilpotent groups and the discrete groups occurring with the automorphic integrals, if there are some.

We are very grateful to Prof. Dr. Derek J. S. Robinson for some helpful discussions about nilpotent groups.

2. Euler Product Decomposition

Definition 1. Let G be a finitely generated group. Then

$$\zeta_G(s) := \sum_{n=1}^{\infty} a(n) n^{-s}$$

with

$$a(n) := |\{H \leq G \mid |G : H| = n\}|$$

is called the zeta function of G. σ_G denotes the abscissa of convergence of the Dirichlet series.

Moreover we define

$$\zeta_G^N(s) := \sum_{n=1}^{\infty} b(n) n^{-s}$$

where

$$b(n) := |\{H \trianglelefteq G \mid |G : H| = n\}|$$

and write σ_G^N for the corresponding abscissa of convergence.

The coefficients $a(n)$ and $b(n)$ are finite since G is finitely generated and thus it contains only finitely many subgroups of given index n.

Definition 2. If G is a finitely generated group and p a prime then

$$\zeta_G^p(s) := \sum_{n=0}^{\infty} a(p^n) p^{-ns} \text{ with } a(p^n) = |\{H \leq G \mid |G : H| = p^n\}|$$

and

$$\zeta_G^{N,P}(s) := \sum_{n=0}^{\infty} b(p^n) p^{-ns} \quad \text{with } b(p^n) = |\{H \trianglelefteq G \mid |G : H| = p^n\}|$$

are the p-Dirichlet series associated with $\zeta_G(s)$ and $\zeta_G^N(s)$ respectively.

For the proof of the main theorem we still need the following:

Definition 3. Let Π be a nonempty subset of the set of primes. Then a positive integer whose prime divisors belong to Π is called a Π-number.

We now show the mentioned result which emphasizes an important property of finitely generated nilpotent groups.

Theorem 1. *If G is a finitely generated nilpotent group then*

$$\zeta_G(s) = \prod_p \zeta_G^P(s) \quad \text{(for $\sigma > \sigma_G$)}$$

and

$$\zeta_G^N(s) = \prod_p \zeta_G^{N,P}(s) \quad \text{(for $\sigma > \sigma_G^N$)}$$

where p runs over all primes.

Proof. a) Proof of the first formula.

Of course we have

$$a(1) = 1 = \prod_p a(p^0)$$

so that it is enough to show

$$a(n) = \prod_{i=1}^{k} a(p_i^{l_i})$$

for an integer $n \geq 2$, with the prime decomposition $n = p_1^{l_1} p_2^{l_2} \cdots p_k^{l_k}$ ($p_1, p_2, \ldots, p_k$ pairwise distinct, $l_1, l_2, \ldots, l_k \in \mathbb{N}$). Without loss of generality we can assume $k \geq 2$. Let H be a subgroup of G with $|G : H| = n < \infty$, $n \geq 2$, $n = p_1^{l_1} p_2^{l_2} \cdots p_k^{l_k}$ as above. If Π is a nonempty subset of $\{p_1, p_2, \ldots, p_k\}$ then we define

$$G_\Pi := \{g \in G \mid g^m \in H \text{ for a } \Pi\text{-number } m \in \mathbb{N}\}$$

and write shortly G_{p_i} instead of $G_{\{p_i\}}$.

1st step: G_Π is a subgroup of G with $H \leq G_\Pi$ (see [6], Th. 3.25).

2nd step: If $\Pi = \{p_{i_1}, p_{i_2}, \ldots, p_{i_r}\}$ then $G_\Pi = G_{p_{i_1}} G_{p_{i_2}} \cdots G_{p_{i_r}}$.

Proof. This is obvious for $r = 1$, so consider the case $r \geq 2$. We have $G_{p_{i_j}} \subset G_\Pi$ ($j \in$ $\{1, 2, \ldots, r\}$) by definition and therefore $G_{p_{i_1}} G_{p_{i_2}} \cdots G_{p_{i_r}} \subset G_\Pi$. Conversely, let $g \in G_\Pi$ with $g^m \in H$ for a Π-number $m = p_{i_1}^{d_1} p_{i_2}^{d_2} \cdots p_{i_r}^{d_r} \in \mathbb{N}$. Put

$$m_j := \frac{m}{p_{i_j}^{d_j}} \quad (j \in \{1, 2, \ldots, r\}).$$

Then there exist integers $c_1, c_2, \ldots, c_r$ with

$$c_1 m_1 + c_2 m_2 + \cdots + c_r m_r = 1$$

so that

$$g = g^{c_1 m_1 + c_2 m_2 + \cdots + c_r m_r} \in G_{p_{i_1}} G_{p_{i_2}} \cdots G_{p_{i_r}}$$

because

$$\left(g^{c_j m_j}\right)^{p_{i_j}^{d_j}} = (g^m)^{c_j} \in H$$

for all $j \in \{1, 2, \ldots, r\}$.

Note that the order of the $G_{p_{i_j}}$ is irrelevant, i.e. $G_{p_{i_j}} G_{p_{i_k}} = G_{p_{i_k}} G_{p_{i_j}}$ for $i_j, i_k \in$ $\{i_1, i_2, \ldots, i_r\}$.

In the following we take $\Pi = \{p_1, p_2, \ldots, p_k\}$.

3rd step: $G_{p_i} \cap \prod_{j \neq i} G_{p_j} = H$ for all $i \in \{1, 2, \ldots, k\}$.

Proof. "$\supset$" is clear while "$\subset$" can be shown thus: Let $g_i \in G_{p_i} \cap \prod_{j \neq i} G_{p_j}$ ($i \in$ $\{1, 2, \ldots, k\}$). $g_i \in G_{p_i}$ implies $g^{p_i^{c_i}} \in H$ for some c_i. Now $g_i \in \prod_{j \neq i} G_{p_j} \overset{2.}{=}$ $G_{\{p_1, \ldots, p_{i-1}, p_{i+1}, \ldots, p_k\}}$ implies $g_i^{\prod_{j \neq i} p_j^{c_j}} \in H$ for some c_j. Then we find integers λ, μ with

$$1 = \lambda p_i^{c_i} + \mu \prod_{j \neq i} p_j^{c_j}$$

and

$$g_i = (g_i^{p_i^{c_i}})^\lambda (g_i^{\prod_{j \neq i} p_j^{c_j}})^\mu \in H.$$

4th step: $G = G_{p_1} G_{p_2} \cdots G_{p_k}$.

Proof. Because of step 2 we already know $G_{p_1} G_{p_2} \cdots G_{p_k} = G_\Pi \subset G$, so we have to show $G \subset G_\Pi$. As a subgroup of the nilpotent group G, H is subnormal in G, i.e., there exist subgroups $H_0, H_1, \ldots, H_e$ of G with

$$H = H_0 \leq H_1 \leq \cdots \leq H_e = G$$

and

$$H_i \trianglelefteq H_{i+1} \quad \text{for } i \in \{0, 1, \ldots, e-1\}.$$

From

$$n = |G : H|$$
$$= |G/H_{e-1}| \cdot |H_{e-1}/H_{e-2}| \cdot \ldots \cdot |H_1/H|$$

we deduce that $g^n \in H$ for arbitrary $g \in G$ and therefore $G \subset G_\Pi$.

5th step: $|G_{p_i} : H| = p_i^{l_i}$ for all $i \in \{1, 2, \ldots, k\}$.

Proof. $G_{p_1} G_{p_2} \cdots G_{p_k} = (G_{p_2} \cdots G_{p_k}) G_{p_1} = G_{p_1}(G_{p_2} \cdots G_{p_k}) \Longrightarrow |G_{p_1} \cdots G_{p_k} : G_{p_2} \cdots G_{p_k}| = |G_{p_1} : ((G_{p_2} \cdots G_{p_k}) \cap G_{p_1})| \overset{3.}{=} |G_{p_1} : H|$. Then observe that

$$p_1^{l_1} \cdot \ldots \cdot p_k^{l_k} = n = |G : H|$$
$$\overset{4.}{=} |G_{p_1} \cdots G_{p_k} : H|$$
$$= |G_{p_1} \cdots G_{p_k} : G_{p_2} \cdots G_{p_k}| \cdot |G_{p_2} \cdots G_{p_k} : H|$$
$$= |G_{p_1} : H| \cdot |G_{p_2} \cdots G_{p_k} : H|$$
$$\vdots$$
$$= |G_{p_1} : H| \cdot \ldots \cdot |G_{p_k} : H|$$

so $|G_{p_i} : H| = p_i^{l_i}$ for all $i \in \{1, 2, \ldots, k\}$.

6th step: Put

$$R_i := \prod_{j \neq i} G_{p_j} \quad (i \in \{1, 2, \ldots, k\}).$$

Then $G \overset{4.}{=} G_{p_i} R_i = R_i G_{p_i}$ so that

$$|G : R_i| = |G_{p_i} : R_i \cap G_{p_i}| \overset{3.}{=} |G_{p_i} : H| \overset{5.}{=} p_i^{l_i}$$

and

$$\bigcap_{i=1}^{k} R_i = H.$$

Note that if subgroups $S_1, S_2, \ldots, S_k$ of G are given satisfying $|G : S_i| = p_i^{l_i}$ for $i \in \{1, 2, \ldots, k\}$ and $\bigcap_{i=1}^{k} S_i = H$ then $S_i = R_i$, hence the R_i are uniquely determined by H.

Thus we have seen that to every subgroup H of G with

$$|G : H| = n = p_1^{l_1} p_2^{l_2} \cdots p_k^{l_k} < \infty$$

there exist uniquely determined subgroups $R_1, R_2, \ldots, R_k$ of G with

$$|G : R_i| = p_i^{l_i} \quad (i \in \{1, 2, \ldots k\}) \quad \text{and} \quad \bigcap_{i=1}^{k} R_i = H.$$

7th step: Conversely, if $R_1 \leq G, R_2 \leq G, \ldots, R_k \leq G$ and

$$|G : R_i| = p_i^{l_i} \quad \text{for } i \in \{1, 2, \ldots, k\}$$

then

$$H := \bigcap_{i=1}^{k} R_i$$

is a subgroup of G and

$$|G : H| = p_1^{l_1} p_2^{l_2} \cdots p_k^{l_k} = n$$

by the Poincaré theorem.

The convergence of the infinite product is clear so that the proof is complete.

b) For the second formula the proof is identical. Note that $H \trianglelefteq G$ if and only if $G_{p_i} \trianglelefteq G$ for all $i \in \{1, 2, \ldots, k\}$.

Examples

The theorem can be widely used to compute the zeta functions of finitely generated nilpotent groups by looking at the corresponding p-Dirichlet series. Two well-known torsion-free examples are the following.

Example 1. $G = \mathbb{Z}^r$, the free abelian group of rank r.
 In this case we get

$$\zeta_{\mathbb{Z}^r}(s) = \zeta_{\mathbb{Z}^r}^N(s) = \zeta(s) \cdot \zeta(s-1) \cdot \ldots \cdot \zeta(s-r+1) \quad (\sigma > r)$$

where ζ is the Riemann ζ-function. There are various proofs for this formula, see e.g. [1], [2], [3], [5], which may also be found in [4] altogether. Another possibility is to calculate the p-Dirichlet series for all primes p by explicitly determining the number of subgroups of $\mathbb{Z}^r$ of p-power index so that finally the Euler product decomposition theorem is applicable.

Example 2.

$$G = \left\{ \begin{pmatrix} 1 & a & b \\ 0 & 1 & c \\ 0 & 0 & 1 \end{pmatrix} \mid a, b, c \in \mathbb{Z} \right\}, \quad \text{the discrete Heisenberg group.}$$

Smith ([5]) computes the Euler factors of the zeta function of G for each prime p and obtains the following results using the Euler product formula:

$$\zeta_G(s) = \frac{\zeta(s)\zeta(s-1)\zeta(2s-2)\zeta(2s-3)}{\zeta(3s-3)} \quad (\sigma > 2)$$

and

$$\zeta_G^N(s) = \zeta(s)\zeta(s-1)\zeta(3s-2) \quad (\sigma > 2).$$

Example 3. Here we study the abelian group $G = \mathbb{Z} \oplus \mathbb{Z}_q$, q a prime. If G is generated by y and z, z of order q, and U a subgroup of G of finite index then we have either

$$U = \langle y^n z^l \rangle \quad (n \in \mathbb{N}, \, l \in \{0, 1, \ldots, q-1\})$$

of index qn in G or

$$U = \langle y^n, z \rangle \quad (n \in \mathbb{N})$$

of index n in G.

Easy counting yields:
$a(p^0) = 1$ for each prime p,
$a(p^\alpha) = 1$ for each prime p, $p \neq q (\alpha \geq 1$ an integer $)$,
$a(q^\alpha) = q + 1 (\alpha \geq 1$ an integer $)$
so that

$$\zeta_G^q(s) = (1 + q^{-s+1})\zeta^q(s) \quad (\sigma > 1)$$

and

$$\zeta_G^p(s) = \zeta^p(s) \quad (\sigma > 1)$$

if $p \neq q$, p a prime. Using the Euler product decomposition theorem we get

$$\zeta_G(s) = \zeta_G^N(s) = (1 + q^{-s+1})\zeta(s) \quad (\sigma > 1).$$

Example 4. $G = \langle x, y, z \mid z \text{ central }, z^q = 1, x^q = 1, xyx^{-1} = yz \rangle$, q a prime, G is nilpotent of class 2. $H := \mathbb{Z} \oplus \mathbb{Z}_q = \langle y, z \mid yz = zy, z^q = 1 \rangle$ may be embedded in G and therefore we can take it as a subgroup of G with $|G : H| = q$.

At first we calculate $\zeta_G^N(s)$. Let $N \lhd G$ with $|G : N| = t < \infty$ and $K := N \cap H$. Then $K \lhd G$ and also $K \lhd N$. We find $|H : K| < \infty$ because $|G : N| < \infty$ and $H \leq G$ imply:

$$|H : K| = |H : N \cap H| \leq |G : N| < \infty.$$

It follows that there are two possible types for K, namely
$$\text{type 1:} \quad K = \langle y^n z^l \rangle \quad (n \in \mathbb{N}, l \in \{0, 1, \dots, q-1\}),$$
$$\text{type 2:} \quad K = \langle Y^n, z \rangle \quad (n \in \mathbb{N}).$$

The condition $K \trianglelefteq G$ yields no restrictions for type 2 but if $K = \langle y^n z^l \rangle$ of type 1 then we have especially

$$x y^n z^l x^{-1} = (xyx^{-1})^n z^l = y^n z^{n+l} \in K$$

so that $z^n = 1$, i.e, $n = qu$ for some $u \in \mathbb{N}$ and thus

$$K = \langle y^{qu} z^l \rangle \quad (u \in \mathbb{N}, l \in \{0, 1, \dots, q-1\}) \text{ with } |H : K| = q^2 u.$$

If K is of type 2 then of course $|H : K| = n$. Moreover $x \in N$ or $x \notin N$, i.e. $|N : K| = q$ or $|N : K| = 1$. Then observe that

$$|G : N| \cdot |N : K| = |G : K| - |G : H| \cdot |H : K|$$

$$\implies |G : N| = \frac{|G : H| \cdot |H : K|}{|N : K|}$$

$$= c \cdot |H : K| = \begin{cases} cq^2 u, & K \text{ of type 1} \\ cn, & K \text{ of type 2} \end{cases}$$

where

$$c = \begin{cases} 1, & x \in N \\ q, & x \notin N. \end{cases}$$

From $yxy^{-1} = xz^{-1}$ we deduce that $x \notin N$ if K is of type 1, i.e. $c = q$. To sum up we have shown that for $N \trianglelefteq G$ of index t we have either

$$(*) \quad t = q^3 u, \quad u \in \mathbb{N} \quad (K \text{ of type 1})$$

or

$$(**) \quad t = cn, \quad n \in \mathbb{N}, \ c = 1 \vee c = q \quad (K \text{ of type 2}).$$

The representation is uniquely determined because the normal subgroup K of G and c are uniquely determined by N.

Conversely if a factorization $t = q^3 u$ of t is given as in $(*)$ then let K be one of the subgroups $\langle y^{qu} z^l \rangle$ with $l \in \{0, 1, \dots, q-1\}$ of H of type 1 and put $N := K$. We have $N \trianglelefteq G$ and

$$q^3 u = \frac{q}{1} \cdot |H : K| = \frac{|G : H| \cdot |H : K|}{|N : K|} = |G : N|.$$

If a factorization $t = cn$ of t is given as in $(**)$ then choose the subgroup $K = \langle y^n, z \rangle$ of H of type 2 and put

$$N := K \cup x^c K \cup \cdots \cup x^{c(d-1)} K \quad (d := \tfrac{q}{c}).$$

It is easy to show that in both cases $(c = 1,\ c = q)$ N is in fact a normal subgroup of G with

$$cn = \frac{q}{d} \cdot |H : K| = \frac{|G : H| \cdot |H : K|}{|N : K|} = |G : N|.$$

Altogether it is proved that the number of normal subgroups of G of finite index $t \in \mathbb{N}$ is equal to the q-fold number of representations of t in the form $t = q^3 u$ with $u \in \mathbb{N}$ plus the number of representations of t in the form $t = cn$ with $n \in \mathbb{N}$ and $c = 1$ or $c = q$.

Now let p be a prime and $\alpha \in \mathbb{N}_0$.

The representation $p^\alpha = q^3 u$ for a $u \in \mathbb{N}$ is only possible if $p = q$, $\alpha \geq 3$ and $u = q^{\alpha-3}$.

The representation $p^\alpha = cn$ for $n \in \mathbb{N}$ and $c = 1$ or $c = q$ is only possible thus:

$$p \neq q : c = 1, n = p^\alpha \ (\alpha \geq 0)$$
$$p = q : c = 1, n = p^0 \ (\alpha = 0)$$
$$c = 1, n = p^\alpha \vee c = p, n = p^{\alpha-1} \ (\alpha \geq 1).$$

So we get the following results:

$b(p^\alpha) = 1$ for each prime p, $p \neq q(\alpha \geq 0$ an integer);
$b(q^0) = 1, b(q^1) = b(q^2) = 2,\ b(q^\alpha) = q + 2\ (\alpha \geq 3$ an integer) if $p = q$.
This yields

$$\zeta_G^{N,p}(s) = \sum_{n=0}^{\infty} 1 \cdot p^{-ns} = \zeta^p(s) \quad (\sigma > 1)$$

if $p \neq q$ and

$$\zeta_G^{N,q}(s) = 1 + 2q^{-s} + 2q^{-2s} + \sum_{n=3}^{\infty}(q+2)q^{-ns}$$
$$= -1 - q - q^{-s+1} - q^{-2s+1} + (q+2)\zeta^q(s)$$
$$= (1 + q^{-s} + q^{-3s+1})\zeta^q(s) \quad (\sigma > 1)$$

so that

$$\zeta_G^N(s) = (1 + q^{-s} + q^{-3s+1})\zeta(s) \quad (\sigma > 1)$$

by the Euler product decomposition theorem. Finally we calculate $\zeta_G(s)$. Let $S \leq G$ with $|G : S| = t < \infty$ and $K := S \cap H$. Then $K \leq G$ and, as above, $K \leq H$ of finite index so that K is either of type 1 or of type 2. We have $x \in S$ or $x \notin S$, i.e. $|S : K| = q$ or $|S : K| = 1$. Moreover $K \trianglelefteq S$ which is clear if K is of type 2 or if K

is of type 1 with $x \notin S$. Now let $x \in S$ and K of type 1. Then

$$K \trianglelefteq S \Longleftrightarrow x(y^n z^l) x^{-1} \in K$$
$$\Longleftrightarrow y^n z^l z^n \in K$$
$$\Longleftrightarrow q|n .$$

This yields

$$|G : S| \cdot |S : K| = |G : K| = |G : H| \cdot |H : K|$$

$$\Longrightarrow |G : S| = \frac{|G : H| \cdot |H : K|}{|S : K|} = c \cdot |H : K|$$

$$= \begin{cases} cqn \text{ with } q|n, & K \text{ of type 1} \\ cn, & K \text{ of type 2} \end{cases}$$

where

$$c = \begin{cases} 1, & x \in S \\ q, & x \notin S. \end{cases}$$

To sum up we have shown that for $S \leq G$ of index t we have either

$$(*) \quad t = cqn, \; n \in \mathbb{N}, \; c = 1 \vee c = q, \; q|cn \quad (K \text{ of type 1})$$

or

$$(**) \quad t = cn, \; n \in \mathbb{N}, \; c = 1 \vee c = q \quad (K \text{ of type 2}).$$

The representation is uniquely determined because the subgroup K of G and c are uniquely determined by S.

Conversely if a factorization $t = cqn$ of t is given as in $(*)$ then let K be one of the subgroups $\langle y^n z^l \rangle$ with $l \in \{0, 1, \ldots , q - 1\}$ of H of type 1 and put

$$S := K \cup x^c K \cup \cdots \cup x^{c(d-1)} K \quad (d := \frac{q}{c}).$$

One checks that $S \leq G$ with

$$cqn = \frac{q}{d} \cdot |H : K| = \frac{|G : H| \cdot |H : K|}{|S : K|} = |G : S|.$$

If a factorization $t = cn$ of t is given as in $(**)$ then choose the subgroup $K = \langle y^n, z \rangle$ of H of type 2 and put

$$S := K \cup x^c K \cup \cdots \cup x^{c(d-1)} K \quad (d := \frac{q}{c}).$$

In both cases $(c = 1, \; c = q)$ S is a subgroup of G with

$$cn = \frac{q}{d} \cdot |H : K| = \frac{|G : H| \cdot |H : K|}{|S : K|} = |G : S|.$$

Altogether it is proved that the number of subgroups of G of finite index $t \in \mathbb{N}$ is equal to the q-fold number of representations of t in the form $t = cqn$ with $n \in \mathbb{N}$, $c = 1$ or $c = q$ and $q|cn$ plus the number of representations of t in the form $t = cn$ with $n \in \mathbb{N}$ and $c = 1$ or $c = q$.

Now let p be a prime and $\alpha \in \mathbb{N}_0$.
The representation $p^\alpha = cqn$ as in $(*)$ is only possible if $p = q$, $\alpha \geq 2$ and we have either

$$c = 1, \; n = q^{\alpha-1}$$

or

$$c = q, \; n = q^{\alpha-2}.$$

The case $p^\alpha = cn$ can be discussed as in the first section of this example.

Thus we get:
$a(p^\alpha) = 1$ for each prime p, $p \neq q$ ($\alpha \geq 0$ an integer),
$a(q^0) = 1, a(q^1) = 2, a(q^\alpha) = 2q + 2$ ($\alpha \geq 2$ an integer) if $p = q$, so that

$$\zeta_G^p(s) = \sum_{n=0}^{\infty} 1 \cdot p^{-ns} = \zeta^p(s) \quad (\sigma > 1)$$

if $p \neq q$,

$$\zeta_G^q(s) = 1 + 2q^{-s} + (2q + 2) \cdot \sum_{n=2}^{\infty} q^{-ns}$$
$$= (2q + 2)\zeta^q(s) - 1 - 2q - 2q^{-s+1}$$
$$= (1 + q^{-s} + 2q^{-2s+1})\zeta^q(s) \quad (\sigma > 1)$$

and finally

$$\zeta_G(s) = (1 + q^{-s} + 2q^{-2s+1})\zeta(s) \quad (\sigma > 1)$$

by the Euler product decomposition theorem.

References

[1] Bushnell, Colin J., and Irving Reiner, Zeta functions of arithmetic orders and Solomon's conjectures, Math. Z. 173 (1980), 135–161.

[2] Grunewald, Fritz J., Daniel Segal and Geoffrey C. Smith, Subgroups of finite index in nilpotent groups, Invent. Math. 93 (1988), 185–223.

[3] Ilani, I., Counting finite index subgroups and the P. Hall enumeration principle, Israel J. Math. 68 (1989), 18–26.

[4] Lubotzky, Alexander, Subgroup Growth, preliminary version No. 1 (June 1993).

[5] Smith, Geoffrey C., Zeta-functions of torsion-free finitely generated nilpotent groups, Ph.D. Thesis, Manchester 1983.

[6] Warfield, Robert B., Nilpotent Groups, Lecture Notes in Math. 513, Springer-Verlag, New York-Heidelberg-Berlin 1976.

Cyclic Presentations and 3-Manifolds

M. J. Dunwoody

1. Introduction

Let F_n be the free group on free generators $x_0, \ldots, x_{n-1}$. Let $\theta : F_n \to F_n$ be the automorphism for which

$$\theta(x_i) = x_{i+1}, \ i = 0, 1, \ldots, n-2; \ \theta(x_{n-1}) = x_0.$$

As in [5], for $w \in F$, define $G_n(w) = F/R$ where R is the normal closure in F of the set $\{w, \theta(w), \ldots, \theta^{n-1}(w)\}$. A group G is said to have a *cyclic presentation* if $G \cong G_n(w)$ for some n and w. Let $A_n(w) = G_n(w)^{ab}$. The *polynomial associated with* the cyclically presented group $G = G_n(w)$ is given by

$$f(t) = \sum_{i=0}^{n-1} a_i t^i,$$

where a_i is the exponent sum of x_i in w, $1 \le i \le n$. It is shown in [5] that $|A_n(w)| = \prod_{\xi^n=1} f(\xi)$ (with the convention that $0 = \infty$), and that $A_n(w)$ is trivial if and only if $f(t)$ is a unit in the ring $\mathbb{Z}[t]/(t^n - 1)$. If $n = 2, 3, 4$ or 6, the only units in this ring are cosets containing elements of the form $\pm t^i$. Clearly if w is a conjugate of x_i or x_i^{-1}, then $G_n(w)$ is trivial. It is quite hard to find other examples for which this occurs. If $n = 2$ or 3, and $w = x_0^{-1} x_1^{-1} x_0 x_1^2$ then $G_n(w)$ is trivial. This is proved in [4]. I do not know if there are any non-trivial cyclic presentations of the trivial group for $n > 3$. I make the following conjecture.

Conjecture. If $G_n(w)$ has one element, then the polynomial associated with the presentation is of the form $\pm t^i$.

Thus the conjecture is true for $n = 1, 2, 3, 4$ or 6.

Possibly the simplest example of a non-trivial unit in $\mathbb{Z}[t]/(t^n - 1)$ is $1 - t + t^2$, which is a unit if $n \equiv \pm 1 \pmod 6$. One might hope to find a counterexample to the conjecture, therefore, for $w = x_0 x_1^{-1} x_2$. However none of the groups $G_n(w)$ is trivial for $n > 1$. In fact $G_5(w) \cong SL(2, 5)$, and if $n \equiv \pm 1 \pmod 6$ then $G_n(w) \cong \langle a, b \mid a^2 = b^3 = (ab)^n \rangle$.

Consider now the general case. Clearly the automorphism θ of F_n induces an automorphism (also denoted θ) of $G_n(w)$. This automorphism has order dividing n and so we can form the group H_n which is the split extension of $G_n(w)$ by a cyclic group of order n. The group $H_n = H_n(v)$ always has a 2-generator, 2-relator presentation of the form

$$\langle t, x \mid t^n = v(t, x) = 1 \rangle,$$

where $v = v(t, x)$ lies in the normal closure of t^n and x. Conversely any group with such a presentation is the split extension of a $G_n(w)$ for some w. Thus if $w = x_0 x_1^{-1} x_2$, then $v = x t^{-1} x^{-1} t^{-1} x t^2$ and

$$H_n(v) = \langle t, x | t^n = x t^{-1} x^{-1} t^{-1} x t^2 = 1 \rangle.$$

In general $G_n(w) = 1$ if and only if $H_n(v)$ is cyclic of order n and $G_n(w)$ is finite if and only if $H_n(v)$ is finite. If $w = x_0 x_1 x_2^{-1}$ then the groups $G_n(w)$ are the Fibonacci groups (see [5]). It proved very difficult to decide which Fibonacci groups were finite although this was eventually achieved. If $a \geq 3$, and $w = x_1^{-1} x_0 x_1 x_0^a$, then the Mennicke group $M(a, a, a) := G_3(w)$ is finite [6]. Thus there are infinitely many cyclic presentations (with w cyclically reduced) of finite groups.

In [7] Neuwirth describes an algorithm for deciding if a group presentation with n generators and n relations corresponds to the spine of a closed compact 3-manifold. To determine which cyclic presentations correspond to the spines of 3-manifolds it is much more efficient to enumerate the spines (or equivalently the Heegaard diagrams) with the necessary cyclic symmetry. This process is described in Section 2. Some of the results obtained are presented in Section 3. The 3-manifolds obtained appear to be cyclic branched covers of knots. I have been unable to prove this, nor have I been able to determine what is the class of knots that arises.

Three manifolds with a different sort of cyclic symmetry have been considered by H. Helling, A. C. Kim and J. Mennicke [2].

2. Heegaard Diagrams with Cyclic Symmetry

Consider graphs Γ with the following properties.

(1) Γ is a plane graph. (It has a specific embedding in the plane.)
(2) Γ is d-regular, i.e. every vertex has degree d.
(3) Γ has no loops, i.e. every edge has 2 vertices. However there may be more than one edge with the same pair of vertices.
(4) Γ has $2n$ vertices.
(5) Γ has an automorphism θ acting as a regular permutation of order n both on the edges and the vertices of Γ.
(6) There is an edge joining two distinct vertices in the same orbit of θ.
(7) There is an edge joining two distinct vertices in the two different orbits of θ.

Let (a, b, c) be a triple of non-negative integers, such that $ab \neq 0$. Let $\tilde{\Gamma}(a, b, c)$ be the graph shown in Figure 1. This is an infinite graph with an automorphism θ such that $\theta(u_n) = u_{n+1}$ and $\theta(v_n) = v_{n+1}$. The labelling indicates the number of edges joining a pair of vertices. Thus there are a edges joining u_1 and u_2. We see that $\tilde{\Gamma}(a, b, c)$ is d-regular where $d = 2a + b + c$. Let $\Gamma_n = \Gamma_n(a, b, c)$ denote the graph obtained from $\tilde{\Gamma}(a, b, c)$ by identifying all edges and vertices in each orbit of θ^n. Thus Γ_n has $2n$ vertices.

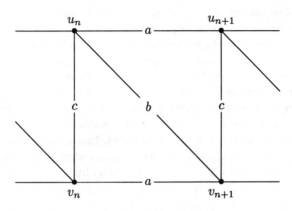

Figure 1

Theorem 1. *A graph Γ satisfying conditions (1)–(7) is isomorphic to $\Gamma_n(a, b, c)$ for some a, b, c.*

Proof. Suppose that Γ satisfies conditions (1)–(7). Let u, v be adjacent vertices in the two orbits of θ. Such vertices exist by (7). Let $u_i = \theta^i(u)$, $v_i = \theta^i(v)$. By (5) and (6) it can be assumed that there is an edge joining u to u_k for some k. By considering the action of θ on this edge, it follows from (5) that k is coprime to n. Replacing θ by θ^k, it can therefore be assumed that $u = u_0, u_1, \ldots, u_{n-1}$ are connected by edges in a cyclic manner. Since the degree of v is d, and u is joined to u_1, it follows that v is joined to v_m for some m. Again m is coprime to n and $v_0, v_m, v_{2m}, \ldots, v_{(n-1)m}$ are joined in a cyclic manner. Since $v_0, v_1, \ldots, v_{n-1}$ are the vertices of a connected subgraph of Γ, it can be assumed that, in the plane embedding of Γ, these vertices all lie in the region of the plane bounded by edges $u_0u_1, u_1u_2, \ldots, u_{n-1}u_0$. Now there is an edge joining u_i and v_i for $i = 0, 1, \ldots, n - 1$. If $m \neq 1$ or $n - 1$, then v_1 (which is adjacent to u_1) lies in the region bounded by edges $v_0v_m, v_mu_m, u_mu_{m-1}, u_{m-1}u_{m-2}, \ldots, u_1u_0, u_0v_0$, whereas v_{m+1} lies outside this region. But v_1 and v_{m+1} are adjacent. Hence $m = 1$ or $n - 1$.

A similar argument shows that apart from u_1 and u_{n-1} the only vertices that can be joined to u_0 by an edge are v_{n-1}, v_0 and v_1. If there is an edge u_0v_1 then there is no edge u_0v_{n-1} or u_1v_0. By relabelling vertices if necessary we can assume that there is no edge u_0v_{n-1} or u_1v_0. We see then that $\Gamma \cong \Gamma_n(a, b, c)$ for some a, b, c. $\qquad\square$

In [7] Neuwirth describes an algorithm for deciding if a group presentation with n generators and n relations corresponds to the spine of a closed compact 3-manifold. Such a spine C is a 2-dimensional cell complex with one 0-cell O, n 1-cells and n 2-cells. As in [7] let R be a regular neighbourhood of O in some triangulation of M. If C corresponds to a cyclic presentation, then $C \cap \delta R$ will be a 1-dimensional cell complex Γ satisfying properties (1)–(5) above. If in addition $C \cap \delta R$ satisfies both (6) and (7) then it must be isomorphic to $\Gamma_n(a, b, c)$ for some a, b, c.

An alternative but equivalent way of obtaining Γ is from a Heegaard diagram for the manifold M. Every closed compact 3-manifold M has a Heegaard decomposition as a union of two handlebodies of genus n. Thus $M = U \cup V$, and $S = U \cap V = \delta U = \delta V$ is a closed surface of genus n. Let $a_1, a_2, \ldots, a_n$ and $b_1, b_2, \ldots, b_n$ be the two families of disjoint closed curves in S which comprise the corresponding Heegaard diagram (see [3]). Thus $a_1, a_2, \ldots, a_n$ span discs in U, and $b_1, b_2, \ldots, b_n$ span discs in V and cutting S along either the a_i's or b_i's produces a 2-sphere with $2n$-holes. After cutting along the a_i's the b_i's which intersect at least one a_i form a collection of arcs joining the boundary circles. The 1-complex Γ is obtained by replacing each boundary circle by a vertex.

The two families of curves in S give two presentations for $\pi_1(M)$, obtained in the following way. Orient the curves in the two families. Passing round the curve b_i in the positive direction, one obtains a word $w_i(a_1, a_2, \ldots, a_n)$ corresponding to the intersections of b_i with the family of curves $a_1, a_2, \ldots, a_n$; write a_j for a crossing by a_j from left to right, a_j^{-1} for a crossing from right to left. These words give a presentation of $\pi_1(M)$ (for arbitrary base point). The **dual presentation** is obtained by passing round the a_i's and recording the intersections with the b_i's.

A typical Heegaard diagram (with cyclic symmetry) is shown in Figure 2. What is shown here is the surface S after cutting along the a_i curves, which is a 2-sphere with $2n$ holes. To recover S and retain cyclic symmetry we need to identify the boundary of the hole labelled i with the boundary of the hole labelled $-(i+s)$ (where $i+s$ is reduced modulo n). Of course the identifications are carried out so that the end point of line segments match up. The end points of lines on the hole labelled i are labelled $1, 2, \ldots, d$ while the endpoints on the hole labelled $-(i + s)$ are labelled $-1, \ldots, -d$. In order to maintain the cyclic symmetry of the presentation, there is a fixed integer r (independent of i), such that the endpoint labelled j is identified with $-j + r$ (or $-j + r - d$ if $-j + r \geq 0$). After carrying out the identifications in this way the line segments will form simple closed curves on the surface S. Since the diagram together with the prescribed identifications is the Heegaard diagram of a 3-manifold, the number of simple closed curves resulting is n. Conversely if we are given such a diagram (given by parameters (a, b, c, n)) and we are given a further pair of parameters r, s which determine how the boundaries of the holes are identified, then the resulting set of simple closed curves determines a Heegaard decomposition if (a) there are n curves, and (b) removing the curves from S does not disconnect S.

We say that the 6-tuple (a, b, c, r, s, n) has property $\mathcal{M}$ if it corresponds to the Heegaard diagram of a 3-manifold. An algorithm for determining which 6-tuples have property $\mathcal{M}$ is now described. Put $d = 2a + b + c$ and let

$$X = \{-d, -d + 1, \ldots, -1, 1, 2, \ldots, d\}.$$

Let α, β be the permutations of X defined as follows:

$$\alpha = (1, d)(2, d - 1) \ldots (a, d - a + 1)(a + 1, -a - c - 1)$$
$$(a + 2, -a - c - 2) \ldots (a + b, -a - c - b)(a + b + 1, -a - 1)$$
$$(a + b + 2, -a - 2) \ldots (a + b + c, -a - c)(-1, -d)$$
$$(-2, -d + 1) \ldots (-a, -d + a - 1),$$

$$j\beta = \begin{cases} -j+r & \text{if } j > 0 \text{ and } -j+r < 0 \text{ or if } j < 0 \text{ and } -j+r \leq d, \\ -j+r-d & \text{otherwise.} \end{cases}$$

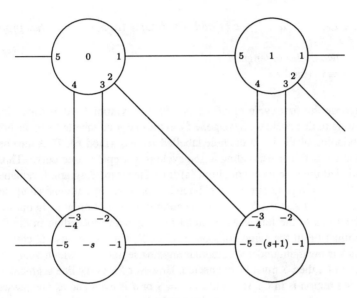

Figure 2

Thus for example in the case illustrated in Figure 2

$$\alpha = (1,5)(2,-4)(3,-2)(4,-3)(-5,-1),$$
$$\beta = (1,-4)(2,-5)(3,-1)(4,-2)(5,-3).$$

The cycles of α, which are all 2-cycles, correspond to the end points of line segments in the Heegaard diagram. Each cycle of β corresponds to a pair of endpoints which is identified in forming the surface S. In this example

$$\alpha\beta = (1,-3,-2,-1,2)(3,4,5,-4,-5).$$

In general (as in this case) if the 6-tuple (a,b,c,r,s,n) has property $\mathcal{M}$, then $\alpha\beta$ is the product of two disjoint cycles of length d. This is because a cycle of $\alpha\beta$ represents the initial points of line segments in an oriented simple closed curve resulting from the identification specified by β. The two cycles represent the same curve with different orientations. Conversely, given values of a, b, c, r we can write down the permutations α and β as above, and calculate the product $\alpha\beta$. If we obtain two disjoint cycle of length $d = 2a+b+c$, then it is possible to determine a simple linear condition satisfied by s and n for the 6-tuple (a,b,c,n,r,s) to have property $\mathcal{M}$. To see how this condition comes about let us consider the example of Figure 2. The entries in the first cycle of $\alpha\beta$ contain one vertex from each line segment of the diagram. Regard these vertices as initial points of the line segments and place arrows on the diagram accordingly. Let p be the number of arrows pointing down the page minus the number of arrows pointing up. In this case

$p = -1$. Let q be the number of arrows pointing from left to right minus the number pointing from right to left, so that in the case considered $q = 3$. In general we have the following result.

Theorem 2. *Let $d = 2a + b + c$ be odd. The 6-tuple (a, b, c, r, s, n) has property $\mathcal{M}$ if and only if*
(i) $\alpha\beta$ *has two cycles of length d, and*
(ii) $ps + q \equiv 0 \pmod{n}$.

Proof. Suppose the first cycle of $\alpha\beta$ starts with the symbol 1. After traversing the path corresponding to this cycle we have gone from the endpoint labelled 1 in the hole labelled 0 to the endpoint labelled 1 in the hole labelled $ps + q \pmod{n}$. Thus the condition (ii) ensures that the path corresponding to the cycle is a simple closed curve. Thus if (i) and (ii) are satisfied then we have condition (a) for a Heegaard diagram. It remains to verify condition (b), i.e. if $\hat{S}$ is the surface with boundary obtained by cutting S along the n curves, then $\hat{S}$ is connected. Let $\theta : S \to S$ be the automorphism of order n. In our diagram any region which lies to the left of an oriented line segment proceeding in the direction of the arrow must lie in the same component C of $\hat{S}$. This is because C contains the top region which is invariant under θ. If there is any line segment for which both neighbouring regions are in C, then $\hat{S}$ must be connected. However for every line segment at least one neighbouring region is in C. Thus either $C = \hat{S}$ or d is even and as one passes around a hole the regions which are in C occur alternately. However we are assuming d is odd, and so $\hat{S}$ must be connected. $\qquad\qquad\qquad\qquad\qquad\qquad\qquad\qquad\qquad\qquad\qquad\qquad\qquad\qquad\qquad\quad\Box$

Theorem 2 is probably true without the restriction that d be odd. However the most interesting cases occur when d is odd.

In the example of Figure 2, we have seen that (i) is satisfied, $p = -1$, $q = 3$. Thus $(1, 1, 2, 3, s, n)$ has property $\mathcal{M}$ if and only if $-s + 3 \equiv 0 \pmod{n}$, i.e. if $s = 3$ and n takes any value. The first cycle of $\alpha\beta$ and the value of s can also be used to calculate the word w of the corresponding cyclic presentation. In our example $w = x_0 x_1^{-1} x_{-2}^{-1} x_{-5}^{-1} x_{-4}$. Thus for example the fourth letter of w is x_{-5}^{-1}. The suffix -5 occurs because if we look at the arrows in our diagram corresponding to the first three terms in the cycle, namely 1, -3, -2, we have $p_3 = -2$ arrows pointing down the page and $q_3 = 1$ arrows pointing from left to right. The suffix of the fourth symbol is $p_3 s + q_3 = -5$. The index -1 occurs because the fourth term in the cycle is negative.

Operating on w by θ^5 to make the smallest suffix 0, we have $\theta^5(w) = x_5 x_6^{-1} x_3^{-1} x_0^{-1} x_1$ and the corresponding polynomial is

$$f(t) = -1 + t - t^3 + t^5 - t^6 = -\frac{(1 - t^{12})(1 - t)}{(1 - t^3)(1 - t^4)},$$

which is the Alexander polynomial of the torus knot $K_{3,4}$.

Let us find a different presentation for $G_n(w)$ in this case. In $G_n(w)$,

$$x_0 x_3 x_6 = x_1 x_5, \quad x_1 x_4 x_7 = x_2 x_6,$$

and so on. Hence

$$z = x_0 x_3 x_6 x_9 = x_1 x_5 x_9 = x_1 x_4 x_7 x_9 = \theta(z).$$

Also

$$z = x_0 x_3 x_6 x_9 = \theta^3(z) = x_3 x_6 x_9 x_{12}.$$

Hence $z x_{12} = x_0 z$ and in general $x_i z = z x_{i+12}$. Thus z^n is central, and if $(n, 12) = 1$, then $G_n(w)$ is generated by x_0 and z. Suppose, for instance $n = 12k + 1$. Let $u = x_0 z^{3k}$ and $v = x_0 z^{4k}$, then

$$u^4 = x_0 z^{3k} x_0 z^{3k} x_0 z^{3k} x_0 z^{3k} = z^{12k} x_{144k} x_{108k} x_{72k} x_{36k}.$$

But $36k \equiv -3 \pmod{n}$. Hence

$$u^4 = z^{12k} x_{-12} x_{-9} x_{-6} x_{-3} = z^{12k+1} = z^n.$$

Similarly $v^3 = z^n$, and in fact $G_{12k+1}(w)$ has presentation

$$G_{12k+1}(w) = \langle x, z \mid (xz^{3k})^4 = (xz^{4k})^3 = z^{12k+1} \rangle = \langle u, z \mid u^4 = (uz^k)^3 = z^{12k+1} \rangle,$$

(putting $u = xz^{3k}$). This group is an infinite perfect group with a triangle group as a homomorphic image.

3. Results

The algorithm described above was implemented on a small computer. All the 4-tuples (a, b, c, r) for which $d = 2a + b + c$ is odd and less than 10 and $0 \le r \le d$, were successively enumerated. The 4-tuples determine the permutations α and β, and so it is possible to work out for which 4-tuples condition (i) of Theorem 2 is satisfied. It is also possible, in each case, to calculate the values of p and q defined above. Condition (ii) of Theorem 2 then provides a simple condition for the 6-tuple (a, b, c, r, s, n) to satisfy condition $\mathcal{M}$. The case when $p = \pm 1$ is of particular interest. In this case, if we put $s = pq$, then (a, b, c, r, s, n) has property $\mathcal{M}$ for every value of $n > 0$. In Table 1 all such 5-tuples (a, b, c, r, s) are listed for $d = 2a + b + c \le 10$. Also shown is the corresponding word w and polynomial $f(t)$. In every case it can be seen that $f(t)$ is the Alexander polynomial of a knot, since, after multiplication by a suitable power of t, $f(t)$ has even degree $2m$ and if $f(t) = \sum_{i=0}^{2m} a_i t^i$, then $f(1) = \pm 1$ and $a_{2m-i} = a_i$, $i = 0, 1, \ldots, m - 1$. It is well known (see [1]) that any polynomial with these properties is the Alexander polynomial of a knot. It seems very plausible that the manifolds listed are all n-fold branched cyclic coverings of a knot. I have not been able to prove this, nor to see what is the class of knots which occur.

M. J. Dunwoody

Table 1

d, a, b, c, r, s	w	$f(t)$
$(3, 1, 0, 1, 1, -2)$	$x_1^{-1}x_2x_0$	$1 - t + t^2$
$(3, 1, 0, 1, 2, 2)$	$x_2^{-1}x_0^{-1}x_1$	$-1 + t - t^2$
$(5, 1, 0, 3, 2, 2)$	$x_4^{-1}x_2^{-1}x_0^{-1}x_1x_3$	$-1 + t - t^2 + t^3 - t^4$
$(5, 1, 0, 3, 3, -2)$	$x_1^{-1}x_3^{-1}x_4x_2x_0$	$1 - t + t^2 - t^3 + t^4$
$(5, 1, 1, 2, 3, 3)$	$x_6^{-1}x_3^{-1}x_0^{-1}x_1x_5$	$-1 + t - t^3 + t^5 - t^6$
$(5, 2, 0, 1, 1, 0)$	$x_1^{-1}x_2x_1^{-2}x_0$	$1 - 3t + t^2$
$(5, 2, 0, 1, 2, -4)$	$x_1^{-1}x_2x_3^{-1}x_4x_0$	$1 - t + t^2 - t^3 + t^4$
$(5, 2, 0, 1, 3, 4)$	$x_4^{-1}x_0^{-1}x_1x_2^{-1}x_3$	$-1 + t - t^2 + t^3 - t^4$
$(5, 2, 0, 1, 4, 0)$	$x_2^{-1}x_1^2x_0^{-1}x_1$	$-1 + 3t - t^2$
$(7, 1, 0, 5, 2, 2)$	$x_6^{-1}x_4^{-1}x_2^{-1}x_0^{-1}x_1x_3x_5$	$-1 + t - t^2 + t^3 - t^4$ $+ t^5 - t^6$
$(7, 1, 0, 5, 5, -2)$	$x_1^{-1}x_3^{-1}x_5^{-1}x_6x_4x_2x_0$	$1 - t + t^2 - t^3 + t^4 - t^5 + t^6$
$(7, 1, 1, 4, 2, -3)$	$x_1^{-1}x_4^{-1}x_7^{-1}x_8x_5x_3x_0$	$1 - t + t^3 - t^4 + t^5 - t^7 + t^8$
$(7, 1, 2, 3, 4, 4)$	$x_{12}^{-1}x_8^{-1}x_4^{-1}x_0^{-1}x_1x_6x_{11}$	$-1 + t - t^4 + t^6 - t^8$ $+ t^{11} - t^{12}$
$(7, 2, 0, 3, 3, 0)$	$x_2^{-2}x_1x_0^{-2}x_1^2$	$-2 + 3t - 2t^2$
$(7, 2, 0, 3, 4, 0)$	$x_1^{-2}x_2^2x_1^{-1}x_0^2$	$2 - 3t + 2t^2$
$(7, 2, 1, 2, 4, 5)$	$x_8^{-1}x_3^{-1}x_4x_5^{-1}x_0^{-1}x_1x_7$	$-1 + t - t^3 + t^4 - t^5$ $+ t^7 - t^8$
$(7, 3, 0, 1, 1, -2)$	$x_1^{-1}x_2x_1^{-1}x_2x_0x_1^{-1}x_0$	$2 - 3t + 2t^2$
$(7, 3, 0, 1, 2, -2)$	$x_1^{-1}x_2x_0x_1^{-1}x_2x_1^{-1}x_0$	$2 - 3t + 2t^2$
$(7, 3, 0, 1, 3, -6)$	$x_1^{-1}x_2x_3^{-1}x_4x_5^{-1}x_6x_0$	$1 - t + t^2 - t^3 + t^4 - t^5 + t^6$
$(7, 3, 0, 1, 4, 6)$	$x_6^{-1}x_0^{-1}x_1x_2^{-1}x_3x_4^{-1}x_5$	$-1 + t - t^2 + t^3 - t^4$ $+ t^5 - t^6$
$(7, 3, 0, 1, 5, 2)$	$x_2^{-1}x_1x_0^{-1}x_1x_2^{-1}x_0^{-1}x_1$	$-2 + 3t - 2t^2$
$(7, 3, 0, 1, 6, 2)$	$x_2^{-1}x_1x_2^{-1}x_0^{-1}x_1x_0^{-1}x_1$	$-2 + 3t - 2t^2$
$(9, 1, 0, 7, 2, 2)$	$x_8^{-1}x_6^{-1}x_4^{-1}x_2^{-1}x_0^{-1}x_1x_3x_5x_7$	$-1 + t - t^2 + t^3 - t^4 + t^5$ $- t^6 + t^7 - t^8$
$(9, 1, 0, 7, 7, -2)$	$x_1^{-1}x_3^{-1}x_5^{-1}x_7^{-1}x_8x_6x_4x_2x_0$	$1 - t + t^2 - t^3 + t^4 - t^5 + t^6$ $- t^7 + t^8$
$(9, 1, 1, 6, 3, 3)$	$x_{12}^{-1}x_9^{-1}x_6^{-1}x_3^{-1}x_0^{-1}x_1x_4x_8x_{11}$	$-1 + t - t^3 + t^4 - t^6 + t^8 - t^9$ $+ t^{11} - t^{12}$
$(9, 1, 3, 4, 2, -3)$	$x_1^{-1}x_4^{-1}x_6^{-1}x_9^{-1}x_{10}x_7x_5x_3x_0$	$1 - t + t^3 - t^4 + t^5 - t^6 + t^7$ $- t^9 + t^{10}$
$(9, 1, 3, 4, 5, 5)$	$x_{20}^{-1}x_{15}^{-1}x_{10}^{-1}x_5^{-1}x_0^{-1}x_1x_7x_{13}x_{19}$	$-1 + t - t^5 + t^7 - t^{10} + t^{13}$ $- t^{15} + t^{19} - t^{20}$
$(9, 2, 0, 5, 4, 0)$	$x_2^{-2}x_1^3x_0^{-2}x_1^2$	$-2 + 5t - 2t^2$
$(9, 2, 0, 5, 5, 0)$	$x_1^{-2}x_2^2x_1^{-3}x_0^2$	$2 - 5t + 2t^2$

$(9, 2, 1, 4, 3, 3)$	$x_{10}^{-1}x_7^{-1}x_3^{-1}x_0^{-1}x_1x_4x_5^{-1}x_6x_9$	$-1 + t - t^3 + t^4 - t^5 + t^6 - t^7$ $+t^9 - t^{10}$
$(9, 2, 1, 4, 5, -5)$	$x_1^{-1}x_6^{-1}x_7x_3x_4^{-1}x_9^{-1}x_{10}x_5x_0$	$1 - t + t^3 - t^4 + t^5 - t^6 + t^7$ $-t^9 + t^{10}$
$(9, 2, 2, 3, 4, -2)$	$x_1^{-1}x_3^{-1}x_4x_2x_1^{-1}x_2^{-1}x_3^{-1}x_2x_0$	$1 - 2t + t^2 - 2t^3 + t^4$
$(9, 2, 2, 3, 5, 6)$	$x_{12}^{-1}x_6^{-1}x_0^{-1}x_1x_8x_9^{-1}x_3^{-1}x_4x_{11}$	$-1 + t - t^3 + t^4 - t^6 + t^8$ $-t^9 + t^{11} - t^{12}$
$(9, 3, 0, 3, 2, -2)$	$x_3^{-1}x_4x_2x_0x_1^{-1}x_2x_1^{-1}x_3^{-1}x_2$	$1 - 2t + 3t^2 - 2t^3 + t^4$
$(9, 3, 0, 3, 4, -2)$	$x_2^{-1}x_4^{-1}x_3x_2^{-1}x_1x_0^{-1}x_2^{-1}x_3x_1$	$-1 + 2t - 3t^2 + 2t^3 - t^4$
$(9, 3, 0, 3, 5, 2)$	$x_3^{-1}x_1^{-1}x_2x_4x_3^{-1}x_2x_1^{-1}x_0x_2$	$1 - 2t + 3t^2 - 2t^3 + t^4$
$(9, 3, 0, 3, 7, 2)$	$x_2^{-1}x_1x_3x_2^{-1}x_3x_4^{-1}x_2^{-1}x_0^{-1}x_1$	$-1 + 2t - 3t^2 + 2t^3 - t^4$
$(9, 3, 1, 2, 5, 7)$	$x_{10}^{-1}x_3x_4x_5^{-1}x_6x_7^{-1}x_0^{-1}x_1x_9$	$-1 + t - t^3 + t^4 - t^5 + t^6 - t^7$ $+t^9 - t^{10}$
$(9, 3, 1, 2, 7, -1)$	$x_4^{-1}x_3x_2x_1x_0^{-1}x_1x_2^{-2}x_3$	$-1 + 2t - t^2 + 2t^3 - t^4$
$(9, 4, 0, 1, 1, 0)$	$x_1^{-1}x_2x_1^{-1}x_2x_1^{-2}x_0x_1^{-1}x_0$	$2 - 5t + 2t^2$
$(9, 4, 0, 1, 2, 0)$	$x_1^{-1}x_2x_1^{-1}x_0x_1^{-1}x_2x_1^{-2}x_0$	$2 - 5t + 2t^2$
$(9, 4, 0, 1, 4, -8)$	$x_1^{-1}x_2x_3^{-1}x_4x_5^{-1}x_6x_7^{-1}x_8x_0$	$1 - t + t^2 - t^3 - t^4 - t^5 + t^6$ $-t^7 + t^8$
$(9, 4, 0, 1, 5, 8)$	$x_8^{-1}x_0^{-1}x_1x_2^{-1}x_3x_4^{-1}x_5x_6^{-1}x_7$	$-1 + t - t^2 + t^3 - t^4 + t^5 - t^6$ $+t^7 - t^8$
$(9, 4, 0, 1, 7, 0)$	$x_2^{-1}x_1^2x_0^{-1}x_1x_2^{-1}x_1x_0^{-1}x_1$	$-2 + 5t - 2t^2$
$(9, 4, 0, 1, 8, 0)$	$x_2^{-1}x_1x_2^{-1}x_1^2x_0^{-1}x_1x_0^{-1}x_1$	$-2 + 5t - 2t^2$

References

[1] Burde, G., Zieschang, H., Knots, de Gruyter Stud. Math. 5, Walter de Gruyter, Berlin-New York 1985.

[2] Helling, H., Kim, A. C., Mennicke, J., A geometric study of Fibonacci groups, preprint, University of Bielefeld.

[3] Hempel, J., 3-Manifolds, Annals of Math. Studies 86, Princeton University Press, 1976.

[4] Higman, G., A finitely generated infinite simple group, J. London Math. Soc. 26 (1951), 59–61.

[5] Johnson, D., Topics in the Theory of Group Presentations, London Math. Soc. Lecture Note Series 42, Cambridge University Press, 1980.

[6] Mennicke J., Einige endliche Gruppen mit drei Erzeugenden und drei Relationen, Arch. Math. 10 (1959), 409–418.

[7] Neuwirth, L. An algorithm for the construction of 3-manifolds from 2-complexes, Proc. Camb. Phil. Soc. 64 (1968), 603–613.

Combing Lattices in Semisimple Lie Groups

Benson Farb

Abstract. Let Γ be an irreducible lattice in a semisimple linear Lie group $G \neq SO(n, 1)$. Then Γ is bicombable if and only if Γ is uniform[1]. If $\mathrm{rank}(G) = 1$ then Γ is automatic if and only if Γ is uniform.

1991 Mathematics Subject Classification: 22E40, 20F32

1. Introduction

The connection between the curvature of a manifold and combinatorial properties of its fundamental group goes back to the work of Dehn at the start of this century [De]). Today, notions of curvature in groups are central to the theory of infinite discrete groups. Many manifestations of negative curvature are captured by Gromov's theory of hyperbolic groups [Gr], [GdlH]. The main candidate for the property of nonpositive curvature in groups has been Thurston's (bi)combable, or "semi-hyperbolic" groups; these groups have been studied extensively by Alonso–Bridson and others [AB].

It has been an open problem to determine precisely which lattices in Lie groups are (bi)combable or (bi)automatic (for precise definitions, see below). Thurston–Epstein have shown [E et al.] that $SL_n(\mathbb{Z})$ is not combable for $n > 2$. Using completely different techniques, Gersten–Short have shown [GS], [Sh] that $SL_2(\mathcal{O})$ is not bicombable, where $\mathcal{O} = \mathcal{O}(\sqrt{d})$ is the ring of algebraic integers in the real quadratic number field $\mathbb{Q}(\sqrt{d})$. The groups $SL_2(\mathcal{O})$ and $SL_n(\mathbb{Z})$ are nonuniform, irreducible lattices in semisimple linear Lie groups of (real) rank 2 and $n - 1$, respectively. In this paper we give a complete solution to the above problem for irreducible lattices in semisimple linear Lie groups. The main result is the following:

Theorem 1.1. *Let Γ be an irreducible lattice in a semisimple linear Lie group $G \neq SO(n, 1)$. Then Γ is bicombable if and only if Γ is uniform.*

Remark. Our results easily extend to irreducible lattices in (not necessarily linear) semisimple Lie groups with finitely many connected components and finite center, as

---

1   Recall that a lattice in a Lie group is ***uniform*** if it is cocompact.

the kernel of the adjoint representation of G is the center of G (also see Lemma 2.1 below). Further, by results in [Ra], one can easily deal with reducible lattices in these semisimple Lie groups.

The proof of Theorem 1.1 falls into two cases, depending on rank:

Theorem 1.2. *Let Γ be a lattice in a rank one Lie group $G \neq SO(n, 1)$. Then the following are equivalent:*
(1) Γ *is word-hyperbolic.*
(2) Γ *is (bi)combable.*
(3) Γ *is (bi)automatic.*
(4) Γ *is uniform.*

It should be noted that every lattice in $SO(n, 1)$ is biautomatic. This follows from Epstein's result that geometrically finite hyperbolic groups are biautomatic [E et al.], and from the fact that lattices in $SO(n, 1)$ are geometrically finite [Bo].

The fact that (3) and (4) are equivalent in Theorem 1.2 can be used to give an algorithm to determine when a lattice in a rank 1 Lie group is uniform.

Theorem 1.3. *Let Γ be an irreducible lattice in a semisimple linear Lie group G with* rank$(G) \geq 2$. *Then Γ is bicombable if and only if Γ is uniform.*

Note that the lattices $SL_2(\mathcal{O})$ and $SL_n(\mathbb{Z})$ are covered by Theorem 1.3. The essential ingredient in the proof of Theorem 1.3 is the recent proof by Lubotzky–Mozes–Raghunathan [LMR] of the Kazhdan Conjecture, which provides so-called "U-elements" for nonuniform, irreducible lattices in semisimple linear groups of rank ≥ 2. By definition, U-elements generate exponentially distorted (in the lattice) cyclic subgroups. On the other hand, by showing that every element of infinite order in a bicombable group has positive translation number, one sees that cyclic subgroups of bicombable groups embed quasi-isometrically, hence have linear distortion.

I am extremely grateful to Jim Lewis and Amie Wilkinson for many useful discussions, and to Steve Gersten for his corrections to and comments on an earlier version of this paper.

2. Combings and Bicombings

In this section we review the definition of (bi)combings and (bi)automatic structures, and recall a few basic facts which we shall need in the proof of the main theorem.

Let Γ be a finitely generated group with generating set X. Let $\mathcal{A} = X \cup X^{-1}$, and let $\mathcal{A}^*$ denote the free monoid on $\mathcal{A}$. For a word $w \in \mathcal{A}^*$, we denote by $\overline{w}$ the image of w under the natural map $\mathcal{A}^* \to \Gamma$. For $g \in \Gamma$, we let $||g||_\Gamma = d_\Gamma(1, g)$, where d_Γ denotes the word metric in Γ (with respect to X). We identify Γ with its Cayley graph (with respect to X), and think of each word $w \in \mathcal{A}^*$ as a path $w : [0, \infty) \to \Gamma$ which is eventually constant.

The group Γ is *word-hyperbolic* if there exists $\delta > 0$ so that geodesic triangles in Γ are uniformly δ-thin; that is, the δ-neighborhood of any two sides of a geodesic triangle contains the third side. For an introduction to word-hyperbolic groups, see [GdlH].

A *combing* is a section $\sigma : \Gamma \to \mathcal{A}^*$ of the natural map $\mathcal{A}^* \to \Gamma$ so that

$\sigma(g)$ is a quasi-geodesic in Γ for all $g \in \Gamma$; and

There is a constant $K > 0$ so that $d_\Gamma(\sigma(g)(t), \sigma(g \cdot x)(t)) \leq K$ for all $g \in \Gamma, x \in X, t > 0$.

In other words, a combing on Γ is a choice of paths from $1 \in \Gamma$ to each element of Γ, so that paths which end close stay uniformly close together. A *bicombing* is a combing $\sigma : \Gamma \to \mathcal{A}^*$ which is "almost-equivariant" in the following sense: there exists $K' > 0$ so that

$$d_\Gamma(x \cdot \sigma(g)(t), \sigma(x \cdot g)(t)) \leq K'$$

for all $g \in \Gamma, x \in X, t > 0$. Note that the $x \cdot \sigma(g)$ refers to the action (on the left) of Γ on the set of paths in its Cayley graph. The group Γ is said to be *(bi)combable* if there exists a (bi)combing on Γ. The group Γ is *(bi)automatic* if there is a (bi)combing $\sigma : \Gamma \to \mathcal{A}^*$ such that $\sigma(\Gamma)$ is a regular language.

(Bi)combable and (bi)automatic groups are among the central examples studied in geometric group theory. We refer the reader to [E et al.], [Fa1], [Sh] for background and elementary properties of these groups.

Warning: Different authors mean different things by "(bi)combing"; in particular what we call "combing" certain others call "bounded width combing by quasi-geodesics". What we call a bicombable group, Alonso–Bridson call a "semi-hyperbolic group" [AB].

Note the following list of implications:

$$
\begin{array}{ccc}
\text{biautomatic} & \Longrightarrow & \text{automatic} \\
\Downarrow & & \Downarrow \\
\text{bicombable} & \Longrightarrow & \text{combable}
\end{array}
$$

It is not known whether or not any of the arrows is reversible, although it is widely believed that there are combable groups which are not automatic. Word-hyperbolic groups are biautomatic [E et al.]. Groups which act cocompactly and properly by isometries on a $CAT(0)$ space (e.g., a complete, simply connected manifold of non-positive curvature) are bicombable [AB]; indeed such groups were the motivation behind the definition of bicombable.

We will need the following property of bicombable groups, which is proved in [AB]:

Lemma 2.1. *Let* $1 \to A \to G \to C \to 1$ *be a short exact sequence of finitely generated groups. If* A *is finite and* C *is bicombable then* G *is bicombable.*

Definition. Let (X, d) and (X', d') be metric spaces with metrics d, d'. A K-quasi-isometry is a (not necessarily continuous) map $f : X \to X'$ such that every point of X' lies in a K-neighborhood of the image of f, and so that

$$\frac{1}{K} d(x, y) - K \leq d'(f(x), f(y)) \leq K d(x, y) + K$$

for all $x, y \in X$.

It is a fundamental observation (made by Svarc and Milnor [Mi]) that if M is a closed Riemannian manifold and $m \in M$, then the map $\gamma \mapsto \gamma \cdot m$ is a quasi-isometry from $\pi_1(M)$ to the universal cover $\tilde{M}$. Milnor's observation holds more generally: let X be a proper geodesic metric space and let Γ be a discrete group of isometries acting properly discontinuously on X. If $\Gamma \backslash X$ is compact, then Γ is finitely generated and is quasi-isometric to X.

A path $\alpha : [0, p] \to X$ is a *K-quasi-geodesic* if it is a K-quasi-isometric embedding of $[0, p]$ into X.

3. Quasiconvexity and Translation Numbers

All of the results in this section are due to Gersten and Short, and can be found in [GS] and [Sh]. We include them here for completeness, as they will be used in the proofs of Theorems 1.2 and 1.3.

Let $\sigma : \Gamma \to \mathcal{A}^*$ be a combing, and let $L = \sigma(\Gamma)$. A subgroup N of Γ is *L-quasiconvex* if there exists a $C > 0$ so that for all $g \in N$, $\sigma(g)$ lies in a C-neighborhood of N in Γ. We will need the following:

Theorem 3.1. *If* $\sigma : \Gamma \to L$ *is a (bi)combing, and if* N *is an* L-*quasiconvex subgroup of* Γ, *then* N *is (bi)combable. Furthermore, if this (bi)combing is the (bi)combing of a (bi)automatic structure, then* N *is (bi)automatic.*

The *translation number* of an element $g \in \Gamma$ is defined to be

$$\tau(g) = \liminf_{n \to \infty} \frac{\|g^n\|_\Gamma}{n}.$$

Although τ depends on the choice of generators for Γ, the vanishing of $\tau(g)$ for a given element $g \in \Gamma$ does not depend on this choice, since changing generators gives a Lipschitz equivalent metric on Γ. We will need the following two facts:

Proposition 3.2. *If* Γ *is bicombable and* $g \in \Gamma$ *has infinite order then* $\tau(g) > 0$.

Proposition 3.2 is proved by showing that the center of the centralizer of g in Γ is quasiconvex in Γ; note that this subgroup is virtually free abelian. It is this first fact that uses the "almost-equivariance" of the bicombing. It is then shown that for any quasiconvex subgroup N of Γ, an element $g \in N$ has positive translation number in N if and only if g has positive translation number in Γ. The argument concludes by noting that every nontrivial element of a free abelian group has nonzero translation number. For details, see [Sh].

Proposition 3.3. *Let* Γ *be a finitely generated group, and let* $g \in \Gamma$ *have positive translation number. Then* $\|g^n\|_\Gamma = O(n)$.

Proof. Clearly $\|g^n\|_\Gamma \le \|g\|_\Gamma \cdot n$. For the reverse inequality, note that since $\tau(g)$ is a lim inf, there exists $N > 0$ so that $\|g^n\|_\Gamma / n \ge \tau(g)/2$ for all $n \ge N$; that is, $\|g^n\|_\Gamma \ge (\tau(g)/2) \cdot n$ for $n \ge N$, and we are done since $\tau(g) > 0$. □

4. Proofs

In this section we present proofs of Theorem 1.2 and Theorem 1.3, from which Theorem 1.1 immediately follows.

Our proof of Theorem 1.2 requires the following proposition concerning the geometry of what has come to be known as "the neutered space".

Proposition 4.1. *Let* H *be a complete, simply connected Riemannian manifold with sectional curvatures satisfying* $-\infty < -\delta^2 \le K(H) \le -1$, *where* $\delta \ge 1$. *Let* Γ *be a discrete group of isometries of* H, *and let* X *be the result of removing the interiors of a* Γ-*invariant set of closed disjoint horoballs from* H. *Give* X *the path metric* d_X. *Then* d_X *is at most exponentially distorted with respect to* d_H; *namely,*

$$d_H(x, y) \le d_X(x, y) \le \sinh(\delta \cdot d_H(x, y))$$

for all $x, y \in X$.

We will only apply Proposition 4.1 to globally symmetric spaces H, so that the metric is known explicitly and the proof is a direct computation. We provide a proof in the more general case here since this proposition is basic to the study of fundamental groups of finite volume, (not necessarily symmetric) negatively curved manifolds.

Proof. The first inequality is clear since any path in X is a path in H. To prove the second inequality, let $x, y \in X$ be given, and let α be the H-geodesic from x to y. To bound the X-distance from x to y, do the following replacement: if α penetrates a horoball S_i at $\alpha(t_i)$ and exits at $\alpha(s_i)$, replace the subsegment $\alpha([t_i, s_i])$ by a geodesic

α_i' on the horosphere from $\alpha(t_i)$ to $\alpha(s_i)$. Do this for each horoball that α penetrates, and call the resulting path α' (see Figure 1).

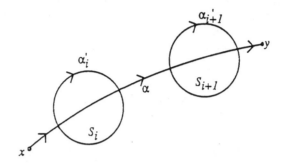

Figure 1. A path α' in X of length $\sinh(\delta \cdot d_H(x, y))$ between $x, y \in X$.

Let β be the length of that part of α (and of α') which lies outside every horoball. Now

$$\sinh(x + y) = \sinh(x)\cosh(y) + \cosh(x)\sinh(y)$$
$$\geq \sinh(x) + \sinh(y)$$

as $\cosh(x) \geq 1$ for all $x \geq 0$. Furthermore, if $h(p, q)$ denotes the distance between two points p, q on a horosphere, then

$$h(p, q) \leq \frac{2}{\delta} \sinh(\frac{\delta}{2} \cdot d_H(p, q))$$

for any such p, q ([HI], Theorem 4.6).

From these two facts it follows that

$$d_X(x, y) \leq \ell(\alpha')$$
$$= \beta + \sum \ell(\alpha_i')$$
$$\leq \beta + \sum \frac{2}{\delta} \sinh(\frac{\delta}{2} \cdot \ell(\alpha_i))$$
$$\leq \sinh(\beta) + \frac{2}{\delta} \sinh(\frac{\delta}{2} \cdot \sum \ell(\alpha_i))$$
$$\leq 2 \sinh(\beta + \frac{\delta}{2} \cdot \sum \ell(\alpha_i))$$
$$\leq 2 \sinh(\frac{\delta}{2} \cdot d_H(x, y))$$
$$\leq \sinh(\delta \cdot d_H(x, y))$$

since $\delta \geq 1$, completing the proof. $\square$

Proof of Theorem 1.2. Let Γ be a lattice in a rank 1 Lie group $G \neq SO(n, 1)$; in other words $G = SU(n, 1)$, $SP(n, 1)$, or F_4^{-20}. We will show that a uniform lattice in G is word-hyperbolic and that a nonuniform lattice in G is not combable. By the discussion in Section 2, this suffices to prove the theorem.

By Selberg's Lemma, Γ contains a torsion-free subgroup Γ' of finite index. Since Γ is word-hyperbolic (resp. combable) iff Γ' is word-hyperbolic (resp. combable) [GdlH], [E et al.], we assume without loss of generality that Γ is torsion free. Then Γ is the fundamental group of a finite volume Riemannian manifold M whose sectional curvatures are pinched between -1 and -4; M is the quotient of an exotic hyperbolic space (the symmetric space associated to G) by Γ acting freely as a discrete group of isometries. If Γ is uniform then M is a compact, negatively curved manifold. It follows that Γ is word-hyperbolic [GdlH].

Now suppose that Γ is a nonuniform lattice in G, so that M is a finite volume, (pinched) negatively curved, locally symmetric manifold with cusps; the universal cover $\widetilde{M}$ is either complex or quaternionic hyperbolic space or the hyperbolic Cayley plane, according to whether $G = SU(n, 1)$, $SP(n, 1)$, or F_4^{-20}. Now Γ acts freely by isometries on $\widetilde{M}$ with finitely many orbits of parabolic fixed points. Choose a Γ-invariant set of disjoint horoballs centered on the parabolic fixed points; these horoballs can be thought of as lifts of the cusps of M. By shrinking these horoballs sufficiently (i.e., looking at lifts of the cusps further and further out), we may assume that any two translates of any of the horoballs are a distance of at least 1 from each other. Let X be the space formed by deleting the interiors of the Γ-orbits of these horoballs, and give X the path metric. This makes each component of the boundary of X a totally geodesic horosphere. Choose some component S of the boundary of X.

Now Γ acts freely and cocompactly by isometries on X, there is a maximal parabolic subgroup N of Γ which preserves S, and N acts on S cocompactly. Picking a basepoint $s \in S$, we see that the map $f : \Gamma \to X$ defined by $f(\gamma) = \gamma \cdot s$ is a K-quasi-isometry for some $K > 0$, and that the restriction of f to N is a quasi-isometry of N and S. Let $\sigma : \Gamma \to \mathcal{A}^*$ be a combing of Γ, and let $L = \sigma(\Gamma)$. Now there exists $C > 0$ so that for all $g \in N$, $\sigma(g)$ is a C-quasi-geodesic in Γ which begins and ends in N. Hence $f \circ \sigma(g)$ is a KC-quasigeodesic in X which begins and ends in S.

We now show that there exists a $K' > 0$ so that $f \circ \sigma(g)$ lies in a K'-neighborhood of S; this argument is similar to the familiar proof that quasigeodesics track geodesics in a simply connected manifold of (pinched) negative curvature. It will then follow that $\sigma(g)$ lies in a KK'-neighborhood of N in Γ, so that N is L-quasiconvex in Γ.

So choose a number P so big that $\cosh(2P) > KC$. Suppose that $\alpha : [0, 1] \to X$ is a segment of $f \circ \sigma(g)$ which lies completely outside a P-neighborhood of S, and let ℓ be the length of α. Let x be the point at which α exits the P-neighborhood of S, and let y be the point at which α returns to the P-neighborhood of S. We bound ℓ universally.

Let x' (resp. y') be the point in S which is the image of x (resp. y) under orthogonal projection onto S (see Figure 2).

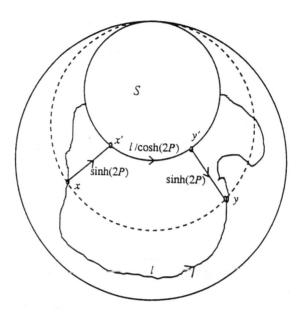

Figure 2. Why quasi-geodesics beginning and ending on a horosphere must stay uniformly close to the horosphere.

Since the universal cover $H = \tilde{M}$ is a simply connected Riemannian manifold with sectional curvatures pinched between -1 and -4, Proposition 4.1 gives

$$d_H(x, y) \leq d_X(x, y) \leq \sinh(2 \cdot d_H(x, y))$$

for all $x, y \in X$. So we have

$$d_H(x, x') = d_H(x, S) \leq d_X(x, S) = P$$

so that $d_X(x, x') \leq \sinh(2P)$; similarly $d_X(y, y') \leq \sinh(2P)$.

Now ignore all of the deleted horospheres. Orthogonal projection onto S shows that $d_X(x', y') \leq \ell/\cosh(2P)$. Hence there is a path in X from x to y of length $2\sinh(2P) + \ell/\cosh(2P)$. Since $f \circ \sigma(g)$, hence α, is a KC-quasi-geodesic, this gives :

$$\ell \leq KC(2\sinh(2P) + \ell/\cosh(2P))$$

which gives

$$\ell \leq \frac{2KC\sinh(2P)}{(1 - KC/\cosh(2P))}$$

which gives a (universal) bound on ℓ. Note that $1 - KC/\cosh(2P) > 0$ by our choice of P. Now let

$$K' = 2\sinh(2P) + \frac{KC\sinh(2P)}{(1 - KC/\cosh(2P))}.$$

Then $f \circ \sigma(g)$ lies in a K'-neighborhood of S, as required.

Since N is L-quasiconvex, it follows from Theorem 3.1 that the combing $\sigma : \Gamma \rightarrow L$ descends to a combing on N. But $G = SU(n, 1)$, $SP(n, 1)$, or F_4^{-20}, so that N is not virtually abelian. This gives a contradiction, for Thurston has shown that these groups are not combable [Th].   □

Remark. There are two alternatives to using Thurston's unpublished result in the proof of Theorem 1.2; each of these gives a slightly weaker conclusion. By our same argument, any automatic structure on Γ descends to an automatic structure on N; but it is a result of Epstein–Holt that nilpotent groups are not automatic [E et al.]. Hence Γ is not automatic. It is a result of Gersten–Short [GS,Sh] that any nilpotent subgroup of a bicombable group is virtually abelian; hence Γ is not bicombable. It is this latter fact that we really use in the proof of Theorem 1.1.

Remark. Although many of the nonuniform lattices in rank 1 Lie groups are not hyperbolic, automatic, or bicombable, much can still be said. Every such lattice is hyperbolic (and automatic) relative to its cusp subgroups; this gives precise combinatorial information about these lattices [Fa2].

The rank$(G) \geq 2$ case rests wholly on the following result of Lubotzky–Mozes–Raghunathan:

Theorem 4.2 [LMR]. *Let Γ be an irreducible lattice in a semisimple linear Lie group G, and let $\gamma \in \Gamma$ be an element of infinite order. Then $\|\gamma^n\|_\Gamma = O(\log(n))$ if and only if γ is virtually unipotent and* rank$(G) \geq 2$.

The proof of Theorem 4.2 makes an essential use of the Margulis arithmeticity theorem, which states that irreducible lattices are arithmetic for rank$(G) \geq 2$. Lubotzky–Mozes–Raghunathan apply Theorem 4.2 to prove a conjecture of Kazhdan: for semisimple linear Lie groups G with rank$(G) \geq 2$, the word metric on an irreducible lattice Γ in G is Lipschitz equivalent to the restriction to Γ of the left invariant Riemannian metric on G.

Proof of Theorem 1.3. Let G be a semisimple linear Lie group with rank$(G) \geq 2$, and let Γ be an irreducible lattice in G. Suppose that Γ is uniform. Let U be a maximal connected, normal, compact subgroup of G. Then the short exact sequence

$$1 \rightarrow U \rightarrow G \rightarrow G/U \rightarrow 1$$

gives a short exact sequence

$$1 \rightarrow A \rightarrow \Gamma \rightarrow \Gamma' \rightarrow 1$$

where A is finite and Γ' is a uniform lattice in G/U. By Lemma 2.1, to show that Γ is bicombable, it suffices to show that Γ' is bicombable. Hence without loss of generality, we assume that G is a semisimple linear Lie group whose symmetric space is a product of irreducible symmetric spaces of noncompact type.

Let K be a maximal compact subgroup of G. Since the symmetric space G/K associated to G is a product of irreducible symmetric spaces of noncompact type, we may apply the classical result of Cartan (see [He], Theorem 3.1) which states that G/K has nonpositive curvature. Since Γ acts properly and cocompactly by isometries on G/K, it follows that Γ' is bicombable.

Now suppose that Γ is nonuniform and bicombable. Let $\gamma \in \Gamma$ be a unipotent element of Γ of infinite order. The existence of such an element for irreducible lattices in linear semisimple Lie groups G with rank$(G) \geq 2$ follows from the Compactness Criterion for arithmetic lattices ([Ra], Theorem 10.19); for arbitrary irreducible lattices, this follows from the Margulis Arithmeticity Theorem ([Ma]). By Theorem 4.2, γ satisfies $||\gamma^n||_\Gamma = O(\log(n))$. Since Γ is bicombable and $\gamma \in \Gamma$ has infinite order, it follows from Proposition 3.2 that γ has nonzero translation number. Hence $||\gamma^n||_\Gamma = O(n)$ by Proposition 3.3, which gives a contradiction. □

Note. The arguments which show that an element of infinite order in a bicombable group generates a quasi-isometrically embedded cyclic subgroup depend heavily on the "almost-equivariance" of the bicombing; hence these arguments do not extend to combable groups. It is not known whether infinite cyclic subgroups of combable groups must embed quasi-isometrically.

References

[AB]   J. Alonso and M. Bridson, Semi-hyperbolic groups, preprint.

[Bo]   B. Bowditch, Geometrically finite hyperbolic groups, University of Warwick, preprint.

[De]   M. Dehn, Papers on Group Theory and Topology, translated by J. Stillwell, Springer-Verlag, 1987.

[E et al.] D. B. A. Epstein, J. Cannon, D. F. Holt, S. Levy, M. S. Patterson and W. P. Thurston, Word Processing in Groups, Jones and Bartlett, 1992.

[Fa1]   B. Farb, Automatic groups: a guided tour, Enseign. Math. 38 (1992), 291–313.

[Fa2]   B. Farb, Relatively hyperbolic groups and automatic groups with applications to finite volume negatively curved manifolds, thesis, Princeton University, June 1994.

[GdlH]   E. Ghys and P. de la Harpe, editors, Sur les Groupes Hyperboliques d'aprés Mikhael Gromov, Progress in Mathematics 83, Birkhäuser, 1990.

[Gr]   M. Gromov, Hyperbolic groups, Essays in Group Theory, Math. Sci. Res. Inst. Publ. 8, S. M. Gersten, ed., Springer-Verlag, 1987, 75–263.

[GS]   S. M. Gersten and H. Short, Rational subgroups of biautomatic groups, Annals of Math. 134 (1991), 125–158.

[HI]   E. Heintze and H. Im Hof, Geometry of horospheres, J. Differential Geom. 12 (1977), 481–491.

[He]   S. Helgason, Differential Geometry, Lie Groups, and Symmetric Spaces, Academic Press, New York, 1978.

[LMR]   R A. Lubotzky, S. Mozes and M. S. Raghunathan, Cyclic subgroups of exponential growth and metrics on discrete groups, preprint.

[Ma]   G.A. Margulis, Discrete Subgroups of Semisimple Lie Groups, Springer-Verlag, 1991.

[Mi]   J. Milnor, A note on curvature and fundamental group, J. Differential Geom. 2, 1–7.

[Ra]   M.S. Raghunathan, Discrete Subgroups of Lie Groups, Springer-Verlag, New York, 1972.

[Sh]   H. Short, Groups and combings, preprint, Ecole Normale Supérieure de Lyon, July 1990.

[Th]   W. P. Thurston, unpublished.

Nielsen Transformations
and Applications: A Survey

Benjamin Fine, Gerhard Rosenberger and Michael Stille

Contents

1. Introduction

One of the main tools in the study of free groups and related constructions involving infinite groups is the linear cancellation method using Nielsen transformations. Introduced originally by Nielsen [N] to prove the subgroup theorem for free groups they can be considered as the basic transformations in moving from one generating system of any finitely generated group to another generating system. Along these lines Nielsen also proved that the Nielsen transformations generate the automorphism group for any free group of finite rank. Nielsen transformations can be considered as the non-commutative analogs of row reduction of matrices and have proved to be indispensable in the theory of free groups.

The theory of Nielsen transformations has been extended to free products with amalgamation by Zieschang [Z1], and then refined a bit by Rosenberger [R1], [R6], [R8], Kalia and Rosenberger [K-R] and Collins and Zieschang [C-Z2] and has been extended to HNN groups by Peczynski and Reiwer [P-R]. Using the theory many subgroup results on these constructions can be deduced. In addition several important results, such as the Grushko theorem and the Kurosh theorem, can be reproved using the Nielsen technique.

The purpose of this paper is to present a comprehensive look at Nielsen transformation techniques in group amalgams and then describe some important applications. Most of these applications have appeared in scattered other locations however some are new. In Section 2 we discuss the basic notations and theory in free groups. As an application we present a theorem due to Rosenberger [R3] which can be applied to the study of both surface groups and co-compact Fuchsian groups. In Section 3 we describe the Nielsen reduction method in amalgamated free products. To do this we must introduce a length function on such groups. In Section 4 we use the theory in free products with amalgamation to give a solution, due to Rosenberger [R10], of the isomorphism problem for cyclically pinched one-relator groups. As another application of the method in free products with amalgamation in Section 5 we consider certain systems of equations in such constructions. As a by-product of this we recover the Kurosh theorem and the Grushko–Neumann theorem. In Section 6 we give the Nielsen theory as extended to HNN groups by Peczynski and Reiwer [P-R]. As applications, we present certain subgroup free-ness results in HNN groups due to Fine, Rohl and Rosenberger [F-R-R1, 2]. As an outgrowth of these results we can answer a question of G. Baumslag concerning a class of special para-free groups. Finally as application we discuss a class of groups generalizing surface groups.

We would like to thank Nicole Isermann for reading this paper and giving some very helpful comments and suggestions.

2. Nielsen Transformations in Free Groups

In this section we describe the basic notation of Nielsen transformations for free groups. We then give some applications related to the theory of surface groups. For our purposes we consider all groups G to be countable and only consider finite subsets $\{x_1, \ldots, x_n\}$, $n \geq 1$, in G. For a finite subset $\{x_1, \ldots, x_n\}$, $n \geq 1$, in G we define an *elementary Nielsen transformation* as a transformation of one of the following five types:

(N1) replace $\{x_1, x_2, x_3, \ldots, x_n\}$ by $\{x_2, x_1, x_3, \ldots, x_n\}$
(N2) replace $\{x_1, \ldots, x_{n-1}, x_n\}$ by $\{x_n, x_1, \ldots, x_{n-1}\}$
(N3) replace $\{x_1, x_2, \ldots, x_n\}$ by $\{x_1^{-1}, x_2, \ldots, x_n\}$
(N4) replace $\{x_1, x_2, \ldots, x_n\}$ by $\{x_1 x_2, x_2, \ldots, x_n\}$
(N5) delete some x_i where $x_i = 1$, $1 \leq i \leq n$.

A *Nielsen transformation* is a finite product of elementary Nielsen transformations. It is a *regular* Nielsen transformation if there is no factor of type (N5) otherwise it is *singular*. Each elementary Nielsen transformation of type (N1), (N2), (N3) or (N4) has an inverse which is a regular Nielsen transformation. It follows then that the regular Nielsen transformations form a group which contains every permutation of the set $\{x_1, \ldots, x_n\}$. For a subset $\{x_1, \ldots, x_n\}$ we let $\langle x_1, \ldots, x_n \rangle$ denote the subgroup of G that they generate. If $\{x_1, \ldots, x_n\}$ is carried by a Nielsen transformation into $\{y_1, \ldots, y_m\}$, $1 \le m \le n$, then clearly $\langle x_1, \ldots, x_n \rangle = \langle y_1, \ldots, y_m \rangle$, that is they generate the same subgroup of G. If this Nielsen transformation is regular then we must have that $m = n$ in which case we say that $\{x_1, \ldots, x_n\}$ is **Nielsen equivalent** to $\{y_1, \ldots, y_m\}$.

Now let F be a free group with fixed basis A. The length $L(w)$ of an element from F is the length of the reduced word for w. This clearly depends on the basis A. We write $u_1 \cdots u_q \equiv v_1 \cdots v_m$ for the equality together with the fact that $L(v_1 \cdots v_m) = L(v_1) + L(v_2) + \cdots + L(v_m)$, all $u_i, v_j \in F$.

Let $X = \{x_1, \ldots, x_n\}$, $n \ge 1$, be a finite subset of the free group F (with $X \cap X^{-1} = \emptyset$). X is called **Nielsen reduced** if for all triples of elements u, v, w from $X^{\pm 1}$ of the form $x_i^{\epsilon_i}$, $\epsilon_i = \pm 1$, the following conditions hold:

(R1) $u \ne 1$

(R2) $uv \ne 1$ implies $L(uv) \ge L(u), L(v)$

(R3) $uv \ne 1$ and $vw \ne 1$ implies that

$$L(uvw) > L(u) - L(v) + L(w).$$

Being Nielsen reduced implies that there is not too much cancellation in multiplying elements from X.

Using in F any fixed order relative to the basis A which does not distinguish between inverse elements, if $X = \{x_1, \ldots, x_n\}$ is finite then X can be carried by a Nielsen transformation into some $Y = \{y_1, \ldots, y_m\}$, $1 \le m \le n$, with Y Nielsen reduced.

If $X = \{x_1, \ldots, x_n\}$, $n \ge 1$, is Nielsen reduced then for each $u \in X^{\pm 1}$ there are words p_u, q_u, k_u with $k_u \ne 1$ such that $u \equiv p_u k_u q_u$ is reduced and such that if $w = u_1 \cdots u_q$, $q \ge 0$, $u_i \in X^{\pm 1}$ all $u_i u_{i+1} \ne 1$, then $k_{u_1}, \ldots, k_{u_q}$ remain uncancelled in the reduced form of w and $L(w) \ge q$. From this it is straightforward that if X is Nielsen reduced then $\langle X \rangle$ is free with X as a basis. Hence every finitely generated subgroup of a free group is free. This was Nielsen's original proof. It can be extended to remove the finitely generated restriction {see [L-S]}. Further if F has finite rank m and $X = \{x_1, \ldots, x_n\}$, $1 \le n \le m$, is Nielsen reduced and generates F then $m = n$ and X is a basis for F. If $X = \{x_1, \ldots, x_n\}$ is Nielsen reduced then $x_1, \ldots, x_n$ are the shortest generators of $\langle X \rangle$ which exist: that is if $y_1, \ldots, y_n$ are other free generators of $\langle X \rangle$ and both sets are ordered according to the length L then $L(y_i) \ge L(x_i)$, $i = 1, \ldots, n$.

It is clear that a Nielsen transformation applied to a generating system of a finitely generated free group defines an automorphism. Nielsen also proved that the Nielsen transformations generate the automorphism group of a free group of finite rank.

We now describe some applications which we used frequently. These applications are related to the study of surface groups and co-compact Fuchsian groups. Recall that a surface group has a presentation

$$\langle a_1, b_1, \ldots, a_g, b_g \, ; \, [a_1, b_1] \cdots [a_g, b_g] = 1 \rangle \text{ in the orientable case}$$

or

$$\langle a_1, \ldots, a_g \; ; \; a_1^2 \cdots a_g^2 = 1 \rangle \text{ in the non-orientable case.}$$

Any automorphism of a surface group moves the relator to a Nielsen equivalent word. What is of interest then is how free group words of the form of the surface group relator behave under Nielsen transformations. This was studied by Rosenberger [R2], [R3] and what was obtained was the following:

Theorem 2.1 ([R2], [R3]). *Let F be the free group on $a_1, \ldots, a_n$ and*

$$P(a_1, \ldots, a_n) = a_1^{\alpha_1} \cdots a_p^{\alpha_p} [a_{p+1}, a_{p+2}] \cdots [a_{n-1}, a_n] \in F$$

with $0 \leq p \leq n$, $n - p$ even and $\alpha_i \geq 1$ for $i = 1, \ldots, n$. Let $X = \{x_1, \ldots, x_m\}$ be any finite system in F and let $H = \langle X \rangle$. Suppose that H contains some conjugate of $P(a_1, \ldots, a_n)$. Then:
(a) *$\{x_1, \ldots, x_m\}$ can be carried by a Nielsen transformation into a free basis for H which contains a conjugate of $P(a_1, \ldots, a_n)$; or*
(b) *$\{x_1, \ldots, x_m\}$ can be carried by a Nielsen transformation into a free basis $\{y_1, \ldots, y_k\}$ for H with $m \geq k \geq n$, $y_i = za_i^{\gamma_i} z^{-1}$, $1 \leq \gamma_i < \alpha_i$, $\gamma_i | \alpha_i$ for $i = 1, \ldots, p$, $y_j = za_j z^{-1}$ for $j = p+1, \ldots, n$ and $z \in F$.*

Proof. First of all we may assume that $P(a_1, \ldots, a_n) \in H = \langle X \rangle$ and that $\{x_1, \ldots, x_m\}$ is Nielsen reduced. In particular, as a freely reduced word in $a_1, \ldots, a_n$, each x_i^ϵ, $\epsilon = \pm 1$, contains an uncancellable symbol a_j^ν, $\nu = \pm 1$, which remains unchanged in each freely reduced word in $x_1, \ldots, x_m$ at that place where x_i^ϵ occurs. For each x_i^ϵ, $\epsilon = \pm 1$, choose such a stable symbol a_j^ν, $\nu = \pm 1$, and take the inverse symbol for $x_i^{-\epsilon}$. We may choose the stable symbol in such a way that $x_i^\epsilon \equiv ya_j^\nu z$ with $L(x_i) = L(y) + L(z) + 1$ and either $L(y) = L(z)$ or $|L(y) - L(z)| = 1$. It is now more convenient to write x_i^ϵ, $\epsilon = \pm 1$, in the freely reduced form $x_i^\epsilon \equiv ua_j^\beta v$ with $\beta \in \mathbb{Z}$, $\beta \neq 0$, and the reduced form of u does not end and the reduced form of v does not start with a power of a_j. In particular then, $L(x_i^\epsilon) = L(u) + |\beta| + L(v)|$.

For the Nielsen reduced system $\{x_1, \ldots, x_m\}$ we have an equation

$$\prod_{j=1}^{q} x_{\nu_j}^{\epsilon_j} = U(x_1, \ldots, x_m) \equiv P(a_1, \ldots, a_n) \tag{1}$$

with $\epsilon_j = \pm 1$, $\epsilon_j = \epsilon_{j+1}$ if $\nu_j = \nu_{j+1}$. Among the equations as in (1) there is one for which q is minimal, and let us assume that this is the case in equation (1). We may also assume that each x_i occurs in (1). If one x_i occurs only once in (1) as either x_i or x_i^{-1} then case (a) holds, that is $\{x_1, \ldots, x_m\}$ can be carried by a Nielsen transformation into a free basis for H which also contains $P(a_1, \ldots, a_n)$. Hence we may assume that each x_i either occurs twice in (1) with the same exponent $\epsilon = \pm 1$ or occurs in (1) exactly once with exponent $+1$ and once with exponent -1.

If x_i occurs in (1) twice with the same exponent $\epsilon = \pm 1$ and has stable symbol a_j^ν, $\nu = \pm 1$, then we must have $1 \le j \le p$ and $\nu = 1$ because of the special form of $P(a_1, \ldots, a_n)$ and because all $\alpha_i \ge 1$. Hence, in addition $x_i^{-\epsilon}$ cannot also occur in (1). Therefore if x_i occurs twice in (1) with the same exponent $\epsilon = \pm 1$ we may assume that $\epsilon = +1$ and that x_i^{-1} does not occur in (1). If x_i and x_k both occur twice in (1) with the same exponent and if both have the same stable symbol a_j, then there is no x_h^ϵ, $\epsilon = \pm 1$, in (1) between x_i and x_k which has a stable symbol different from a_j again because of the special form of $P(a_1, \ldots, a_n)$. In this situation it could be the case that $i \ne k$ and that at least two x_i have the same stable symbol a_j, $1 \le j \le p$. Hence if again write $x_i^\epsilon \equiv u a_j^\beta v$, $\epsilon = \pm 1$, then we must consider subwords in (1) of the type

$$u_1 a_j^{\beta_1} v_1 u_i a_j^{\beta_i} v_i \cdots u_1 a_j^{\beta_1} v_1 u_k a_j^{\beta_k} v_k \quad \text{and} \tag{2}$$

$$u_i a_j^{\beta_i} v_i u_1 a_j^{\beta_1} v_1 \cdots u_k a_j^{\beta_k} v_k u_1 a_j^{\beta_1} v_1 \tag{3}$$

where $1 \le j \le p, 1 \le i, k \le q$. Recall that each x_h with a stable symbol a_t, $1 \le t \le p$, occurs twice in equation (1). We first consider the subword (2). Then $v_1 = u_i^{-1} = u_k^{-1}$. Let $z_1 \equiv u_i a_j^{\beta_i} v_i$ and $z_2 \equiv u_k a_j^{\beta_k} v_k \equiv u_i a_j^{\beta_k} v_k$. We have $\beta_i, \beta_k \ge 1$. If $z_1 \ne z_2$ then $z_1^{-1} z_2 = v_i^{-1} a_j^{\beta_k - \beta_i} v_k \ne 1$ and $z_2^{-1} z_1 = v_k^{-1} a_j^{\beta_i - \beta_k} v_i \ne 1$ which contradicts the fact that $\{x_1, \ldots, x_m\}$ is Nielsen reduced. Hence we must have $z_1 = z_2$.

For the subword (3) we get that $u_1 = v_i^{-1} = v_k^{-1}$ and analogously $u_i a_j^{\beta_i} v_i \equiv u_k a_j^{\beta_k} v_k$. Altogether we get that a maximal subword $\prod x_{v_i}^{\epsilon_i}$ of (1) in which each x_{v_i} has a_j, $1 \le j \le p$, as stable symbol (recall that here $\epsilon_i = +1$) must have the reduced form

$$(x_{v_{i_0}} \cdots x_{v_{i_0 + j_0}})^\delta w,$$

where $w = 1$ or $w = x_{v_{i_0}} \cdots x_{v_{i_0 + j}}$ and with $\delta \ge 2, 0 \le j \le j_0 - 1, x_{v_{i_0 + r}} \ne x_{v_{i_0 + s}}$ for $r \ne s$ and $0 \le r, s \le j_0$.

If $j_0 = 0$ then $x_{v_{i_0}}$ is conjugate to a power of a_j. Now let $j_0 \ne 0$.

If $w \ne 1$ then we replace $x_{v_{i_0 + j_0}}$ by $x_0 = x_{v_{i_0}} \cdots x_{v_{i_0 + j_0}}$ which defines a Nielsen transformation if the other x_i are left fixed. In this situation then case (a) holds.

Now let $w = 1$. Without loss of generality we may assume that $v_{i_0} < v_{i_0 + j_0}$. Then there is a Nielsen transformation from $\{x_1, \ldots, x_m\}$ to the system

$$\{x_1, \ldots, x_{v_{i_0} - 1}, x_{v_{i_0}} \cdots x_{v_{i_0 + j_0}}, x_{v_{i_0} + 1}, \ldots, x_m\},$$

and then $P(a_1, \ldots, a_n)$ is already contained in the proper subgroup of $H = \langle X \rangle$, generated by

$$x_1, \ldots, x_{v_{i_0} - 1}, x_{v_{i_0}} \cdots x_{v_{i_0 + j_0}}, x_{v_{i_0} + 1}, \ldots, x_{v_{i_0 + j_0} - 1}, x_{v_{i_0 + j_0} + 1}, \ldots, x_m.$$

Hence there is a Nielsen transformation from $\{x_1, \ldots, x_m\}$ to a system $\{y_1, \ldots, y_m\}$ with $P(a_1, \ldots, a_n) \in \langle y_1, \ldots, y_{m-1} \rangle$. Now we start with the system $\{y_1, \ldots, y_{m-1}\}$

and argue as above. Therefore without loss of generality we may assume from the very beginning that there is no Nielsen transformation from $\{x_1, \ldots, x_m\}$ to a system $\{y_1, \ldots, y_m\}$ with $P(a_1, \ldots, a_n) \in \langle y_1, \ldots, y_{m-1}\rangle$, that is m is minimal with respect to this property. We also assume now that case (a) does not hold.

If we now argue as above, with $\{x_1, \ldots, x_m\}$ Nielsen reduced and each x_i occurs in equation (1) we get: If x_i occurs twice in (1) with the same exponent $\epsilon = \pm 1$ then x_i is conjugate to a power of some a_j, $1 \leq j \leq p$, and no x_k with $k \neq i$ is also conjugate to a power of a_j. On the other hand, for each a_j, $1 \leq j \leq p$, there is an x_i which is conjugate to a power of a_j for otherwise in the abelianized group F^{ab} there is some power a_j^ν, $\nu \neq 0$, $1 \leq j \leq p$, in the subgroup of F^{ab} generated by $a_1, \ldots, a_{j-1}, a_{j+1}, \ldots, a_n$. Recall that if x_i does not occur twice in (1) with the same exponent then it occurs in (1) exactly once with exponent $+1$ and once with exponent -1. Also the centralizer of a non-trivial word in F is cyclic. Therefore without loss of generality we may now assume that $x_i = a_i^{\gamma_i}$, $1 \leq \gamma_i < \alpha_i$, $\gamma_i | \alpha_i$ for $i = 1, \ldots, p$, possibly after a suitable Nielsen transformation.

Therefore we have reduced the proof to the following situation: For $x_{p+1}, \ldots, x_m$ we have an equation

$$\prod_{j=1}^{t} x_{v_j}^{\epsilon_j} \equiv [a_{p+1}, a_{p+2}] \cdots [a_{n-1}, a_n], \tag{4}$$

with $\epsilon_j = \pm 1$, $\epsilon_j = \epsilon_{j+1}$ if $v_j = v_{j+1}$, and each x_i, $p+1 \leq i \leq n$ occurs in (4) exactly once with exponent $+1$ and once with exponent -1. Now we argue analogously as in Chapter 5 of [Z-V-C] to obtain that here $m = n$ and a Nielsen transformation from $\{a_1^{\gamma_1}, \ldots, a_p^{\gamma_p}, x_{p+1}, \ldots, x_m\}$ to $\{a_1^{\gamma_1}, \ldots, a_p^{\gamma_p}, a_{p+1}, \ldots, a_n\}$.

This theorem has some very nice consequences and corollaries.

Corollary 2.1 ([R2]). *Let* F, $P(a_1, \ldots, a_n)$, $X = \{x_1, \ldots, x_m\}$ *and* $H = \langle X \rangle$ *be as in the theorem. Suppose that there are* $u_1, \ldots, u_k \in H$, $k \geq 1$ *with*

$$u_1^{\beta_1} \cdots u_q^{\beta_q} [u_{q+1}, u_{q+2}] \cdots [u_{k-1}, u_k] = P(a_1, \ldots, a_n)$$

where $0 \leq q \leq k$, $2 \leq \beta_j$ *for* $j = 1, \ldots, q$ *and* $\gcd(\beta_1, \ldots, \beta_q) \geq 2$ *if* $q \geq 1$. *Then* $\{x_1, \ldots, x_m\}$ *can be carried by a Nielsen transformation into a free basis* $\{y_1, \ldots, y_l\}$ *for* H *with* $m \geq l \geq n$, $y_i = a_i^{\gamma_i}$, $1 \leq \gamma_i < \alpha_i$, $\gamma_i | \alpha_i$ *for* $i = 1, \ldots, p$, $y_j = a_j$ *for* $j = p+1, \ldots, n$. *In particular we have* $k \geq n$.

If in addition each α_i, $1 \leq i \leq p$, is a prime number then $H = F$ and if further $m = n$ then X is a free generating system for F.

Corollary 2.2 ([R2]). *Let* F *be a the free group on* $a_1, \ldots, a_n$ *and*

$$P(a_1, \ldots, a_n) = a_1^{\alpha_1} \cdots a_p^{\alpha_p} [a_{p+1}, a_{p+2}] \cdots [a_{n-1}, a_n] \in F$$

with $0 \le p \le n$, $n - p$ even, $\alpha_i \ge 1$ for $i = 1, \ldots, n$ and $\gcd(\alpha_1, \ldots, \alpha_p) \ge 2$ if $p \ge 1$. If $\phi : F \to F$ is an endomoprhism which fixes $P(a_1, \ldots, a_n)$, that is $\phi(P(a_1, \ldots, a_n)) \equiv P(a_1, \ldots, a_n)$, then ϕ is already an automorphism of F.

For both corollaries all that is necessary is that in a free group F a primitive element cannot be contained in the subgroup $F^t F'$, $t \ge 2$, generated by the t-th powers and the commutators. This excludes case (a) of the theorem and the corollaries follow.

It is also of interest to consider subgroups of the free group F which contain proper powers of the word $P(a_1, \ldots, a_n)$ as in the theorem. As an example of the results we can obtain here is the following:

Theorem 2.2 ([R4]). *Let F be a free group on $a_1, b_1, \ldots, a_g, b_g$, $g \ge 1$ and let $P(a_1, b_1, \ldots, a_g, b_g) = [a_1, b_1] \cdots [a_g, b_g]$. Let $X = \{x_1, \ldots, x_m\}$ be any finite system in F and let $H = \langle X \rangle$. Suppose that H contains some conjugate of some non-trivial power of $P(a_1, b_1, \ldots, a_g, b_g)$ and let β be the smallest positive integer such that some conjugate of $P^\beta(a_1, b_1, \ldots, a_g, b_g)$ lies in H. Then:*
(a) *$\{x_1, \ldots, x_m\}$ can be carried by a Nielsen transformation into a free basis for H which contains a conjugate of $P^\beta(a_1, b_1, \ldots, a_g, b_g)$; or*
(b) *the index of H in F is β and in this case $\{1, P(a_1, b_1, \ldots, a_g, b_g), \ldots, P^{\beta-1}(a_1, b_1, \ldots, a_g, b_g)\}$ form a set of coset representatives for H in F.*

In addition, if some conjugate of $P^\beta(a_1, b_1, \ldots, a_g, b_g)$ is contained in the commutator subgroup H' of H then (a) does not occur and β is odd in (b).

A weaker version of this theorem is used in [R5] by Rosenberger to classify the congruence subgroups of the Modular group which can be generated by elements of finite order.

For further results with restrictions on m, such as $1 \le m \le n$ see [R3]. For example:

Theorem 2.3 ([R3]). *Let F be a the free group on $a_1, \ldots, a_n$ and*

$$P(a_1, \ldots, a_n) = a_1^{\alpha_1} \cdots a_p^{\alpha_p} [a_{p+1}, a_{p+2}] \cdots [a_{n-1}, a_n] \in F$$

with $0 \le p \le n$, $n - p$ even and $\alpha_i \ge 1$ for $i = 1, \ldots, n$. Let $X = \{x_1, \ldots, x_m\}$ with $1 \le m \le n$ be any finite system in F and let $H = \langle X \rangle$. Suppose that H contains some $y^{-1} P^\alpha(a_1, \ldots, a_n) y$ with $\alpha \ne 0$ and $y \in F$. Then
(a) *$\{x_1, \ldots, x_m\}$ can be carried by a Nielsen tra sformation into a free basis $\{y_1, \ldots, y_k\}$ with $y_1 = z P^\beta(a_1, \ldots, a_n) z^{-1}$, $\beta \ge 1$ and $z \in F$. Further β is the smallest positive integer for which a relation $y^{-1} P^\beta(a_1, \ldots, a_n) y \in X$ holds for some $y \in F$; or*
(b) *we have $m = n$ and $\{x_1, \ldots, x_m\}$ is Nielsen equivalent to a system $\{y_1, \ldots, y_m\}$ with $y_i = z a_i^{\gamma_i} z^{-1}$, $1 \le \gamma_i < \alpha_i$, $\gamma_i | \alpha_i$ for $i = 1, \ldots, p$ and $y_j = z a_j z^{-1}$ for $j = p + 1, \ldots, n$ and $z \in F$.*

3. Nielsen Reduction in Amalgamated Free Products

In this section we describe the Nielsen cancellation method in free products with amalgamation as developed by H. Zieschang [Z1] and refined a bit by G. Rosenberger [R1], [R6], [R8] and R. N. Kalia and Rosenberger [K-R]. A further refinement of this technique was given by D. Collins and H. Zieschang in [C-Z2] which we do not consider here in detail. We restrict ourselves in this paper to the free product of two groups with an amalgamated subgroup, although the method works more generally.

Let $G = H_1 \star_A H_2$, $H_1 \neq A \neq H_2$, denote the non-trivial free product of the groups H_1 and H_2 with the amalgamated subgroup $A = H_1 \cap H_2$. If $A = \{1\}$ then G is just the free product $G = H_1 \star H_2$ of H_1 and H_2. We choose in each H_i, $i = 1, 2$, a system L_i of left coset representatives of A in H_i, normalized by taking 1 to represent A. Each $x \in G$ has a unique representation $x = h_1 \cdots h_n a$ with $a \in A$, $1 \neq h_j \in L_1 \cup L_2$ and $h_{j+1} \notin L_i$ if $h_j \in L_i$. The **length** of x denoted $L(x)$ is then defined to be n and G is then (partially) ordered by length.

In order to obtain results analogous to those in free groups it is found that the ordering defined by the length L is too coarse. Therefore, as in the free group case, we need a finer pre-ordering for G. For this purpose we define a symmetric normal form for elements $x \in G$. Take the inverses L_i^{-1} of the left coset representatives as a system of right coset representatives. Then each $x \in G$ has a unique representation

$$x = l_1 \cdots l_m k_x r_m \cdots r_1$$

with $m \geq 0$, $k_x \in H_1 \cup H_2$, $1 \neq l_j \in L_1 \cup L_2$, $1 \neq r_j \in L_1^{-1} \cup L_2^{-1}$ and $l_{j+1} \notin L_i$ if $l_j \in L_i$, $r_{j+1} \notin L_i^{-1}$ if $r_j \in L_i^{-1}$. Further if $k_x \in A$ then l_m and r_m belong to different H_i (if $m \geq 1$), and if $k_x \in H_i \setminus A$ then $l_m \notin H_i$, $r_m \notin H_i$ (if $m \geq 1$).

We then have $L(x) = 2m$ if $k_x \in A$ and $L(x) = 2m + 1$ if $k_x \notin A$. We call $l_1 \cdots l_m$ the **leading half**, $r_m \cdots r_1$ the **rear half** and k_x the **kernel** of x. One advantage of this symmetric normal form is that in forming products, cancellations can usually be reduced to free cancellations.

We now introduce an ordering $\leq$ on G. We assume that for each H_i, the system L_i of left coset representatives has a strict total order. For our applications we may assume that the groups are countable. This is no restriction if one considers a given finitely generated subgroup of G or a given finite system in G. If G is countable then just enumerate the system L_i and order it correspondingly. Let the elements of L_1 precede those of L_2. Then we order for each m the product $l_1 \cdots l_m$ of left coset representatives (where $1 \neq l_j \in L_1 \cup L_2$ and $l_{j+1} \notin L_i$ if $l_j \in L_i$), first by length and second lexicographically. Hence if $l_1 \cdots l_m < l_1' \cdots l_m'$ then for any permitted l_{m+1}, l_{m+1}' we have $l_1 \cdots l_{m+1} < l_1' \cdots l_{m+1}'$. If G is countable then further each product $l_1 \cdots l_m$ has only finitely many predecessors of the form $l_1 \cdots l_{m-1} l_m'$ (where $l_m' \in L_i$ if $l_m \in L_i$). Thus without loss of generality we assume that G is countable if we consider a given finitely generated subgroup of G or a given finite system in G.

We define an ordering on the products of right coset representatives in the L_i^{-1} by taking inverses. We now extend this ordering to the set of pairs $\{g, g^{-1}\}$, $g \in G$, where the notation is chosen so that the leading half of g precedes that of g^{-1} with respect to ordering $\leq$. Then we set $\{g, g^{-1}\} < \{h, h^{-1}\}$ if either $L(g) < L(h)$ or $L(g) = L(h)$ and the leading half of g strictly precedes that of h, or $L(g) = L(h)$, the leading halves of g and h coincide, and the leading half of g^{-1} precedes that of h^{-1}.

Thus if $\{g, g^{-1}\} < \{h, h^{-1}\}$ and $\{h, h^{-1}\} < \{g, g^{-1}\}$ then g and h differ only in the kernel; since this can occur with $g \neq h$ then $<$ is only a pre-order.

For $g \in G$ let the leading half of $g^{\epsilon(g)}$, $\epsilon(g) = \pm 1$, precede that of $g^{-\epsilon(g)}$. A finite system $\{g_1, \ldots, g_m\}$ in G is called **shorter** than a system $\{h_1, \ldots, h_m\}$ if $\{g_i^{\epsilon(g_i)}, g_i^{-\epsilon(g_i)}\} < \{h_i^{\epsilon(h_i)}, h_i^{-\epsilon(h_i)}\}$ holds for all $i \in \{1, \ldots, m\}$ and for at least one $i \in \{1, \ldots, m\}$, $\{h_i^{\epsilon(h_i)}, h_i^{-\epsilon(h_i)}\} < \{g_i^{\epsilon(g_i)}, g_i^{-\epsilon(g_i)}\}$ fails to hold.

A finite system $\{g_1, \ldots, g_m\}$ in G is said to be **Nielsen reduced** or **minimal** with respect to $<$ if $\{g_1, \ldots, g_m\}$ cannot be carried by a Nielsen transformation into a system $\{h_1, \ldots, h_m\}$ with $h_i = 1$ for some $i \in \{1, \ldots, m\}$ or there is no system Nielsen equivalent to $\{g_1, \ldots, g_m\}$ which is shorter. If G is countable then every finite system can be carried by a Nielsen transformation into a minimal system. In general, as already mentioned, for a given finite system, a suitable order can always be chosen such that this finite system can be carried by a Nielsen transformation into a minimal system.

We now make the following conventions on notation. We write $u_1 \cdots u_q \equiv v_1 \cdots v_n$ to stand for the equality together with the fact that $L(v_1 \cdots v_n) = L(v_1) + \cdots + L(v_n)$ for all $u_i, v_j \in G$. If $x \in G$ is given in its symmetric normal form

$$x = l_1 \cdots l_m k_x r_m \cdots r_1$$

then we write $x \equiv p_x k_x r_x$ where $p_x = l_1 \cdots l_m$ is the leading half and $q_x = r_m \cdots r_1$ is the rear half of x. Further $x = p_x k_x r_x$ refers to the symmetric normal form as above.

The Nielsen reduction method in G now refers to Nielsen transformations from given systems to shorter systems and the resulting investigation of minimal systems.

An analysis of the result of H. Zieschang [Z1] for G {see also Rosenberger [R8]} produces the following result.

Theorem 3.1. *Let* $G = H_1 \star_A H_2$. *If* $\{x_1, \ldots, x_m\}$ *is a finite system of elements in* G *then there is a Nielsen transformation from* $\{x_1, \ldots, x_m\}$ *to a system* $\{y_1, \ldots, y_m\}$ *for which one of the following cases hold:*

(i) $y_i = 1$ *for some* $i \in \{1, \ldots, m\}$.

(ii) *Each* $w \in \langle y_1, \ldots, y_m \rangle$ *can be written as* $w = \prod_{i=1}^q y_{v_i}^{\epsilon_i}$, $\epsilon_i = \pm 1$, $\epsilon_i = \epsilon_{i+1}$ *if* $v_i = v_{i+1}$ *with* $L(y_{v_i}) \leq L(w)$ *for* $i = 1, \ldots, q$.

(iii) *There is a product* $a = \prod_{i=1}^q y_{v_i}^{\epsilon_i}$, $a \neq 1$ *with* $y_{v_i} \in A$ $(i = 1, .., q)$ *and in one of the factors* H_j *there is an element* $x \notin A$ *with* $x^{-1}ax \in A$.

(iv) *There is a* $g \in G$ *such that for some* $i \in \{1, \ldots, m\}$ *we have* $y_i \notin gAg^{-1}$, *but for a suitable natural number* k *we have* $y_i^k \in gAg^{-1}$.

(v) *Of the* y_i *there are* $p \geq 1$ *contained in a subgroup of* G *conjugate to* H_1 *or* H_2 *and a certain product of them is conjugate to a non-trivial element of* A.

The Nielsen transformation can be chosen so that $\{y_1, \ldots, y_m\}$ is shorter (with respect to the length and a suitable order) than $\{x_1, \ldots, x_m\}$ or the lengths of the elements of $\{x_1, \ldots, x_m\}$ are preserved. Further if $\{x_1, \ldots, x_m\}$ is a generating system of G then in case (v) we find $p \geq 2$ for in this case conjugations determine a Nielsen transformation.

If we are interested in the combinatorial description of $\langle x_1, \ldots, x_m \rangle$ in terms of generators and relations we find again that $p \geq 2$ in case (v), possibly after suitable conjugations.

The main tool in the proof of Theorem 3.1 is the following:

Lemma 3.1. *Suppose* $x, y, z \in G$, $1 \neq x, 1 \neq z$ *with* $L(xy) \geq L(x), L(y)$ *and* $L(yz) \geq L(y), L(z)$. *Suppose further that neither* xy *precedes* x *nor* yz *precedes* z. *Then the following holds.*
(a) *If* $L(xyz) \leq L(x) - L(y) + L(z)$ *then* y *is conjugate to an element of* H_1 *or* H_2.
(b) *If* $L(x), L(z) > L(y)$ *and* $L(xyz) < L(x) - L(y) + L(z)$ *then* y *is conjugate to an element of* A.
(c) *If* $V \subset G$ *is a finite Nielsen reduced system,* $x, y, z \in V^{\pm 1}$ *and* $L(xyz) < L(x) - L(y) + L(z)$ *then* y *is conjugate to an element of* A *or* $x = y = z$.

Proof. Let $x \equiv p_x k_x q_x$, $y \equiv p_y k_y q_y$, $z \equiv p_z k_z q_z$ be the symmetric normal forms for x, y, z.

(a) Let $L(xyz) \leq L(x) - L(y) + L(z)$. If $y \in H_1$ or $y \in H_2$ then there is nothing to prove. Hence suppose $p_y \neq 1$, $q_y \neq 1$. Since $L(xyz) \leq L(x) - L(y) + L(z)$, $L(xy) \geq L(x)$, $L(y)$ and $L(yz) \geq L(y), L(z)$ we must have $q_x \equiv r_x p_y^{-1}$ and $p_z \equiv q_y^{-1} l_y$. Since neither xy precedes x nor yz precedes z we must have $p_y^{-1} \leq q_y$ and $q_y^{-1} \leq p_y$, that is $p_y = q_y^{-1}$.

(b) Let $L(x), L(z) > L(y)$ and $L(xyz) < L(x) - L(y) + L(z)$. If $y \in A$ there is nothing to prove. Hence suppose that $L(y) \geq 1$. We know from part (a) that y is conjugate to an element of H_1 or H_2 and therefore $k_y \notin A$.

Now we must have $q_x \equiv r_x r p_y^{-1}, r \in L_i^{-1}, r \neq 1$, and $p_z \equiv q_y^{-1} l l_y, l \in L_i, l \neq 1$, $p_y = q_y^{-1}$ and $r k_y l \in A$ ($i = 1$ or 2). Let r' be the right coset representative of $r k_y$ and l' the left coset representative of $k_y l$. Since xy does not precede x we must have $r' \geq r$. Analogously $l' \geq l$. Hence $r' = l^{-1}$, $l' = r^{-1}$ and $r^{-1} = l$ since $r k_y l \in A$. Therefore k_y is conjugate to an element of A and hence also y.

(c) Let $x, y, z \in V^{\pm 1}$ with $V \subset G$ a finite Nielsen reduced system and suppose $L(xyz) < L(x) - L(y) + L(z)$. If $y \in A$ then there is nothing to prove so suppose that $y \notin A$. Then from part (a) $k_y \notin A$ and $q_x \equiv r_x p_y^{-1}$, $p_z \equiv q_y^{-1} l_y$. If $r_x \neq 1, l_y \neq 1$ then the statement follows from part (b). Now suppose $r_x = 1$ or $l_y = 1$. Then $r_x = l_y = 1$ since V is Nielsen reduced. Since $k_y \notin A$ we have $L(xyz) < L(x), L(y), L(z)$. This implies $x = y = z$ since V is Nielsen reduced.

We note that if $L(g^n) < L(g)$ for $g \in G$ and $n \in \mathbb{N}$ then g is conjugate to an element of H_1 or H_2 and g^n is conjugate to an element of A.

In a finite system $\{x_1, \ldots, x_m\}$, $m \geq 1$, it can happen that some of the x_i are conjugate to elements of H_1 or H_2. In applying Nielsen reduction to results like the Kurosh theorem it is important to preserve this property if one tries to replace $\{x_1, \ldots, x_m\}$ by a shorter system. This leads to the following ideas developed by Kalia and Rosenberger [K-R].

Definition 3.1. A finite subset $X = \{x_1, \ldots, x_m\}$, $m \geq 1$, in G is called an **E-set** if it is composed as follows:

(a) $X = X_1 \cup X_2$, $X_1 \cap X_2 = \emptyset$,

(b) each $x_j \in X_1$ is conjugate to an element of H_1 or H_2, and

(c) each $x_j \in X_2$ is not conjugate to an element of H_1 or H_2.

Definition 3.2. On an E-set $X = \{x_1, \ldots, x_m\}$ with partition $X = X_1 \cup X_2$ as above we define the following types of transformations which produce an E-set Y with partition $Y = Y_1 \cup Y_2$:

(E1) replace some $x_j \in X_1$ by $x_j' = x_k^\epsilon x_j x_k^{-\epsilon}$, $k \neq j$, $\epsilon = \pm 1$, and leave the remaining x_i, $i \neq j$ fixed;

(E2) replace some $x_j \in X_2$ by $x_j' = x_k^\epsilon x_j$, or $x_j' = x_j x_k^\epsilon$, $k \neq j$, $\epsilon = \pm 1$, and leave the remaining x_i, $i \neq j$, fixed;

(E3) replace some $x_j \in X$ by $x_j' = x_j^{-1}$ and leave the remaining x_i, $i \neq j$, fixed;

(E4) permute in X_1 and leave X_2 fixed;

(E5) permute in X_2 and leave X_1 fixed;

(E6) delete some $x_j \in X$ where $x_j = 1$, $1 \leq j \leq m$.

We call these the **elementary E-transformations**. An **E-transformation** is a finite product of elementary E-transformations. An E-set Y is **derivable** from an E-set X if there is an E-transformation from X to Y. An E-set $X = \{x_1, \ldots, x_m\} \subset G$ is **E-reduced** if there is no E-set $Y = \{y_1, \ldots, y_m\}$ derivable from X, such that one of the following holds:

(a) $y_i = 1$ for some $i \in \{1, \ldots, m\}$

(b) $y_i \neq 1$ for all $i \in \{1, \ldots, m\}$ and Y is shorter than X.

We note that using the elementary E-transformation (E2) we can go from an E-set X to an E-set Y which has more elements conjugate to an element of H_1 or H_2: hence in the corresponding partitions $X = X_1 \cup X_2$, $Y = Y_1 \cup Y_2$ the set Y_1 contains more elements than X_1. Also we note that in $X = X_1 \cup X_2$ we could have $X_1 = \emptyset$ or $X_2 = \emptyset$.

Further we have the following two direct observations:

(a) If the E-set Y is derivable from the E-set X then $\langle Y \rangle = \langle X \rangle$.

(b) Suppose $X = \{x_1, \ldots, x_m\} \subset G$, $m \geq 1$, is an E-set and suppose G is countable. Then there exists an E-reduced E-set Y which is derivable from X.

As already mentioned if we consider a given finite system in G we may assume that G is countable.

Analogously to Theorem 3.1 we obtain

Theorem 3.2. *Let $G = H_1 \star_A H_2$. If $\{x_1, \ldots, x_m\}$, $m \geq 1$, is a finite system of elements in G then there is an E-transformation from $\{x_1, \ldots, x_m\}$ to a system $\{y_1, \ldots, y_m\}$ for which one of the following cases holds:*

(i)   *$y_i = 1$ for some $i \in \{1, \ldots, m\}$;*

(ii) *Each $w \in \langle y_1, \ldots, y_m \rangle$ can be written as $w = \prod_{i=1}^{q} y_{v_i}^{\epsilon_i}$, $\epsilon_i = \pm 1$, $\epsilon_i = \epsilon_{i+1}$ if $v_i = v_{i+1}$ with $L(y_{v_i}) \leq L(w)$ for $i = 1, \ldots, q$;*

(iii) *Of the y_i there are p, $p \geq 1$ contained in a subgroup of G conjugate to H_1 or H_2 and a certain product of them is conjugate to a non-trivial element of A.*

As stated this theorem is weaker than Theorem 3.1. However we note that it can be refined in a manner similar to the earlier theorem (see [K-R]).

Two of the main tools for the proof of Theorem 3.2 are the following lemmas.

Lemma 3.2 ([K-R]). *Suppose $x, y \in G$ with $L(y) \leq L(x)$ and $y \equiv p_y k_y q_y$, $x \equiv p_x k_x p_x^{-1}$, $k_x \in H_i \setminus A$, $i = 1$ or 2, in symmetric normal form. If $L(xy^\epsilon) < L(x)$ or $L(y^{-\epsilon}x) < L(x)$, $\epsilon = \pm 1$, then one of the following cases occurs:*

(a) *$L(y^{-\epsilon}xy^\epsilon) < L(x)$;*

(b) *$q_y = p_y^{-1} = p_x^{-1}$, and $k_x k_y^\epsilon$ or $k_y^{-\epsilon} k_x$ is contained in A;*

(c) *$q_y \neq p_y^{-1}$, $k_y \in H_i \setminus A$, $L(x) = L(y)$, $L(y^{-\epsilon}xy^\epsilon) = L(x)$, and $k_x k_y^\epsilon$ or $k_y^{-\epsilon} k_x$ is contained in A; especially $L(xy^\epsilon) < L(y)$ or $L(y^{-\epsilon}x < L(y)$.*

Proof. Suppose $L(xy) < L(x)$. Since $k_x \in H_i \setminus A$ we have $L(x) = 2L(p_x) + 1$. From $L(y) \leq L(x)$ and $L(xy) < L(x)$ we get that $p_x^{-1} \equiv r_x^{-1} p_y^{-1}$.

If $r_x \neq 1$ then necessarily $L(r_x^{-1} k_y q_y) < L(r_x) + L(q_y) + 1$ and case (a) holds.

Suppose now that $p_x^{-1} = p_y^{-1}$. If $k_y \in A$ then necessarily $L(k_x k_y q_y) < L(q_y) + 1$ and $L(y^{-1}xy) = L(q_y^{-1} k_y^{-1} k_x k_y q_y) < L(x)$ and case (a) holds.

Now suppose $k_y \notin A$. Then $L(x) = L(y)$, $xy = p_x k_x k_y q_y$ and $k_x k_y \in A$ since $L(xy) < L(x)$. If $q_y = p_x^{-1}$ then case (b) holds while if $q_y \neq p_x^{-1}$ case (c) holds.

Lemma 3.3 ([K-R]). *Suppose $x, y \in G$ with $L(y) \leq L(x)$ and $x \equiv p_x k_x p_x^{-1}$, $k_x \in H_i \setminus A$, $i = 1$ or 2, in symmetric normal form. If $L(xy^\epsilon) = L(x)$ or $L(y^{-\epsilon}x) = L(x)$, $\epsilon = \pm 1$, then $L(y^{-\epsilon}xy^\epsilon) \leq L(x)$.*

Proof. Let $L(xy) = L(x)$ and $y \equiv p_y k_y q_y$ in symmetric normal form. If $p_x = 1$ then also $p_y = 1$ and $x, y \in H_i$ since $L(y) \leq L(x) = L(xy)$, and hence $L(y^{-1}xy) \leq L(x)$.

Suppose $p_x \neq 1$. We must have $p_x^{-1} \equiv r_x^{-1} p_y^{-1}$ because $L(xy) = L(x)$. Then $xy = p_x k_x r_x^{-1} k_y q_y$. If $r_x \neq 1$ then $L(r_x^{-1} k_y q_y) = L(p_x)$ and hence $L(y^{-1}xy) = L(x)$ since $L(y) \leq L(x)$.

Now suppose $r_x = 1$, that is, $xy = p_x k_x k_y q_y$. Then $L(k_x k_y) = 1$ since $L(xy) = L(x)$ and $L(y^{-1}xy) = L(q_y^{-1} k_y^{-1} k_x k_y q_y) \leq L(x)$ since $L(y) \leq L(x)$.

There are many applications of the above theorems on cancellations in amalgamated free products. To handle these one must consider in detail the "bad situation" where some elements are in a conjugate of a factor H_1 or H_2 and a product of these elements is in a conjugate of the amalgamated subgroup A. The first set of applications are to **co-compact Fuchsian groups** which we present here. In the next two sections we survey other applications: first to a solution of the isomorphism problem for cyclically pinched one-relator groups and then to equations in free products with amalgamation. Recall that geometrically a co-compact Fuchsian group is a non-elementary discrete group of isometries of the hyperbolic plane (see [K]). Combinatorially these groups are finitely generated with presentations of the following form:

$$F = \langle s_1, \ldots, s_m, a_1, b_1, \ldots, a_g, b_g ; s_1^{\alpha_1} = \cdots = s_m^{\alpha_m} = R = 1 \rangle \qquad (3.1)$$

where $R = s_1 \cdots s_m [a_1, b_1] \cdots [a_g, b_g], g \geq 0, m \geq 0, \alpha_i \geq 2, i = 1, \ldots, m$ and $2g - 2 + \sum_{i=1}^m (1 - \frac{1}{\alpha_i}) > 0$.

Theorem 3.3 ([P-R-Z]). *Suppose F is a co-compact Fuchsian group with presentation of the form (3.1). Then F has rank*
(a) *$2g$ if $m = 0$, $g \geq 1$;*
(b) *$m - 2$ if $g = 0$, m is even, and all α_i equal 2 except for one of them which is odd;*
(c) *$2g + m - 1$ in all other cases.*

Here we say that a finitely generated group G has **rank r** if it can be generated by r elements but not by $(r - 1)$ elements.

For co-compact Fuchsian groups F, case (b) of the theorem, where the rank of F is $m - 2$, is unexpected and particularly interesting. Here there is only one Nielsen equivalence class of generating systems $\{x_1, \ldots, x_{m-2}\}$. Even more concretely if m is even, $\alpha_1 = \cdots = \alpha_{m-1} = 2$ and α_m is odd then each generating system $\{x_1, \ldots, x_{m-2}\}$ is Nielsen equivalent to $\{s_1 s_2, \ldots, s_1 s_{m-1}\}$ (see [R3] and [R9]). In the next theorem we extend this application.

Theorem 3.4 ([R3], [R9]). *Suppose F is a co-compact Fuchsian group with presentation of the form (3.1). Let the rank of F be $2g$ if $m = 0$ or $2g + m - 1$ if $m \geq 1$.*
(a) *If $m \leq 1$ then each generating system $x_1, \ldots, x_{2g}$ is Nielsen equivalent to $\{a_1, b_1, \ldots, a_g, b_g\}$.*
(b) *If $m \geq 2$ then each automorphism $\phi : F \rightarrow F$ is induced by some automorphism ϕ^* of the free group $F^* = F^*(S_1, \ldots, S_{m-1}, A_1, B_1, \ldots, A_g, B_g)$ with respect to the epimorphism $F^* \rightarrow F$ given by $S_i \rightarrow s_i$, $A_j \rightarrow a_j$, $B_j \rightarrow b_j$.*
(c) *Suppose $m \geq 2$, and $m \geq 4$ if $g = 0$, and suppose that one of the following holds:*
   (1) *At least three α_i are greater than 2 if $g = 0$.*
   (2) *All α_i are equal to 2 except for two which are both even and greater than two if $g = 0$.*

(3) *m is even and all α_i are equal to 2 except for two of them which are both odd if $g = 0$.*

Assume that $\{x_1, \ldots, x_{2g+m-1}\}$ is a minimal generating system for F. Then $\{x_1, \ldots, x_{2g+m-1}\}$ is Nielsen equivalent to a system $\{s_{v_1}^{\beta_1}, \ldots, s_{v_{m-1}}^{\beta_{m-1}}, a_1, b_1, \ldots, a_g, b_g\}$ with $v_i \in \{1, \ldots, m\}$, $v_1 < v_2 \cdots < v_{m-1}$ if $m \geq 3$ and $1 \leq \beta_i < \alpha_{v_i}$, $\gcd(\beta_i, \alpha_{v_i}) = 1$.

Theorem 3.4 is a summary and extension of several different results. The references [R3], [R9] give this summary but also indicate the original sources. For example statement (a) for $m = 0$, $g \neq 3$ was done by Zieschang [Z1] while the case when $g = 0$, $m = 3$ is easy to handle by changing the decomposition as a free product with amalgamation. For a discussion of the generating pairs of F for $m = 0$, $g = 3$ see the paper of Fine and Rosenberger [F-R2].

Statement (b) is a small extension of another result of Zieschang [Z3]. Another extension in a different direction can be found in the paper of Lustig, Moriah and Rosenberger [L-M-R].

The proof of statement (c) depends on the decomposition of F as an amalgamated free product under the conditions that $m \geq 2$ if $g \geq 1$ and $m \geq 4$ if $g = 0$. For the consideration of the "bad" situations we use the results described in the section on free groups together with the following result.

Lemma 3.4 ([R9]). *Suppose $H = \langle s_1, \ldots, s_m ; s_1^{\alpha_1} = \cdots = s_m^{\alpha_m} = 1 \rangle$, $n \geq 2$, all $\alpha_i \geq 2$ and suppose $\{x_1, \ldots, x_n\} \subset H$, $n \geq 1$. Let X be the subgroup generated by $\{x_1, \ldots, x_n\}$ and suppose $y^{-1}(s_1 \cdots s_m)^\alpha y \in X$ for some $\alpha \neq 0$ and some $y \in H$. Then one of the following cases occurs:*

(a) *There is a Nielsen transformation from $\{x_1, \ldots, x_n\}$ to a system $\{y_1, \ldots, y_n\}$ with $y_1 = z(s_1 \cdots s_m)^\beta z^{-1}$, $\beta \geq 1$, $z \in H$ or $y_1 = zs_i^{\gamma_i} z^{-1}$ for some i, $1 \leq i \leq m$, $2 \leq \gamma_i < \alpha_i$, $\gcd(\gamma_i, \alpha_i) \geq 2$ and $z \in H$.*

(b) *X has finite index $\beta = [H : X]$, where β is the smallest positive integer for which a relation $y^{-1}(s_1 \cdots s_m)^\beta y \in X$ holds for some $y \in H$.*

In fact, for the proof of statement (c) of Theorem 3.3 we need only consider the case where $n \leq m$. If it is already true that $n < m$ then statement (b) of Lemma 3.4 may be replaced by:

(b') *m odd, $n = m - 1$, all $\alpha_i = 2$, $\{x_1, \ldots, x_n\}$ is Nielsen equivalent to $\{s_1 s_2, \ldots, s_1 s_m\}$ and $s_1 \cdots s_m \notin X$ but $(s_1 \cdots s_m)^2 \in X$ (see [P-R-Z]).*

Statement (c) of the theorem can be extended but the extension is quite diffi-cult. This is evident from the following example in which the occurring exponent is not coprime to the respective order. Let $F = \langle s_1, s_2, s_3, s_4 ; s_1^{10} = s_2^2 = s_3^2 = s_4^5 = s_1 s_2 s_3 s_4 = 1 \rangle$. Let $x_1 = s_1^4$, $x_2 = s_1 s_2$, $x_3 = s_1 s_3$ and $X = \langle x_1, x_2, x_3 \rangle$. Now $\gcd(4, 10) = 2 > 1$. Then $x_2 x_3^{-1} x_1^{-2} x_2^{-1} x_3 = (s_1 s_2 s_3)^2 = s_4^{-2} \in X$. Hence $\{s_4, s_1 s_2 s_3, s_3, s_1, s_2\} \subset X$ and therefore $X = F$.

Finally we mention that Theorem 3.3 has been extended by Rosenberger [R3] to a wider class of groups.

4. The Isomorphism Problem for Cyclically Pinched One-Relator Groups

In this section we present some applications of the Nielsen theory in free products with amalgamation to a certain class of one-relator groups. In particular the Nielsen methods allow for the solution of the isomorphism problem for a subclass containing the cyclically pinched one-relator groups. The class of one-relator groups we consider have the following form:

$$G = \langle a_1, \ldots, a_p, b_1, \ldots, b_q ; (WV)^m = 1 \rangle \qquad (4.1)$$

where $m \geq 1$, $1 \leq p$, $1 \leq q$, $1 \neq W = W_1^r$, $r \geq 1$, $W_1 = W_1(a_1, \ldots, a_p)$ is not a proper power in the free group on $a_1, \ldots, a_p$ and $1 \neq V = V_1^s$, $s \geq 1$, $V_1 = V_1(b_1, \ldots, b_q)$ is not a proper power in the free group on $a_1, \ldots, a_p$. Further we suppose that $r \geq 2$ if W_1 is a primitive element in $\langle a_1, \ldots, a_p ; \rangle$ and $s \geq 2$ if V_1 is a primitive element in $\langle b_1, \ldots, b_q ; \rangle$. Recall that if $m = 1$, so that G is the free product of the free groups on $a_1, \ldots, a_p$ and $b_1, \ldots, b_q$ respectively amalgamated over the infinite cyclic groups $\langle W \rangle = \langle V^{-1} \rangle$ then G is called a **cyclically pinched one-relator group** (see [B1]). We get the following result, which gives a solution to the isomorphism problem for cyclically pinched one-relator groups. By this we mean that given a specific cyclically pinched one-relator group G we can determine algorithmically in finitely many steps whether or not an arbitrary one-relator group (given by a specific one-relator presentation) is isomorphic to G.

Theorem 4.1 (see [R10]). *Suppose G is a one-relator group with a presentation of the form* 4.1. *Then:*

(a) *Each automorphism $\phi : G \to G$ is induced by some automorphism ϕ^* of the free group $F^* = F^*(A_1, \ldots, A_p, B_1, \ldots, B_q)$ with respect to the epimorphism $F^* \to G$ given by $A_i \to a_i$, $B_j \to b_j$. In particular the automorphism group of G is finitely generated.*

(b) *If $m \geq 2$ and $\{x_1, \ldots, x_{p+q}\}$ is a generating system of G then $\{x_1, \ldots, x_{p+q}\}$ is Nielsen equivalent to $\{a_1, \ldots, a_p, b_1, \ldots, b_q\}$ and we can decide algorithmically in finitely many steps whether or not an arbitrary one-relator group (given by a specific one-relator presentation) is isomorphic to G.*

(c) *Suppose $m = 1$ and at most one of W_1 or V_1 is primitive. If $\{x_1, \ldots, x_{p+q}\}$ is a generating system of G then one of the following cases occurs:*

(1) *There is a Nielsen transformation from $\{x_1, \ldots, x_{p+q}\}$ to a system $\{a_1, \ldots, a_p, y_1, \ldots, y_q\}$ with $\{y_1, \ldots, y_q\} \subset H_2$ where $H_2 = \langle V, y_1, \ldots, y_q \rangle$.*

(2) *There is a Nielsen transformation from* $\{x_1, \ldots, x_{p+q}\}$ *to a system* $\{y_1, \ldots, y_p, b_1, \ldots, b_q\}$ *with* $\{y_1, \ldots, y_p\} \subset H_1$ *where* $H_1 = \langle W, y_1, \ldots, y_p \rangle$.

For this generating system $\{x_1, \ldots, x_{p+q}\}$, *there is a presentation of G with a single defining relation. Further G has only finitely many Nielsen equivalence classes of minimal generating systems and we can decide algorithmically in finitely many steps whether or not an arbitrary one-relator group (given by a specific one-relator presentation) is isomorphic to G.*

As an example consider the one-relator group (see [R8], [R11])

$$G = \langle c_1, \ldots, c_p, a_1, b_1, \ldots, a_g, b_g \,;\, c_1^{\alpha_1} \cdots c_p^{\alpha_p} [a_1, b_1] \cdots [a_g, b_g] = 1 \rangle$$

where all $\alpha_i \geq 2$, $p \geq 0$, $g \geq 0$, $g \geq 1$ if $p = 0$ and $p \geq 3$ if $g = 0$. If $\{x_1, \ldots, x_{2g+p}\}$ is a generating system for G then there is a Nielsen transformation from $\{x_1, \ldots, x_{2g+p}\}$ to a system $\{c_1, \ldots, c_{i-1}, c_i^{\gamma_i}, c_{i+1}, \ldots, c_p, a_1, b_1, \ldots, a_g, b_g\}$ with $1 \leq i \leq p$, $1 \leq \gamma_i \leq \alpha_i/2$, $\gcd(\alpha_i, \gamma_i) = 1$. To actually prove this we use the theorem together with the results on free groups discussed in Section 2.

We mention that the case where both W_1 and V_1 are primitive was discussed in papers by Rosenberger [R11], Zieschang [Z3] and Collins [C].

In addition to the fact that G is decomposable as a free product with amalgamation the proof of Theorem 4.1 is based on the concept of an **r-stable** Nielsen equivalence class. Let $F = \langle a_1, \ldots, a_n \,;\, \rangle$, $n \geq 2$ be the free group of rank n with basis $\{a_1, \ldots, a_n\}$. Let $r \in F$, $r \neq 1$ be a freely reduced word and suppose that there is no Nielsen transformation from $\{a_1, \ldots, a_n\}$ to a system $\{b_1, \ldots, b_n\}$ with $r \in \langle b_1, \ldots, b_{n-1} \rangle$. We consider generating systems $\{r, x_1, \ldots, x_n\}$ of F and we say that in a Nielsen transformation ϕ from $\{r, x_1, \ldots, x_n\}$ to a system $\{r, y_1, \ldots, y_n\}$, r is **not replaced** if in all the elementary Nielsen transformations comprising ϕ, r either remains unchanged, is replaced by r^{-1} or is put in a different place in the relevant $(n+1)$-tuple. We then refer to a Nielsen transformation and the corresponding Nielsen equivalence class in which r is not replaced as **r-stable**. In the proof of Theorem 4.1 the following result was used.

Lemma 4.1. *Let* $F = \langle a_1, \ldots, a_n \,;\, \rangle$ *be the free group of rank n with basis* $\{a_1, \ldots, a_n\}$ *and let r be as above. Then there are only finitely many r-stable Nielsen equivalence classes of generating systems* $\{r, x_1, \ldots, x_n\}$ *of F.*

The Nielsen techniques used for Theorem 4.1 can also be used to prove certain freeness results for groups of the form (4.1). Recall that a group G is **n-free** if each n-generator subgroup is free (necessarily of rank $\leq n$). In [R13] Rosenberger proved the following:

Theorem 4.2 ([R13]). *Let G have form (4.1) with* $m = r = s = 1$. *Then:*

(a) *G is 3-free; that is every three-generator subgroup* $H \subset G$ *is free of rank* ≤ 3.

(b) *Let $H \subset G$ be a four-generator subgroup. Then one of the following two cases can occur:*

  (1) *H is free of rank ≤ 4;*

  (2) *If $\{x_1, x_2, x_3, x_4\}$ is a generating system of H then there is a Nielsen transformation from $\{x_1, x_2, x_3, x_4\}$ to a system $\{y_1, y_2, y_3, y_4\}$ with $y_1, y_2 \in z\langle a_1, \dots, a_p \rangle z^{-1}$ and $y_3, y_4 \in z\langle b_1, \dots, b_q \rangle^{-1}$ for a suitable $z \in G$. Further for the generating system $\{x_1, x_2, x_3, x_4\}$ of H there is a presentation of H with one defining relation.*

We note that G. Baumslag [B2] showed that when $m = r = s = 1$ the group G is 2-free. This result can be extended to the case where $m = r = 1$, $s \geq 1$. These statements together with Theorem 4.2 are generalizations of facts about surface groups (see [L-S]). Part (a) of Theorem 4.2 was reproved in a different manner by Baumslag and Shalen [B-S]. There are also generalizations by Fine, Rohl and Rosenberger to HNN groups similar in form to (4.1) [F-R-R1, 2]. We will discuss these in Section 6. In connection with questions on logic and the universal theory of free groups the following results were proved by Fine, Gaglione, Rosenberger and Spellman [F-G-R-S1] related to the above.

Theorem 4.3 ([F-G-R-S1]). *Let G be a one-relator group with presentation*

$$G = \langle B_1, \dots, B_n ; \prod_{i=1}^{n} W_i = 1 \rangle$$

where $B_1, \dots, B_n$, $n \geq 1$, are non-empty pairwise disjoint subsets of generators each of size ≥ 2 and for $i = 1, \dots, n$ we have that the $W_i = W_i(B_i) \neq 1$ are non-trivial words neither proper powers nor primitive elements in the free group on B_i. Then G is n-free.

Theorem 4.4 ([F-G-R-S1]). *Let G be a one-relator group with presentation*

$$G = \langle B_1, \dots, B_n ; \prod_{i=1}^{n} V_i^{t_i} = 1 \rangle$$

where $B_1, \dots, B_n$, $n \geq 1$, are non-empty pairwise disjoint subsets of generators and for $i = 1, \dots, n$ we have that the $V_i = V_i(B_i) \neq 1$ are non-trivial words not proper powers in the free group on B_i and $t_i \geq 1$. Then G is $(n-1)$-free.

We note that the result in Theorem 4.4 is the best possible because the non-orientable surface group of genus n has a presentation $G = \langle a_1, \dots, a_n ; a_1^2 \cdots a_n^2 = 1 \rangle$ and is not a free group.

Another straightforward result along the lines of the above is (see [K-S]):

Theorem 4.4. *Suppose* $G = H_1 \star_A H_2$ *with* $H_1 \neq A \neq H_2$ *and* A *is malnormal in both* H_1 *and* H_2. *If* $H = \langle x_1, x_2 \rangle \subset G$ *then* H *is either cyclic or the free product of two cyclic groups or conjugate to a subgroup of* H_1 *or* H_2.

Recall that a subgroup H of a group G is **malnormal** if $H \cap xHx^{-1} = \{1\}$ if $x \notin H$.

5. Equations in Amalgamated Free Products

The final set of applications of the Nielsen theory in free products with amalgamation concerns equations in such constructions. We first recall the basic concepts.

Let H be a group and F a free group of rank $n \geq 1$ with ordered basis $X = \{x_1, \ldots, x_n\}$. We regard the elements of F as reduced words $w = w(x_1, \ldots, x_n)$. The homomorphisms $\phi : F \rightarrow H$ are represented uniquely by ordered sets $\phi(X) = U = \{u_1, \ldots, u_n\}$ of elements of H where $\phi(x_i) = u_i$, $i = 1, \ldots, n$. If $w = w(x_1, \ldots, x_n) \in F$ then $U \subset H$ is a **solution** of the equation $w = 1$ in H if $\phi(w) = w(u_1, \ldots, u_n) = 1$ in H.

If $\alpha : F \rightarrow F$ is an automorphism of F then α is defined by a Nielsen transformation on $\{x_1, \ldots, x_n\}$. If $x_i' = \alpha(x_i) = \alpha_i(x_1, \ldots, x_n)$, $1 \leq i \leq n$, then we write $\alpha(U) = U' = \{u_1', \ldots, u_n'\}$ for the Nielsen equivalent system $\phi\alpha(X) = \{\alpha_1(u_1, \ldots, u_n), \ldots, \alpha_n(u_1, \ldots, u_n)\}$. Suppose U is a solution in H of the equation $w = 1$ and $\alpha : F \rightarrow F$ is an automorphism. If $w' = \alpha(w)$ then $U' = \alpha^{-1}(U)$ is a solution of the equation $w' = 1$. In general we call the pairs (w, U) and (w', U'), where $w' = \alpha(w)$, $U' = \alpha^{-1}(U)$ with α an automorphism of F, **Nielsen equivalent** (induced by the automorphism α). Certainly the study of the solutions of $w = 1$ is equivalent to the study of the solutions of $w' = 1$.

A word $w \in F$ is called **regular** if there is no automorphism $\alpha : F \rightarrow F$ such that $w' = \alpha(w)$, as a word in $x_1, \ldots, x_n$, contains less of the $x_1, \ldots, x_n$ than w itself, that is, W is of minimal rank. Suppose W is not regular. Then if x_i is in w but not in $w' = \alpha(w)$ the element $u_i' = \phi\alpha^{-1}(x_i)$ will appear as an arbitrary parameter in each solution of $w' = 1$ and hence in each solution of $w = 1$.

A word $w = w(x_1, \ldots, x_n) \in F$ is **quadratic** if each x_i which occurs in w occurs exactly twice, each time as x_i or x_i^{-1}. A quadratic word $w = w(x_1, \ldots, x_n) \in F$ is **alternating** if each x_i which occurs in w occurs exactly once as x_i and exactly once as x_i^{-1}.

Now suppose $G = H_1 \star_A H_2$, $H_1 \neq A \neq H_2$, is a free product with amalgamated subgroup A. An ordered set $U = \{u_1, \ldots, u_n\} \subset G$ is **regular** if there is no Nielsen transformation from U to a system $U' = \{u_1', \ldots, u_n'\}$ in which one of the elements is conjugate to an element of A.

If $U \subset G$ is a solution of an equation $w = 1$ and U is not regular, then, for some automorphism $\alpha : F \rightarrow F$, $U' = \alpha^{-1}(U) = \{u_1', \ldots, u_n'\}$ with $u_i' = gag^{-1}, a \in A$,

$g \in G$, is a solution of the equation $w' = 1$ with $w' = \alpha(w)$. If here $a = 1$—this is the case if G were a free product—then U' is essentially a solution $U_0' = \{u_2', \ldots, u_n'\}$ of $w_0 = 1$ where $w_0 = w_0(x_2, \ldots, x_n) = w'(1, x_2, \ldots, x_n)$.

We first have the following result due to Rosenberger [R6].

Theorem 5.1 ([R6]). *Suppose* $G = H_1 \star_A H_2$, $H_1 \neq A \neq H_2$, *is a free product with amalgamated subgroup* A. *Let* $1 \neq w = w(x_1, \ldots, x_n)$ *be a regular word in* F, *with* F *free of rank* $n \geq 1$ *on the ordered basis* $\{x_1, \ldots, x_n\}$. *Further let* $\phi : F \to G$ *be a homomorphism such that* $U = \phi(X)$ *is a Nielsen reduced set and* $\phi(w) = hah^{-1}$, $a \in A$, $h \in G$. *Then there exists a cyclic permutation* $w' \in F$ *of* W, *induced by an automorphism* $\alpha : F \to F$, *such that*
(1) $w' = \alpha(w) = w_1 w_2$ *with* $w_1 \neq 1$;
(2) $\phi \alpha^{-1}(w_1)$ *is conjugate in* G *to an element of* A;
(3) *There is an* $i = 1$ *or* 2 *and a* $g \in G$ *with* $\phi \alpha^{-1}(x_j) \in g H_i g^{-1}$ *for each* x_j *which occurs in* w_1.

If in addition, A is normal in G then we may argue inductively to get the following.

Theorem 5.2 ([R6], [R12]). *Suppose* $G = H_1 \star_A H_2$, $H_1 \neq A \neq H_2$, *with* A *normal in* G. *Let* $1 \neq w = w(x_1, \ldots, x_n)$ *be a regular word in* F, *with* F *free of rank* $n \geq 1$ *on the ordered basis* $\{x_1, \ldots, x_n\}$, *and further let* $\phi : F \to G$ *be a homomorphism such that* $U = \phi(X)$ *is regular and* $\phi(w) = a \in A$. *Then the pair* (w, U) *is Nielsen equivalent to a pair* $(w', U') = (\alpha(w), \alpha^{-1}(U))$ *with* $\alpha : F \to F$ *an automorphism such that:*
(1) $w' = r_1 w_1 r_1^{-1} \cdots r_k w_k r_k^{-1}$ *for some* $k \geq 0$ *where, for* $1 \leq i \leq k$, r_i *and* w_i *are reduced words in* F;
(2) *For each* i, $1 \leq i \leq k$, *we have* $\phi \alpha^{-1}(w_i) \in A$;
(3) *For each* i, $1 \leq i \leq k$, *there exists a* $v_i = 1$ *or* 2 *and a* $g_i \in G$ *with* $\phi \alpha^{-1}(x_j) \in g_i H_{v_i} g_i$ *for each* x_j *which occurs in* w_i.
Moreover if ϕ *is an epimorphism so that* U *is a generating set for* G *then we may assume that* $g_i = 1$ *for all* i, $1 \leq i \leq k$.

Corollary 5.1. *If in addition to all the above the word* $w = w(x_1, \ldots, x_n)$ *is quadratic then statement* (1) *of Theorem 5.2 can be replaced by*
(1') $w' = w_1 \cdots w_k$ *for some* $k \geq 0$, *and where each* w_i, $1 \leq i \leq k$, *is a quadratic word in* F. *Further if* W *is alternating then each* w_i *is also alternating.*

We note that neither the theorem nor the corollary hold without the assumption that A is normal in G. This can be seen from the following example.

Let $F = \langle x_1, x_2, x_3 \rangle$ and let $w = x_1^2 x_2 x_3 x_2^{-1} x_3^{-1}$. Then w is a regular quadratic word in F. Now let G be the group with presentation

$$G = \langle t, s_1, s_2, s_3, s_4 ; s_1^2 = s_2^2 = s_3^2 = s_4^2 = t^2 s_1 s_2 s_3 s_4 = 1 \rangle.$$

G is then a free product with amalgamation $G = H_1 \star_A H_2$ where $H_1 = \langle t; \rangle$, $H_2 = \langle s_1, s_2, s_3, s_4; s_1^2 = s_2^2 = s_3^2 = s_4^2 = 1 \rangle$ and $A = \langle a \rangle = \langle t^2 \rangle = \langle s_1 s_2 s_3 s_4 \rangle$. A is not normal in H_2 and therefore A is not normal in G. Let the homomorphism $\phi : F \to G$ be defined by $\phi(x_1) = u_1 = t$, $\phi(x_2) = u_2 = s_1 s_2$, $\phi(x_3) = u_3 = s_3 s_1$. Then $\phi(X) = U = \{t, s_1 s_2, s_3 s_1\}$. The set U is regular but statement (1') of Corollary 5.1 does not hold. (see Theorem 1 of [P-R-Z]).

If the amalgamated subgroup A is malnormal in both factors then we can obtain the following somewhat weaker result.

Theorem 5.3 ([R12]). *Suppose $G = H_1 \star_A H_2$, $H_1 \neq A \neq H_2$, with A malnormal in both factors H_1 and H_2. Let $1 \neq w = w(x_1, \ldots, x_n)$ be a regular quadratic word in F, with F free of rank n, $1 \leq n \leq 4$, on the ordered basis $\{x_1, \ldots, x_n\}$. Further let $\phi : F \to G$ be a homomorphism such that $U = \phi(X)$ is regular and $\phi(w) = 1$. Then the pair (w, U) is Nielsen equivalent to a pair $(w', U') = (\alpha(w), \alpha^{-1}(U))$ with $\alpha : F \to F$ an automorphism such that:*

(1)   *$w' = w_1 w_2$ where w_1 and w_2 are also quadratic in F;*
(2)   *For $i = 1, 2$ we have that $\phi\alpha^{-1}(w_i)$ is conjugate to an element of A*
(3)   *For $i = 1, 2$ there is a $v_i = 1$ or 2 and a $g_i \in G$ with $\phi\alpha^{-1}(x_j) \in g_i H_{v_i} g_i^{-1}$ for each x_j which occurs in w_i.*

The above theorems on equations in free products with amalgamation have some straightforward consequences. In particular they can be used to recover the theorem of Kurosh–H. Neumann on subgroups of free products with amalgamation and the theorem of Grushko–B. H. Neumann on generating systems of free products with amalgamation.

Corollary 5.2 (Theorem of Kurosh and H. Neumann). *Suppose $G = H_1 \star_A H_2$, $H_1 \neq A \neq H_2$. Let K be a finitely generated subgroup of G with $gKg^{-1} \cap A = \{1\}$ for all $g \in G$. Then*

$$K = F \star (\star_{i \in I} K_i)$$

where F is a free group and each K_i is conjugate to a subgroup of H_1 or H_2.

Notice that if A is trivial, so that G is the free product of H_1 and H_2, we get the standard Kurosh theorem.

Corollary 5.3 (Theorem of Grushko–B. H. Neumann, see also McCool and Pietrowski [Mc-P]). *Suppose $G = H_1 \star_A H_2$, $H_1 \neq A \neq H_2$, with A normal in G and G finitely generated. If $\{x_1, \ldots, x_n\}$ is a generating system of G then there is a Nielsen transformation from $\{x_1, \ldots, x_n\}$ to a system $\{y_1, \ldots, y_n\}$ such that $y_i \in H_1 \cup H_2$ for $i = 1, \ldots, n$ (some of the y_i can be 1).*

Moreover if A is trivial, so that G is the free product of H_1 and H_2, then $\mathrm{rank}(G) = \mathrm{rank}(H_1) + \mathrm{rank}(H_2)$.

We note that Corollary 5.3 does not hold if A is not normal in G. Let $H_1 = \langle s_1, s_2 ; s_1^2 = s_2^2 = 1 \rangle$, $H_2 = \langle s_3, s_4 ; s_3^2 = s_4^m = 1 \rangle$ where $m \geq 3$ and m is odd. Let $A = \langle s_1 s_2 \rangle = \langle (s_3 s_4)^{-1} \rangle$ and let $G = H_1 \star_A H_2$. Then $G = \langle s_1 s_2, s_1 s_3 \rangle$ and the system $\{s_1 s_2, s_1 s_3\}$ is not Nielsen equivalent to a system $\{x, y\}$ with $x \in H_1 \cup H_2$ and at the same time $y \in H_1 \cup H_2$.

For discussions of the proofs of these results together with further references we refer to the book of Lyndon and Schupp [L-S].

The results on equations also yield the following theorem of Shenitzer [Sh].

Corollary 5.4. *Let F be free of rank n, $n \geq 2$, with ordered basis $\{x_1, \ldots, x_n\}$. Let $w = w(x_1, \ldots, x_n\}$ and let N be the normal closure of w in F. Then:*
(1) *If F/N is free then either $w = 1$ or w is primitive in F.*
(2) *If F/N is not free then F/N is decomposable into a non-trivial free product if and only if w is not regular.*

There are many other applications of the Nielsen method in free products with amalgamation. For example we also have the following, due again to Rosenberger.

Theorem 5.4 ([R12]). *Let $G = H_1 \star H_2 \star \ldots \star H_n$, $n \geq 2$. Let $1 \neq a_j \in H_j$ and let p be the number of the a_j which are proper powers in H_j, $1 \leq j \leq n$. Let $\{x_1, \ldots, x_m\} \subset G$, $m \geq 1$, and let $H = \langle x_1, \ldots, x_m \rangle$. If $a_1 \cdots a_n \in H$ then one of the following cases holds:*
(1) *There is a Nielsen transformation from $\{x_1, \ldots, x_m\}$ to a system $\{y_1, \ldots, y_m\}$ with $y_1 = a_1 \cdots a_n$.*
(2) *$m \geq 2n - p$ and there is a Nielsen transformation from $\{x_1, \ldots, x_m\}$ to a system $\{y_1, \ldots, y_m\}$ such that $y_i \in H_j$, $1 \leq j \leq n$, $1 \leq i \leq 2n - p$.*

For further applications of the theory in free products with amalgamation see [K-R] and [C-Z2].

6. The Theory in HNN Groups

The final amalgam construction for infinite groups is that of an HNN group. In this section we describe the Nielsen reduction method for such groups. The theory in HNN groups was carried over by N. Peczynski and W. Reiwer [P-R].

Let

$$K = \langle B, t ; rel\ B, t^{-1} K_1 t = K_{-1} \rangle$$

be an HNN group with **base group B**, **stable letter** or **free part t** and **associated subgroups K_1, K_{-1}** (see [L-S] for additional information on the HNN construction).

For an element $x \in K$ a representation

$$x = h_1 t^{\epsilon_1} h_2 t^{\epsilon_2} \cdots h_n t^{\epsilon_n} h_{n+1}$$

with $\epsilon_i = \pm 1$, $h_i \in B$, is said to be **reduced** if $\epsilon_{i+1} = -\epsilon_i$ implies that $h_{i+1} \notin K_{\epsilon_{i+1}}$.

Choose left transversals R_1 of K_1 and R_{-1} of K_{-1} in B where K_1 and K_{-1} are represented by 1. Then each $x \in K$ may be uniquely represented as

$$x = l_1 t^{\epsilon_1} l_2 t^{\epsilon_2} \cdots l_n t^{\epsilon_n} b$$

with $\epsilon_i = \pm 1$, $b \in B$, $l_i \in R_{\epsilon_i}$ and $\epsilon_i = \epsilon_{i+1}$ whenever $l_{i+1} = 1$. The **length** $L(x)$ of x is then defined to be n.

As right transversals of K_1, K_{-1} we take the inverses R_1^{-1}, R_{-1}^{-1}. Then each $x \in K$ has a reduced representation

$$x = l_1 t^{\epsilon_1} \cdots l_m t^{\epsilon_m} k_x t^{\nu_m} r_m \cdots t^{\nu_1} r_1$$

with $m \geq 0$, ϵ_i, $\nu_i \in \{\pm 1\}$, $l_i \in R_{\epsilon_i}$, $r_i \in R_{-\nu_i}^{-1}$ and $k_x = h_1 t^\epsilon h_2$, $h_1, h_2 \in B$, $\epsilon = \pm 1$, if $L(x)$ is odd or $k_x \in B$ if $L(x)$ is even. In this representation $l_1 t^{\epsilon_1} \cdots l_m t^{\epsilon_m}$ is called the **leading half**, $t^{\nu_m} r_m \cdots t^{\nu_1} r_1$ the **rear half** and k_x the **kernel** of x respectively. The above reduced representation is then called a **symmetric form** for x.

We now introduce, as in the amalgamated free product situation, an ordering on K. For our applications we may assume that the groups are countable. This is no restriction if one considers a given finite system in K, for given a finite system a suitable order can always be chosen so that this system can be carried by a Nielsen transformation into what we will call a Nielsen reduced system. Choose a total order of the transversals R_1, R_{-1}, and order products $l_1 t^{\epsilon_1} \cdots l_m t^{\epsilon_m}$ by using the lexicographic order on the sequences $(l_1, \ldots, l_m)$. Next we extend this order to the set of pairs $\{g, g^{-1}\}$, $g \in K$, where the notation is chosen such that the leading half of g precedes that of g^{-1} with respect to the above ordering. Let $\{g, g^{-1}\} < \{h, h^{-1}\}$ if either $L(g) < L(h)$ or $L(g) = L(h)$ and the leading half of g strictly precedes that of h or $L(g) = L(h)$ and the leading halves of g and h coincide while the leading half of g^{-1} precedes that of h^{-1}. Hence if $\{g, g^{-1}\} < \{h, h^{-1}\}$ and $\{h, h^{-1}\} < \{g, g^{-1}\}$ then at most the kernels of g and h may be different.

For $g \in K$ let the leading half of $g^{\epsilon(g)}$, $\epsilon(g) = \pm 1$, precede that of $g^{-\epsilon(g)}$.

A finite system $\{g_1, \ldots, g_m\}$ in K is called **shorter** than a system $\{h_1, \ldots, h_m\}$ if $\{g_i^{\epsilon(g_i)}, g_i^{-\epsilon(g_i)}\} < \{h_i^{\epsilon(h_i)}, h_i^{-\epsilon(h_i)}\}$ holds for all $i \in \{1, \ldots, m\}$ and at least for one $i \in \{1, \ldots, m\}$, $\{h_i^{\epsilon(h_i)}, h_i^{-\epsilon(h_i)}\} < \{g_i^{\epsilon(g_i)}, g_i^{-\epsilon(g_i)}\}$ fails to hold.

A system $\{g_1, \ldots, g_m\}$ in K is said to be **Nielsen reduced** or **minimal** with respect to $<$ if either $\{g_1, \ldots, g_m\}$ cannot be carried into a system $\{h_1, \ldots, h_m\}$ with $h_i = 1$ for some $i \in \{1, \ldots, m\}$ or there is no system Nielsen equivalent to $\{g_1, \ldots, g_m\}$ which is shorter. If the group K is countable, then each finite system, as in the case of a free product with amalgamation, can be carried by a Nielsen transformation into a minimal system. In general for a given finite system a suitable order can always be chosen so that this finite system can be carried by a Nielsen transformation into a minimal system.

Before giving Peczynski and Reiwer's main result we introduce some further notation. We write $u_1 \cdots u_q \equiv v_1 \cdots v_n$ for the equality together with the fact that $L(v_1 \cdots v_n) = L(v_1) + L(v_2) + \cdots + L(v_n)$ all $u_i, v_j \in K$. If $x \in K$ is given in symmetric normal form

$$x = l_1 t^{\epsilon_1} \cdots l_m t^{\epsilon_m} k_x t^{\nu_m} r_m \cdots t^{\nu_1} r_1$$

then we write $x \equiv p_x k_x q_x$ where p_x is the leading half and q_x is the rear half of x. If we write $x \equiv p_x k_x q_x$ we mean the symmetric normal form as above.

The following theorem is a slightly refined summary of the results of Peczynski and Reiwer [P-R].

Theorem 6.1. *Let* $K = \langle t, B\, ; rel\, B, t^{-1} K_1 t = K_{-1} \rangle$ *be an HNN group. If* $\{x_1, \ldots, x_m\}$ *is a finite system of elements in* K *then there is a Nielsen transformation from* $\{x_1, \ldots, x_m\}$ *to a system* $\{y_1, \ldots, y_m\}$ *for which one of the following cases holds:*

(i) $y_i = 1$ *for some* $i \in \{1, \ldots, m\}$.

(ii) *Each* $w \in \langle y_1, \ldots, y_m \rangle$ *can be written as*

$$w = \prod_{i=1}^{q} y_{v_i}^{\epsilon_i}, \ \epsilon_i = \pm 1, \ and \ \epsilon_i = \epsilon_{i+1}$$

if $v_i = v_{i+1}$ *with* $L(y_{v_i}) \le L(w)$ *for* $i = 1, \ldots, q$.

(iii) *Some subgroup of* B *contains* p, $p \ge 1$, *of the* y_i *and some product of these* y_i *is conjugate to a non-trivial element of* K_1 *or* K_{-1}.

The Nielsen transformation can be chosen in finitely many steps such that $\{y_1, \ldots, y_m\}$ *is shorter than* $\{x_1, \ldots, x_m\}$ *or the lengths of the elements of* $\{x_1, \ldots, x_m\}$ *are preserved.*

The main tool in the proof of this theorem, which is not included in the work of Peczynski and Reiwer is the following lemma. This is proved in an analogous fashion to the corresponding result in the free product with amalgamation case—although the proofs are slightly different.

Lemma 6.1. *Let* $x, y, z \in K$, $x \ne 1 \ne z$, *with* $L(xy) \ge L(x), L(y)$ *and* $L(yz) \ge L(z), L(y)$. *Moreover let neither* xy *precede* x *nor* yz *precede* z. *Then the following hold.*

(a) *If* $L(xyz) \le L(x) - L(y) + L(z)$ *then* y *is conjugate to an element of* B.

(b) *If* $L(x), L(z) > L(y)$ *and* $L(xyz) < L(x) - L(y) + L(z)$ *then* y *is conjugate to an element of* K_ϵ, $\epsilon = \pm 1$.

(c) *If* $V \subset K$ *is a finite Nielsen reduced system,* $x, y, z \in V^{\pm 1}$ *and* $L(xyz) < L(x) - L(y) + L(z)$ *then* y *is conjugate to an element of* $K_\epsilon, \epsilon = \pm 1$, *or* $x = y = z$.

Proof. Let $x \equiv p_x k_x q_x, y \equiv p_y k_y q_y, z \equiv p_z k_z q_z$ be symmetric normal forms for x, y, z.

(a) Suppose $L(xyz) \leq L(x) - L(y) + L(z)$. If $y \in B$ then there is nothing to prove so hence let $p_y \neq 1 \neq q_y$. Since $L(xyz) \leq L(x) - L(y) + L(z), L(xy) \geq L(x), L(y)$ and $L(yz) \geq L(y), L(z)$ we must have $q_x \equiv r_x p_y^{-1}$ and $p_z \equiv q_y^{-1} l_y$. Since neither xy precedes x nor yz precedes z we also must have $p_y^{-1} \leq q_y$ and $q_y^{-1} \leq p_y$, that is $p_y = q_y^{-1}$. Moreover $L(xyz) \leq L(x) - L(y) + L(z)$ implies then that $L(k_y) = 0$ and therefore $k_y \in B$.

(b) Suppose $L(x), L(z) > L(y)$ and $L(xyz) < L(x) - L(y) + L(z)$. If $y \in K_\epsilon$, $\epsilon = \pm 1$, then there is nothing to prove. Therefore suppose $y \notin K_1$ and $y \notin K_{-1}$. (Recall that K_1 is conjugate to K_{-1} in K.) From part (a) we have that y is conjugate to an element of B and therefore $k_y \in B$. Suppose first that $q_x \equiv r_x t^\varsigma r p_y^{-1}, r \in R_{-\varsigma}^{-1}$ and $p_z \equiv q_y^{-1} l t^\epsilon l_y, l \in R_\epsilon$ with $p_y = q_y^{-1}$ $(\epsilon, \varsigma = \pm 1)$.

From $L(xyz) < L(x) - L(y) + L(z)$ we have that $\varsigma = -\epsilon$ and $rk_y l \in K_{-\varsigma}$. Analogously as in the case of free products with amalgamation we get that $r = l^{-1}$ and therefore k_y is conjugate to an element of $K_{-\varsigma}$ and hence, so is y. Now suppose $q_x = p_y^{-1} = q_y$ or $p_z = q_y^{-1} = p_y$. With no loss of generality assume that $q_x = p_y^{-1} = q_y$. Then $k_x = h_1 t^\varsigma h_2$ with $h_1, h_2 \in B, \varsigma = \pm 1$, because $L(x) > L(y)$.

If $p_z \equiv q_y^{-1} l t^\epsilon l_y, l \in R_\epsilon, \epsilon = \pm 1$, then we must have $\varsigma = -\epsilon$ and $h_2 k_y l = ark_y l \in K_\varsigma$, where $h_2 = ar, a \in K_\varsigma$ and $r \in R_{-\varsigma}^{-1}$, that is, $rk_y l \in K_{-\varsigma}$, and we get $r = l^{-1}$ and y is conjugate to an element of K_ς.

If $p_z = q_y^{-1} = p_y$ then we have $k_z = g_1 t^\epsilon g_2$ with $g_1, g_2 \in B, \epsilon = \pm 1$, because $L(z) > L(y)$. In this case we have $\varsigma = -\epsilon$ and $h_2 k_y g_1 = ark_y lb \in K_{-\varsigma}$, where $h_2 = ar, a \in K_{-\varsigma}, r \in R_{-\varsigma}^{-1}$ and $g_1 = lb, b \in K_\epsilon, l \in R_\epsilon$, that is $rk_y l \in K_{-\varsigma}$ since $\varsigma = -\epsilon$. From this we get that $r = l^{-1}$ and y is conjugate to an element of $K_{-\varsigma}$.

(c) Suppose $x, y, z \in V^{\pm 1}$ and $L(xyz) < L(x) - L(y) + L(z)$. If $y \in K_\epsilon, \epsilon = \pm 1$, then there is nothing to prove. Hence suppose $y \notin K_1$ and $y \notin K_{-1}$. If $L(x) > L(y)$ and $L(z) > L(y)$ then the statement follows from part (b). Now suppose then that $L(x) = L(y)$ or $L(z) = L(y)$. Then $L(x) = L(y) = L(z)$ since V is assumed to be Nielsen reduced. Now we have that $L(xyz) < L(x), L(y), L(z)$ and therefore $x = y = z$ since V is Nielsen reduced.

We note that if $L(g^n) < L(g)$ for $g \in K$ and n a positive integer then g is conjugate to an element of the base B and g^n is conjugate in K to an element of K_1 or K_1.

From the above cancellation arguments as in the proof we directly obtain the following results found in the work of S. Pride [P2]. See also [P-R].

Lemma 6.2 ([P2]). *Suppose both K_1 and K_{-1} are malnormal in the base B. Suppose $\langle a, b \rangle$ is a two-generator subgroup of K which is not cyclic. Then $\langle a, b \rangle$ is conjugate to $\langle a', b' \rangle$ satisfying one of the following conditions.*
(1) *$\langle a', b' \rangle$ is a free product of cyclics.*

(2) $\langle a', b' \rangle$ is contained in B.

(3) a' has the form $th_1 \cdots th_n$, $n \geq 1$, $h_j \in B$ for $j = 1, \ldots, n$, $b' \in K_1$ and $a'^{-1}b'a' \in B$.

Corollary 6.1 ([P3], [P-R]). *Suppose K is a two-generator group and both K_1 and K_{-1} are malnormal in B. Then any system $\{x, y\}$ of generators of K is Nielsen equivalent to a system $\{th, b\}$ where $h \in B$ and $b \in K_1$.*

We now present some applications of the theory in HNN groups. In this section we review some of the already known results while in the next section we present some new applications. Most of the applications are concerned with the subgroup structure and the isomorphism problem for one-relator groups

$$H = \langle a_1, \ldots, a_n ; r^\alpha = 1 \rangle, \ n \geq 2, \ \alpha \geq 1.$$

These applications make use of the fact that if the (freely) cyclically reduced form of the word r involves at least two generators then H can be embedded into an HNN group whose base is a one-relator group, the relator of which has free length less then the length of r. This is the basis for both the proof of the Freiheitssatz and the Magnus method for dealing with one-relator groups. (See [B1] and [F-R3].) Along these lines we first present the following due to S. Pride [P1].

Theorem 6.2 ([P1]). *Let $K = \langle t, a ; P^n = 1 \rangle$, $n \geq 2$ with $P \neq 1$ a non-primitive, non proper power element on the free group on t and a. If $\{x, y\}$ is a generating system of K then $\{x, y\}$ is Nielsen equivalent to $\{t, a\}$. Hence the isomorphism problem for K is solvable, that is, we can decide algorithmically in finitely many steps whether or not an arbitrary one-relator group is isomorphic to K.*

We give a sketch of the proof. We note first that the solution to the isomorphism problem uses the Whitehead algorithm to decide if in a free group two elements are congruent via an automorphism α of the free group.

We may assume that P is cyclically reduced. It follows from Lemma 4.1 of the work of Sacerdote and Schupp [S-S] that there is an automorphism ϕ of the free group F on t, a such that $\phi(P)$ has exponent sum zero on t and it is easily seen that $K = \langle a, t ; \phi(P)^n = 1 \rangle$. Hence without loss of generality we may assume that the exponent sum of t in P is zero. Let $a_i = t^{-i}at^i$, $i \in \mathbb{Z}$. Let Q be the word expressing P in terms of the a_i. Let m, M represent respectively the least and greatest value of i for which a_i appears in Q. Then K is an HNN group

$$K = \langle a_m, \ldots, a_M, t ; Q^n = 1, \ t^{-1}a_it = a_{i+1}, \ i = m, \ldots, M - 1 \rangle.$$

Using work of B. B. Newman [Ne] the associated subgroups $K_1 = \langle a_m, \ldots, a_{M-1} \rangle$ and $K_{-1} = \langle a_{m+1}, \ldots, a_M \rangle$ are malnormal in the base $B = \langle a_m, \ldots, a_M ; Q^n = 1 \rangle$. Now applying Corollary 6.1 we see that a generating system $\{x, y\}$ for K is Nielsen equivalent to a system $\{tg, h\}$, $g \in B$, $h \in K_1$. This pair $\{tg, h\}$ is then Nielsen equivalent to $\{t, a\}$. (For the technical details see Pride's paper [P1].) The rest of the theorem then follows from the use of Whiteheads's algorithm.

As a consequence we then have that if K is as in the theorem then K is Hopfian and the automorphism group of K is finitely generated (see [Mc]).

Another result from the cancellation theorem and Lemma 6.1 is the following, also appearing in work of S. Pride's [P3].

Theorem 6.3 ([P3]). *Let* $K = \langle a_1, \ldots, a_m ; P^n = 1 \rangle$, $m \geq 2$, $n \geq 2$, *and* P *is cyclically reduced. Then every two-generator subgroup of K is either a free product of two cyclic groups or is a one-relator group with torsion.*

The situation with two generator, torsion-free one-relator groups is more difficult. In this case it may happen that the group possesses infinitely many Nielsen inequivalent one-relator presentations (see A.M. Brunner [Br]).

To apply the Nielsen cancellation method to one-relator groups with more than two generators is even more complicated. Here we have some results on classes of groups which are of special interest. For the following let

$$K = \langle t, a_1, \ldots, a_n ; t^{-1}Ut = V \rangle \tag{6.1}$$

with $n \geq 1$ and $U = U(a_1, \ldots, a_n)$, $V = V(a_1, \ldots, a_n)$ non-trivial elements of the free group F on $a_1, \ldots, a_n$. Fine, Rohl and Rosenberger [F-R-R1] proved the following.

Theorem 6.4 ([F-R-R1]). *Let* K *be as in* 6.1 *with neither* U *nor* V *a proper power in* F. *Suppose* $\langle x, y \rangle$ *is a two-generator subgroup of K. Then*
(i) $\langle x, y \rangle$ *is free of rank 2; or*
(ii) $\langle x, y \rangle$ *is abelian; or*
(iii) $\langle x, y \rangle$ *is has a presentation* $\langle \alpha, \beta ; \alpha^{-1}\beta\alpha = \beta^{-1} \rangle$ *(isomorphic to a Klein-bottle group).*

This result can be considered as the generalization to HNN groups of the Baumslag result on cyclically pinched one-relator groups mentioned in Section 1. As a corollary we have

Corollary 6.2. *Let* K *be as in Theorem* 6.4 *and suppose further that U is not conjugate in F to* V^{-1}. *Let* $\langle x, y \rangle$ *be a two-generator subgroup of K which is not cyclic. Then either* $\langle x, y \rangle$ *is free abelian of rank 2 or free of rank 2.*

If K is as in Corollary 6.2 then Bass [Ba] points out that K acts freely on a non-archimedean Λ-tree. In [F-R-R1] it posed as a question whether a two-generator group which acts freely on a Λ-tree must be either free or abelian. This was answered positively by Urbanski and Zamboni [U-Z]. Further as a direct consequence of the proof of Theorem 6.4 in [F-R-R1] we have:

Corollary 6.2. *Let K be as in Theorem 6.4 and suppose further that U is not conjugate in F to either V or V^{-1}. Then K is 2-free, that is every two-generator subgroup is free.*

Further results on the three-generator subgroups of groups of the form 6.1 were obtained in [F-R-R2] with restrictions on the subgroup generated by U, V. A two-generator subgroup $N = \langle x, y \rangle$ of a group H is **maximal** if it is not properly contained in any other two-generator subgroup. It is **strongly maximal** if it is maximal and for each $g \in H$ there is an $h \in H$ such that $\langle x, gyg^{-1} \rangle \subset \langle x, hyh^{-1} \rangle$ and $\langle x, hyh^{-1} \rangle$ is a maximal two-generator subgroup of H. From [F-R-R2] we then have the following which is a summary of several results in that paper.

Theorem 6.5. *Let K be as in Theorem 6.4 and suppose further that U is not conjugate in F to either V or V^{-1}. Let $H = \langle x_1, x_2, x_3 \rangle \subset K$. Then H is a free group of rank at most 3 or H has a presentation with one defining relation for the generating system $\{x_1, x_2, x_3\}$. If in addition $\langle U, V \rangle$ is a strongly maximal subgroup of F (the free group on $a_1, \ldots, a_n$) then K is 3-free.*

Using the Nielsen reduction method coupled with the proofs of the above theorems two straightforward corollaries can be obtained concerning the isomorphism problem for certain groups K related to the above. The details are in [F-R-R2]

Corollary 6.4. *Let $K = \langle F, t ; t^{-1}Ut = V \rangle$ where F is free of rank 2 and U, V are non-trivial, non-proper power elements of F which are in the subgroup $F^p F'$ generated by the p-th powers ($p \geq 2$) and commutators of F. Then:*
(i)   *K has rank 3 and for any generating system $\{x_1, x_2, x_3\}$ there is a presentation of K with a single defining relation.*
(ii)   *The isomorphism problem for K is solvable, that is it can be decided algorithmically in finitely many steps whether or not an arbitrary one-relator group is isomorphic to K.*
(iii) *K is Hopfian, every automorphism of K is induced by an automorphism of the free group of rank 3, and Aut K is finitely generated.*

Corollary 6.5. *Let $K = \langle F, t ; t^{-1}Ut = V \rangle$ where F is free of rank n ($n \geq 2$) on $a_1, \ldots, a_n$ and U, V are non-trivial, non-proper power elements of F. Suppose further that there is no Nielsen transformation from $\{a_1, \ldots, a_n\}$ to a system $\{b_1, \ldots, b_n\}$ with $U \in \{b_1, \ldots, b_{n-1}\}$ and there is no Nielsen transformation from $\{a_1, \ldots, a_n\}$ to a system $\{c_1, \ldots, c_n\}$ with $V \in \{c_1, \ldots, c_{n-1}\}$. Then:*
(i)   *K has rank $n + 1$, and for any minimal generating system there is a presentation of K with a single defining relation.*
(ii)   *The isomorphism problem for K is solvable.*
(iii) *K is Hopfian.*

In closing this section we mention that the Nielsen reduction method was used further by the authors in the study of two additional properties of infinite groups—the *restricted Gromov property* and *commutative transitivity*—motivated by corresponding properties in discrete groups. A group G is a *restricted Gromov group* or *RG-group* if for each pair x, y of non-trivial elements in G either $\langle x, y \rangle$ is cyclic or there exists an integer t such that $\langle x^t, y^t \rangle = \langle x^t \rangle \star \langle y^t \rangle$. This class of groups and it closure under certain amalgam constructions was studied in [F-R1]. A group G is *commutative transitive* if G is centerless and commutativity is transitive—that is $[x, y] = [y, z] = 1$ for non-trivial x, y, z implies $[x, z] = 1$. This class is also closed under certain amalgam constructions (see [L-R]) and a study of these groups has been initiated by Fine, Gaglione, Rosenberger and Spellman [F-G-R-S2].

7. On a Class of Para-Free Groups

We now present some new applications of the Nielsen reduction method in HNN groups to two additional classes of groups which are of special interest. The first is the class

$$G_{i,j} = \langle a, b, t \, ; \, a^{-1} = [b^i, a][b^j, t] \rangle. \tag{7.1}$$

This class was introduced by Gilbert Baumslag [B3] (the presentation above is slightly different than Baumslag's presentation since we use $[x, y] = xyx^{-1}y^{-1}$ and he uses $[x, y] = x^{-1}y^{-1}xy$) and is of special interest since they are *para-free*, that is they share many properties with the free group F of rank 2. In particular:

(1) If $\gamma_n G_{i,j}$ are the terms of the lower central series of $G_{i,j}$, then for all n, $G_{i,j}/\gamma_n G_{i,j} \cong F/\gamma_n F$ and further the intersection over all n of the $\gamma_n G_{i,j}$ is $\{1\}$.

(2) $G_{i,j}/G_{i,j}'' \cong F/F''$ where $G_{i,j}''$ respectively F'' represents the second derived subgroup.

The following properties are also true of the $G_{i,j}$ (see [B3] and [Li]).

(3) $G_{i,j}$ has a normal subgroup with an infinite cyclic quotient group.

(4) $G_{i,j}$ is 2-free but not free.

(5) $G_{i,j}$ is residually torsion-free nilpotent.

(6) $G_{i,j}$ is hyperbolic in the sense of Gromov.

Magnus and Chandler [Ch-M] in their *History of Combinatorial Group Theory* mention the class $G_{i,j}$ to demonstrate the difficulty of the isomorphism problem for torsion-free one-relator groups. They remark that as of 1980 there was no proof showing that any of the groups $G_{i,j}$ are non-isomorphic. S. Liriano [Li] used representations of $G_{i,j}$ into $PSL(2, p^k)$, $k \in \mathbb{N}$, to show that $G_{1,1}$ and $G_{30,30}$ are non-isomorphic. In the following theorem we extend these results on $G_{i,j}$, somewhat, by using Nielsen cancellation methods. To apply the Nielsen reduction techniques notice that $G_{i,j}$ can be expressed as an HNN group

$$G_{i,j} = \langle a, b, t \, ; \, t^{-1}a[b^i, a]b^j t = b^j \rangle.$$

If in addition $j = 1$ then $\langle b \rangle$ and $\langle a[b^i, a]b \rangle$ are malnormal in $\langle a, b; \rangle$ and hence we may apply the techniques developed in [F-R-R2] and mentioned in the previous section. In particular we prove the following result.

Theorem 7.1. *Let i be a natural number. Then the isomorphism problem for $G_{i,1}$ is solvable, that is it can be decided algorithmically in finitely many steps whether or not an arbitrary one-relator group is isomorphic to $G_{i,1}$.*

Further if i and k are primes then

$$G_{i,1} \cong G_{k,1} \text{ if and only if } i = k.$$

In addition $G_{i,1}$ is Hopfian, every automorphism of $G_{i,1}$ is induced by an automorphism of the free group $F^* = F^*(A, B, T)$ of rank 3, with respect to the epimorphism $A \to a$, $B \to b$, $T \to t$, and the automorphism group $\text{Aut } G_{i,1}$ is finitely generated.

Proof. We use the notation $F = \langle a, b; \rangle$, $c = a[b^i, a]b$, $d = cb^{-1} = ab^i ab^{-i} a^{-1}$, and recall that in the free group F the conjugacy problem is solvable.

Let $\{x_1, x_2, x_3\}$ be a generating system for $G_{i,1}$. From the results in [F-R-R2] there is a one-relator presentation for $G_{i,1}$ on $\{x_1, x_2, x_3\}$. By analyzing the techniques developed in [F-R-R2] and after replacing t^{-1} by ht^{-1} (or $t^{-1}h$) and b by hbh^{-1} (respectively c by $h^{-1}ch$), $h \in F$, if necessary, we need only consider the cases

(1) $x_1, x_2 \in F$, $c^\alpha \in \langle x_1, x_2 \rangle$, $\alpha \geq 1$, and $x_3 = t$ or

(2) $x_1, x_2 \in F$, $x_2 = b^\alpha$, $\alpha \geq 1$, and $x_3 = t$.

We consider first case (1). The Nielsen cancellation method in free groups easily gives that $c = a[b^i, a]b$ is primitive in each proper two-generator subgroup of F which contains c. Hence by Lemma 2.1 of Rosenberger's 1981 St. Andrews paper [R13] one necessarily has $\langle x_1, x_2 \rangle = F$ and $\{x_1, x_2, t\}$ is Nielsen equivalent to $\{a, b, t\}$, or some c^γ, $\gamma \geq 1$, is primitive in $\langle x_1, x_2 \rangle$. If this latter case holds then we may assume that already $x_2 = c^\gamma$ which reduces the problem to case (2) after a suitable conjugation in $G_{i,1}$ and again applying Nielsen reduction. Hence we now consider case (2).

Let $x_1, x_2 \in F$, $x_2 = b^\alpha$ for some $\alpha \geq 1$ and $x_3 = t$. We have necessarily $F = \langle x_1, x_2, c^\alpha \rangle = \langle x_1, b^\alpha, c^\alpha \rangle$ independently of whether $c \in \langle x_1, x_2 \rangle$ already or not. Further we must then have $\alpha = 1$ since the factor group $\overline{F} = \langle a, b; b^\gamma = (a[b^i, a]b)^\gamma = 1 \rangle$, $\gamma \geq 2$, is not cyclic (see [F-H-R]) Therefore we have $F = \langle x_1, b, c \rangle = \langle x_1, b, d \rangle$. We wish to show that, in a minimizing manner, $\{x_1, b, c\}$ is c-stable Nielsen equivalent to $\{a, b, c\}$, and hence $\{x, b, t\}$ is Nielsen equivalent to $\{a, b, t\}$, or if $i \geq 2$, $\{x_1, b, c\}$ is c-stable Nielsen equivalent to a system $\{b, ab^\beta a^{-1}, c\}$, and hence $\{x_1, b, t\}$ is Nielsen equivalent to $\{b, ab^\beta a^{-1}, t\}$ where $1 \leq \beta < i$, $\beta | i$.

We consider the generating system $\{x_1, b, c\}$ and perform c-stable Nielsen transformations from $\{x_1, b, c\}$ to other systems in a minimizing manner. For this we regard F as the free product $F = \langle a \rangle \star \langle b \rangle$ of two infinite cyclic groups together with the length L and an order with respect to this factorization. The reason for doing this is that in this factorization $L(c) = 6$ whereas the free length of c is $2i + 4$ which can be quite large. We note that $H = \langle b, c \rangle = \langle b, d \rangle$ is free of rank 2, and there is no cancellation between b and $d = ab^i ab^{-i} a^{-1}$ with respect to the above free product factorization. In particular we have that $H \subsetneq F$. We write $x = x_1$ and now use the Nielsen cancellation method in the free product factorization for F to handle $\{x, b, c\}$. For each $\mu, \epsilon = \pm 1$ and for each $r \in H$ we can obtain

$$L(x^\mu r^\epsilon) \geq L(x) \text{ and}$$

$$L(x^\mu r x^\epsilon) > L(x) - L(r) + L(x).$$

In particular the reduced form of x does not end with a power of b.

Consider first the case where $L(x) \geq 5$. Since $F = \langle x, b, d \rangle$ we must have either at least one $L(x^\mu d^\epsilon) < L(d) = 5$ or always $L(x^\mu d^\epsilon) \geq L(d) = 5$, but at least once $L(d^\mu x d^\epsilon) \leq 2L(d) - L(x) = 10 - L(x)$, $\mu, \epsilon = \pm 1$. In the latter situation the left half and the right half of the symmetric normal form of x are cancelled and we obtain $L(x^\gamma d^\beta) < L(x)$ for some $\beta \neq 0$, $\gamma = \pm 1$, which contradicts $L(x^\mu r^\epsilon) \geq L(x)$, $\mu, \epsilon = \pm 1$, $r \in H$. Therefore we have $L(x^\mu d^\epsilon) < 5$ for some $\mu, \epsilon = \pm 1$, and we may replace x by $y = x^\mu d^\epsilon$. Hence from the beginning we may assume that $L(x) < 5$.

Therefore we now consider the case where $L(x) < 5$. Since x neither starts nor ends with a power of b, we must have $L(x) = 1$ or $L(x) = 3$. Consider first $L(x) = 3$. In this case then $x = a^\alpha b^\beta a^\gamma$ for some non-zero integers α, β, γ. Assume that $L(x^\mu d^\epsilon) < L(d) = 5$ for some $\mu, \epsilon = \pm 1$. Suppose for example $L(xd) < 5$. Then we have necessarily $\gamma = -1$, $\beta = -i$ and $xd^{-\alpha} = a^\alpha b^{-i} a^{-1} ab^i a^{-\alpha} b^{-i} a^{-1} = b^{-i} a^{-1}$, that is $L(xd^{-\alpha}) < L(x)$ which contradicts that $L(x^\mu r^\epsilon) \geq L(x)$, $\mu, \epsilon = \pm 1$, $r \in H$. The same type of analysis works for the other possibilities for μ, ϵ and hence we always have $L(x^\mu d^\epsilon) \geq L(d)$. We must have at least once $L(d^\mu x d^\epsilon) \leq 2L(d) - L(x) = 7$ and then we have necessarily that $\alpha = 1$ and $\gamma = -1$. This implies then that $x = ab^\beta a^{-1}$. Without loss of generality let $\beta \geq 1$. We must have $1 \leq \beta \leq i$, $\beta | i$ because for $\beta \geq 2$, $\beta \nmid i$, the factor group $\overline{F} = \langle a, b; b^\beta = 1 \rangle$ cannot be generated by b and $ab^i ab^{-i} a^{-1}$. Let $x = ab^\beta a^{-1}$ with $1 \leq \beta \leq i$, $\beta | i$, and $j = \frac{i}{\beta}$. We then have the relation $a = x^{-j} d x^j$ and hence $G_{i,1} = \langle x, b, t \rangle$. If $\beta = i$ then we have $x^{-1} d = a^2 b^{-i} a^{-1}$ and $L(x^{-1} d) = 3 < L(d) = 5$ which contradicts that $L(x^\mu d^\epsilon) \geq L(d)$, $\mu, \epsilon = \pm 1$, and therefore we may assume that $\beta < i$.

Hence let $i \geq 2$ and $\beta < i$. Here $\{x, b, c\}$ is c-stable Nielsen equivalent, in a minimizing manner, to $\{b, ab^\beta a^{-1}, c\}$ and hence $\{x, b, t\}$ is Nielsen equivalent to $\{b, ab^\beta a^{-1}, t\}$. Using the relation $a = x^{-j} d x^j = x^{-j} t b t^{-1} b^{-1} x^j$, where $x = ab^\beta a^{-1}$ we get for this generating system $\{x, b, t\}$, $x = ab^\beta a^{-1}$, the one-relator presentation

$$G_{i,1} = \langle x, b, t; R = 1 \rangle$$

where $R = x^{-j} b t b^{-1} t^{-1} x t b t^{-1} b^{-1} x^j b^{-\beta}$ and $i \geq 2$, $2 \leq j \leq i$, $j | i$ and $\beta = \frac{i}{j}$. This completes the case when $L(x) = 3$.

Now suppose $L(x) = 1$, that is, $x = a^\alpha$ for some $\alpha \neq 0$. We may assume that $\alpha \geq 1$, but then we must have $\alpha = 1$ for if $\alpha \geq 2$ the factor group $\overline{F} = \langle a, b; a^\alpha = 1 \rangle$ cannot be generated by b and $ab^i ab^{-i} a^{-1}$. Therefore here $\{x, b, c\}$ is, in a minimizing manner, c-stable Nielsen equivalent to $\{a, b, c\}$, and hence, $\{x, b, t\}$ is Nielsen equivalent to $\{a, b, t\}$. We note also that this always holds if $i = 1$, that is in particular we have: If $\{x_1, x_2, x_3\}$ is a generating system for $G_{1,1}$ then $\{x_1, x_2, x_3\}$ is Nielsen equivalent to $\{a, b, t\}$. In all cases the Whitehead method now leads to the solvability of the isomorphism problem for the class $G_{i,1}, i \in \mathbb{N}$. In particular it shows that if i and k are primes then $G_{i,1} \cong G_{k,1}$ if and only if $i = k$ for in the rank 3 free group F_1 with ordered basis $\{X, Y, Z\}$ the words $X[Y^i, X][Y, Z], i \in \mathbb{N}$, and $X^{-j}YZY^{-1}Z^{-1}XZYZ^{-1}Y^{-1}X^jY^{-\beta}, i \geq 2, 2 \leq j \leq i, j|i$ and $\beta = \frac{i}{j}$, have minimal free length among all words which we can obtain from them via automorphisms of the free group F_1 and elementary Whitehead transformations increase the length..

The remaining statements in the theorem are now almost immediate. Let ϕ : $G_{i,1} \to G_{i,1}$ be a homomorphism from $G_{i,1}$ onto $G_{i,1}$, and set $a' = \phi(a), b' = \phi(b)$, $t' = \phi(t)$. Consider the generating system $\{a', b', t'\}$. Applying Nielsen transformations to this system with respect to the equation $t'^{-1}a'[b'^i, a']b't' = b'$ shows that $\{a', b', t'\}$ is Nielsen equivalent to $\{a, b, t\}$ and hence $G_{i,1}$ is Hopfian. Further every automorphism of $G_{i,1}$ is induced by an automorphism of the rank 3 free group. Since the group of inner automorphisms is finitely generated it follows that Aut $G_{i,1}$ is finitely generated (see [Mc]). This completes the theorem.

We note that $G_{1,1}$ has exactly one Nielsen equivalence class of generating triples $\{x_1, x_2, x_3\}$. From the proof of the theorem it follows easily that $G_{i,j}$ with $i \geq 2$ has at least two different Nielsen equivalence classes of generating triples. Hence $G_{1,1}$ is never isomorphic to $G_{i,j}$ with $i \geq 2$.

The second class of groups we consider are motivated by surface groups and have presentations of the following form:

$$K_{\sigma,\alpha} = \langle a_1, \ldots, a_n, t ; t^{-1}a_1^\alpha \cdots a_n^\alpha t = a_{\sigma(1)}^\alpha \cdots a_{\sigma(n)}^\alpha \rangle \qquad (7.2)$$

where $\alpha \in \mathbb{N}$, $n \in \mathbb{N}$ and $\sigma \in S_n$, that is σ is a permutation on $\{1, 2, \ldots, n\}$. Notice that if $\alpha = 1$, n is odd and $\sigma = \begin{pmatrix} 1 & 2 & \cdots & n \\ n & n-1 & \cdots & 1 \end{pmatrix}$ then $K_{\sigma,1}$ is the surface group of genus $\frac{n+1}{2}$. Notice further that if $\alpha = 1$ the relator $t^{-1}a_1 \cdots a_n t a_{\sigma(n)}^{-1} \cdots a_{\sigma(1)}^{-1}$ in $K_{\sigma,1}$ is a quadratic word. We get the following.

Theorem 7.2. Let $K_{\sigma,\alpha}$ be as in 7.2. If $\{x_1, \ldots, x_{n+1}\}$ is a generating system of $K_{\sigma,\alpha}$ then $\{x_1, \ldots, x_{n+1}\}$ is Nielsen equivalent to $\{a_1, \ldots, a_n, t\}$.

Proof. Let $K_{\sigma,\alpha}$ be as in 7.2 and suppose first that $\alpha \geq 2$. We note that the groups $G = \mathbb{Z} \times (\mathbb{Z}_\alpha)^n$ and $G_1 = \mathbb{Z} \times \mathbb{Z}_\beta \times (\mathbb{Z}_\alpha)^{n-1}$ with $\beta \in \mathbb{N}$, $\beta \geq 2$, $\beta|\alpha$ both have rank $n + 1$. Let $\{x_1, \ldots, x_{n+1}\}$ be a generating system for $K_{\sigma,\alpha}$. Then x_1 can neither be conjugate in $K_{\sigma,\alpha}$ to a power of $a_1^\alpha \cdots a_n^\alpha$ nor conjugate to a power a_i^β for some $1 \leq i \leq n$ with $\beta \geq 2$, $\beta|\alpha$. Therefore from the Nielsen cancellation method in free groups and Theorem 2.1 (see Section 2) we get a Nielsen transformation from $\{x_1, \ldots, x_{n+1}\}$ to a system $\{a_1, \ldots, a_n, x\}$. However now we must have that

$\{a_1, \ldots, a_n, x\}$ is Nielsen equivalent to $\{a_1, \ldots, a_n, t\}$ by the standard cancellation arguments in HNN groups. For $n \geq 2$ this follows directly from the techniques in [F-R-R2].

Now suppose $\alpha = 1$. As mentioned before the relator is now a quadratic word. If we apply Nielsen transformations in free groups for quadratic words (see Section 2 and Theorem 5.3 and also [L-S]) then we get that

$$K_{\sigma,1} \cong \langle b_1, \ldots, b_{n+1} ; [b_1, b_2] \cdots [b_{k-1}, b_k] b_{k+1} b_{k+1}^{-1} \cdots b_{n+1} b_{n+1}^{-1} = 1 \rangle$$

$$\cong \langle b_1, \ldots, b_k ; [b_1, b_2] \cdots [b_{k-1}, b_k] = 1 \rangle \star F$$

where $1 \leq k \leq n+1$, k even and $F = \langle b_{k+1}, \ldots, b_{n+1} ; \rangle$. Now if $\{x_1, \ldots, x_{n+1}\}$ is a generating system for $K_{\sigma,1}$ then $\{x_1, \ldots, x_{n+1}\}$ is known to be Nielsen equivalent to $\{b_1, \ldots, b_{n+1}\}$ from the work on minimal generating systems for Fuchsian groups (see Theorem 3.4).

From the Nielsen cancellation method on quadratic words as in the proof of the theorem we directly get the following corollary:

Corollary 7.1. *If $\alpha = 1$ then $K_{\sigma,1}$ is a free product of a free group and a surface group. If n is even $K_{\sigma,1}$ is never a surface group.*

As an example of the corollary consider the case where $n = 3$, $\alpha = 1$. Then

$$K_{\sigma,1} = \langle a_1, a_2, a_3, t ; = t^{-1} a_1 a_2 a_3 t = a_{\sigma(1)} a_{\sigma(2)} a_{\sigma(3)} \rangle.$$

(1) If $\sigma = (132)$ or $\sigma = (123)$ or $\sigma = \{1\}$ then

$$K_{\sigma,1} \cong \langle x, y ; [x, y] = 1 \rangle \star \mathbb{Z} \star \mathbb{Z}.$$

(2) If $\sigma = (13)$ or $\sigma = (12)$ or $\sigma = (23)$ then

$$K_{\sigma,1} \cong \langle u_1, v_1, u_2, v_2 ; [u_1, v_1][u_2, v_2] = 1 \rangle.$$

Finally using the same techniques we may prove the following generalization.

Theorem 7.3. *Let $G_{\sigma,\alpha} = \langle a_1, \ldots, a_n, t ; t^{-1} P(a_1, \ldots, a_n) t = P(a_{\sigma(1)}, \ldots, a_{\sigma(n)}) \rangle$ where $n \geq 1$, $\sigma \in S_n$ and*

$$P(a_1, \ldots, a_n) = a_1^{\alpha_1} \cdots a_p^{\alpha_p} [a_{p+1}, a_{p+1}] \cdots [a_{n-1}, a_n]$$

with $0 \leq p \leq n$, $n - p$ even, $\alpha_i \geq 2$ for $i = 1, \ldots, p$ and $\gcd(\alpha_1, \ldots, \alpha_n) \geq 2$. Then if $\{x_1, \ldots, x_{n+1}\}$ is a generating system for $G_{\sigma,\alpha}$ then $\{x_1, \ldots, x_{n+1}\}$ is Nielsen equivalent to $\{a_1, \ldots, a_n, t\}$.

8. Equations in HNN Groups

As in the free product with amalgamation case the Nielsen cancellation method can be applied to the study of equations in HNN groups. Let $K = \langle B, t\,; \; rel\ B,\ t^{-1}K_1 t = K_{-1}\rangle$ be an HNN group with base B and associated subgroups K_1, K_{-1}. An ordered set $U = \{u_1, \ldots, u_n\} \subset K$ is called *regular* if there is no Nielsen transformation from U to a system $U' = \{u_1', \ldots, u_n'\}$ in which one of the elements is conjugate in K to an element of K_1. The following results are due to Rosenberger [R12].

Theorem 8.1 ([R12]). *Let $K = \langle B, t\,; \; rel\ B, t^{-1}K_{-1}t = K_1\rangle$ be an HNN group as above. Let F be a free group of rank $n \geq 1$ with ordered basis $X = \{x_1, \ldots, x_n\}$. Let $w = w(x_1, \ldots, x_n) \neq 1$ be a regular word in F and let ϕ be a homomorphism $\phi : F \to K$ such that $U = \phi(X)$ is regular and $\phi(w) = hah^{-1}$, $a \in K_1$, $h \in K$.*
(a) *Then the pair (w, U) is Nielsen equivalent to a pair $(w', U') = (\alpha(w), \alpha^{-1}(U))$ with $\alpha : F \to F$ an automorphism such that*
   (1) *$w' = \alpha(w) \equiv w_1 w_2$ with $w_1 \neq 1$;*
   (2) *$\phi\alpha^{-1}(w_1)$ is conjugate in K to an element of K_1;*
   (3) *There is a $g \in K$ with $\phi\alpha^{-1}(x_j) \in gBg^{-1}$ for each x_j which occurs in w_1.*
(b) *If $K_1 = K_{-1}$ and K_1 is normal in K then the pair (w, U) is Nielsen equivalent to a pair $(w', U') = (\alpha(w), \alpha^{-1}(U))$ with $\alpha : F \to F$ an automorphism such that*
   (1) *$w' = r_1 w_1 r_1^{-1} \cdots r_k w_k r_k^{-1}$ for some $k \geq 0$ where $r_i, w_i, 1 \leq i \leq k$, are reduced words in F;*
   (2) *For each i, $1 \leq i \leq k$, we have $\phi\alpha^{-1}(w_i) \in K_1$;*
   (3) *For each i, $1 \leq i \leq k$, there exists a $g_i \in K$ with $\phi\alpha^{-1}(x_j) \in g_i Bg_i^{-1}$ for each x_j which occurs in w_i.*
  If in addition $w = w(x_1, \ldots, x_n)$ is a quadratic word in F then we may replace (1) by
   (1)' *$w' \equiv w_1 \cdots w_k$ for some $k \geq 0$, where each w_i, $1 \leq i \leq k$, is a quadratic word in F. Further if w is alternating then each w_i is also alternating.*
(c) *If $n \leq 4$, both K_1, K_{-1} are malnormal in B, w quadratic and $\phi(w) = 1$ then the pair (w, U) is Nielsen equivalent to a pair $(w', U') = (\alpha(w), \alpha^{-1}(U))$ with $\alpha : F \to F$ an automorphism such that*
   (1) *$w' \equiv w_1 w_2$, where w_1 and w_2 are quadratic in F;*
   (2) *For $i = 1, 2$ we have that $\phi\alpha^{-1}(w_i)$ is conjugate in K to an element of K_1*
   (3) *For $i = 1, 2$ there is a $g_i \in K$ with $\phi\alpha^{-1}(x_j) \in g_i Bg_i^{-1}$ for each x_j which occurs in w_i.*

As a straightforward corollary the following subgroup theorem of H. Neumann on HNN groups can be obtained.

Theorem (H. Neumann, see [L-S]). *Let K be an HNN group as above and H a finitely generated subgroup with $gHg^{-1} \cap K_1 = \{1\}$, for all $g \in K$. Then $H = F \star (\star_{i \in I} G_i)$, where F is a free group and each G_i is conjugate to a subgroup of B.*

9. Extended Nielsen Transformations

Let G be a group and $\{x_1, \ldots, x_n\} \subset G, n \geq 1$, a finite subset. If $x_i, 1 \leq i \leq n$, has finite order $k \geq 2$ then a transformation $\{x_1, \ldots, x_n\} \rightarrow \{x_1, \ldots, x_{i-1}, x_i^m, x_{i+1}, \ldots, x_n\}$ with $1 \leq m < k$ and $\gcd(m, k) = 1$ is called an **elementary extended Nielsen transformation**. An **extended Nielsen transformation** is a finite sequence of Nielsen transformations and elementary extended Nielsen transformations.

If $\{x_1, \ldots, x_n\}$ is carried in to $\{y_1, \ldots, y_m\}, 1 \leq m \leq n$, by an extended Nielsen transformation then $\langle x_1, \ldots, x_n \rangle = \langle y_1, \ldots, y_m \rangle$, and $n = m$ if all Nielsen transformations involved are regular. The concept of extended Nielsen transformations was heavily used in the classification of all generating pairs of co-compact two generator Fuchsian groups. (see [F-R2]) It was also been used in the study of subgroups of the Modular group $PSL_2(\mathbb{Z})$. (see [R5], [R9] and [F])

In an analogous manner to standard Nielsen transformations one may use extended Nielsen transformations in free products with amalgamation and HNN groups with respect to the Nielsen cancellation theory in these constructions. Perhaps this concept can be utilized to give the complete classification of all minimal generating systems for all co-compact Fuchsian groups.

References

[Ba]   H. Bass, Group actions on non-archimedean trees, in: Arboreal Group Theory, Springer-Verlag (1991), 69–131.

[B1]   G. Baumslag, A survey of groups with a single defining relation, in: Proceedings of Groups St. Andrews 1985 (1986), 30–58.

[B2]   G. Baumslag, On generalized free products, Math. Z. 78 (1962) 423–438.

[B3]   G. Baumslag, Groups with the same lower central series as a relatively free group. II. Properties, Trans. Amer. Math. Soc. 142 (1969), 507–538.

[B-S]   G. Baumslag and P. Shalen, Groups whose three generator subgroups are free, Bull. Austral. Math. Soc. 40 (1989), 163–174.

[Br]   A. M. Brunner, A group with an infinite number of Nielsen inequivalent one-relator presentations, J. Algebra 42 (1976), 81–86.

[Co-L]   M. Cohen and M. Lustig, Very small group actions on R-trees and Dehn twist automorphisms, to appear.

[Ch-M]   B. Chandler and W. Magnus, The History of Combinatorial Group Theory, Springer-Verlag (1982).

[C]   D. J. Collins, Presentations of the amalgamated free product of two infinite cyclics, Math. Ann. 237 (1978), 233–241.

[C-Z1]   D. J. Collins and H. Zieschang , Combinatorial Group Theory and Fundamental Groups, Algebra VII, Springer-Verlag (1993), 3–166.

[C-Z2]   D. J. Collins and H. Zieschang , On the Nielsen Method in Free Products with Amalgamated Subgroups, Math. Z. 197 (1987), 97–118.

[F]   B. Fine, Subgroup presentations without coset representatives, in: Topology and Combinatorial Group Theory, Springer-Verlag (1990), 59–74.

[F-H-R]   B. Fine, J. Howie and G. Rosenberger, One-relator quotients and free products of cyclics, Proc. Amer. Math. Soc. 102 (1988), 1–6.

[F-G-R-S1] B. Fine, A. Gaglione, G. Rosenberger and D. Spellman, n-free groups and questions about universally free groups, in: Groups '93 St. Andrews/Galway, London Math. Soc. Lecture Note Ser. 211, Cambridge Univ. Press (1995), 191–204.

[F-G-R-S2] B. Fine, A. Gaglione, G. Rosenberger and D. Spellman, On the commutative transitive kernel, to appear.

[F-R1]   B. Fine and G. Rosenberger, On restricted Gromov groups, Comm. in Algebra 20 (8) (1992), 2171–2182.

[F-R2]   B. Fine and G. Rosenberger, Classfication of all generating pairs of two generator Fuchsian groups, in: Groups '93 St. Andrews/Galway, London Math. Soc. Lecture Note Ser. 211, Cambridge Univ. Press (1995), 205–232.

[F-R3]   B. Fine and G. Rosenberger, The Freiheitssatz of Magnus and its extensions, Contemporary Math. 169 (1994), 213–252.

[F-R-R1]   B. Fine, F. Rohl and G. Rosenberger, Two generator subgroups of certain HNN groups, Contempoary Math. 109 (1990), 19–23.

[F-R-R2]   B. Fine, F. Rohl and G. Rosenberger, On HNN groups whose three-generator subgroups are free, in: Infinite Groups and Group Rings, World Scientific (1993), 13–37.

[K-R]   R. N. Kalia and G. Rosenberger, Über Untergruppen ebener diskontinuierlicher Gruppen, Contemp. Math. 33 (1984), 308–327.

[K-S]   A. Karrass and D. Solitar, The free product of two groups with a malnormal amalgamated subgroup, Canad. J. Math. 23 (1971), 933–959.

[K]   S. Katok, Fuchsian Groups, University of Chicago Press, Chicago (1993).

[L-R]   F. Levin and G. Rosenberger, On power commutative and commutation transitive groups, in: Proc. Groups St. Andrews 1985, Cambridge University Press, Cambridge (1986), 249–253.

[Li]   S. Liriano, The non-isomorphism of two one-relator groups is established, Ph.D. Thesis, New York (1993).

[L-S]   R. C. Lyndon and P. E. Schupp, Combinatorial Group Theory, Springer-Verlag (1977).

[L-M-R]   M. Lustig,Y. Moriah and G. Rosenberger, Automorphisms of Fuchsian groups and their lifts to free groups, Canad. J. Math. XLI (1989), 123–131.

[M-K-S]   W. Magnus, A. Karrass, and D. Solitar, Combinatorial Group Theory, Wiley (1966); Second Edition, Dover Publications, New York (1976),

[Mc]   J. McCool, Some finitely presented subgroups of the automorphism group of a free group, J. Algebra 35 (1975), 205–273.

[Mc-P]   J. McCool and A. Pietrwoski, On free products with amalgamation of two infinite cyclic groups, J. Algebra 18 (1971), 377–383.

[Ne]   B. B. Newman, Some results on one-relator groups, Bull. Amer Math. Soc. 74 (1968), 568–574.

[N]   J. Nielsen, Om Regning med ikke kommutative Faktoren og dens Anvendelse i Gruppeteorien, Mat. Tidsskr. B (1921), 77–94.

[P-R]   N. Peczynski and W. Reiwer, On cancellations in HNN groups, Math. Z. 158 (1978), 79–86.

[P-R-Z]   N. Peczynski , G. Rosenberger and H. Zieschang, Über Erzeugende ebener diskontinuierlicher Gruppen, Invent. Math. 29 (1975), 161–180.

[P1]   S. J. Pride, The isomorphism problem for two generator one-relator groups with torison is solvable, Trans. Amer. Math. Soc. 227 (1977), 109–139.

[P2]   S. J. Pride, The two generator subgroups of one-relator groups with torsion, Trans. Amer. Math. Soc. 234 (1977), 483–496.

[P3]   S. J. Pride, On the generation of one-relator groups, Trans. Amer. Math. Soc. 210 (1975), 331–363.

[R1]   G. Rosenberger, Applications of Nielsen's Reduction Method in the Solution of Combinatorial Problems in Group Theory, London Math. Soc. Lecture Notes 36 (1979), 339–358.

[R2]   G. Rosenberger, Alternierende Produkte in freien Gruppen, Pacific J. Math. 78 (1978), 243–250.

[R3]   G. Rosenberger, Minimal generating systems for plane discontinuous groups and an equation in free groups, in: Proc. Groups Korea 1988, Lecture Notes in Math. 1398, Springer Verlag (1989), 170–186.

[R4]   G. Rosenberger, A property of subgroups of free groups, Bull. Austral. Math. Soc. 43 (1991), 269–272.

[R5]   G. Rosenberger, Über Untergruppen in freien Produkten, Proc. Groups Korea 1983, Lecture Notes in Math. 1098, Springer Verlag (1984), 142–160.

[R6]   G. Rosenberger, Gleichungen in freien Produkten mit Amalgam, Math. Z. 173 (1980), 1–12.

[R7]   G. Rosenberger, Correction to: Gleichungen in freien Produkten mit Amalgam, Math. Z. 178 (1981), 579.

[R8]   G. Rosenberger, Zum Rang und Isomorphieproblem für freie Produkte mit Amalgam, Habilitationsschrift, Hamburg 1974.

[R9]   G. Rosenberger, On subgroups of free products of cyclics, in: Proc. Internat. Conference on Algebra, Contemp. Math. 131 (1989), 315–324.

[R10]   G. Rosenberger, The isomorphism problem for cyclically pinched one-relator groups, J. Pure Appl. Algebra 95 (1994), 75–86.

[R11]   G. Rosenberger, Zum Isomorphieproblem für Gruppen mit einer definierenden Relation, Ill, J. Math. 20 (1976), 614–621.

[R12]   G. Rosenberger, Bemerkungen zu einer Arbeit von R. C. Lyndon, Archiv. Math. 40 (1983), 200–207.

[R13]   G. Rosenberger, On one-relator groups that are free products of two free groups with cyclic amalgamation, in: Groups St. Andrews 1981, (1983), 328–344.

[S-S]   G. S. Sacerdote and P. E. Schupp, SQ-universality in HNN groups and one-relator groups, J. London Math. Soc. 7 (1974), 733–740.

[Sh]   A. Shenitzer, Decomposition of a group with a single defining relation into a free product, Proc. Amer. Math. Soc. 6 (1955), 273–279.

[U-Z]   M. Urbanski and L. Zamboni, On free actions on Λ-trees, Math. Proc. Cambridge. Philos. Soc. 11 (1993), 535–542.

[Z1]   H. Zieschang, Über die Nielsensche Kürzungsmethode in freien Produkten mit Amalgam, Invent. Math. 10 (1970), 4–37.

[Z2]   H. Zieschang, Über Automorphismen ebener discontinuierlicher Gruppen, Math. Ann. 166 (1966), 148–167.

[Z3]   H. Zieschang, Generators of the free product with amalgamation of two infinite cyclic groups, Math. Ann. 227 (1977), 195–221.

[Z-V-C] H. Zieschang, E. Vogt, H. D. Coldeway, Surfaces and Planar Discontinuous Groups, Lecture Notes In Math. 835, Springer-Verlag (1980).

Groups Satisfying the Minimal Condition on Certain Non-Normal Subgroups

Silvana Franciosi and Francesco de Giovanni

In memoriam Brian Hartley

1. Introduction

Let χ be a property pertaining to subgroups. Several authors have studied the structure of groups "many" of whose subgroups have the property χ. In particular, dealing with infinite groups, a natural interpretation of this requirement is that the set of all subgroups not having the property χ satisfies the minimal condition. The investigation of groups with such condition was started by S. N. Chernikov in a series of articles (for a general reference see [5]), and continued by Phillips and Wilson in [12]. In particular, among many other choices for the property χ, groups satisfying the minimal condition either on non-normal or on non-serial subgroups were characterized, under a weak additional condition imposed to avoid Tarski groups and other similar examples. More recently, groups satisfying the minimal condition on non-subnormal subgroups, on non-pronormal subgroups and on subgroups with infinitely many conjugates have been considered in [6], [7] and [8], respectively.

The aim of this article is to study groups satisfying the minimal condition on subgroups having infinite index in their normal closure. The reason of this choice has of course to be found in the well-known theorem of B. H. Neumann [11] proving that every subgroup of the group G has finite index in its normal closure if and only if the commutator subgroup G' of G is finite. It will be proved that a locally finite group satisfies the minimal condition on subgroups having infinite index in their normal closure if and only if either it is a Chernikov group or its commutator subgroup is finite. Moreover, a complete description of non-periodic groups with the same property will be given. These results will be applied in the last section to study groups with finitely many conjugacy classes of subgroups having infinite index in their normal closure. In particular, it will be shown that a locally graded group with such property either has finite commutator subgroup or is a finite extension of an infinite cyclic group. Groups with similar restrictions on non-normal and non-subnormal subgroups have been studied in [3], [6] and [14].

Most of our notation is standard and can be found in [13].

2. The Minimal Condition

Let G be a group, and let H and K be subgroups of G such that the indices $|H^G : H|$ and $|K^G : K|$ are finite; then it is clear that also the subgroups $H \cap K$ and $\langle H, K \rangle$ have finite index in their normal closure in G. This obvious property will be frequently used in our proofs.

Recall that the **FC-centre** of a group G is the subgroup of G consisting of all elements having finitely many conjugates, and G is said to be an **FC-group** if it coincides with its FC-centre.

Lemma 2.1. *Let G be a group, and let x be an element of G such that the index $|\langle x \rangle^G : \langle x \rangle|$ is finite. Then x has finitely many conjugates in G.*

Proof. Let n be a positive integer such that $\left(\langle x \rangle^G \right)^n$ is contained in $\langle x \rangle$. Since $\langle x \rangle^G$ is cyclic-by-finite, the factor group $\langle x \rangle^G / \left(\langle x \rangle^G \right)^n$ is finite, so that $\langle x \rangle$ is almost normal in G, and x has finitely many conjugates in G.

Our first theorem characterizes the groups in which every infinite subgroup has finite index in its normal closure.

Theorem 2.2. *Let G be a group which either is locally finite or non-periodic. Then every infinite subgroup of G has finite index in its normal closure if and only if G satisfies one of the following conditions:*

(a) *The commutator subgroup G' of G is finite.*

(b) *G is a finite extension of a group of type p^∞ for some prime p.*

(c) *G is a finite extension of an infinite cyclic group.*

Proof. The conditions are clearly sufficient, since if G satisfies (b) or (c), then every infinite subgroup of G has finite index. Conversely, suppose that every infinite subgroup of G has finite index in its normal closure, and assume first that G is a locally finite group which is not a Chernikov group. Then G does not satisfy the minimal condition on abelian subgroups by a result of Shunkov [15], and hence it contains an abelian subgroup of the form $A = A_1 \times A_2$, where A_1 and A_2 both are infinite. Then the indices $|A_1^G : A_1|$ and $|A_2^G : A_2|$ are finite, and hence the normal subgroup $A_1^G \cap A_2^G$ of G is finite. Clearly every subgroup of G/A_1^G has finite index in its normal closure, and so the commutator subgroup $G'A_1^G/A_1^G$ of G/A_1^G is finite by a result of

B. H. Neumann (see [13] Part 1, p. 127). Similarly, also the group $G'A_2^G/A_2^G$ is finite. Therefore G' is finite in this case. Suppose now that G is a Chernikov group. If G contains two infinite subgroups with trivial intersection, as above it follows that G' is finite. In the other case, we obtain that G is a finite extension of a group of type p^∞ for some prime p.

Suppose now that G contains an element of infinite order x. The index $|\langle x \rangle^G : \langle x \rangle|$ is finite and every subgroup of $G/\langle x \rangle^G$ has finite index in its normal closure, so that $G'\langle x \rangle^G/\langle x \rangle^G$ is finite, and G' is polycyclic-by-finite. In particular G is a soluble-by-finite group. Let T be the largest periodic normal subgroup of G. If T is infinite, the commutator subgroup $G'T/T$ of G/T is finite, and so also G' is finite. Assume that T is finite. Clearly it is enough to prove that G/T either has finite commutator subgroup or is a finite extension of an infinite cyclic subgroup, so that without loss of generality it can be assumed that $T = 1$ and G has no non-trivial periodic normal subgroups. The Baer radical B of G is contained in the FC-centre of G by Lemma 2.1, so that in particular B is a torsion-free FC-group, and hence is abelian. If B contains two elements of infinite order b_1 and b_2 such that $\langle b_1 \rangle \cap \langle b_2 \rangle = 1$, we obtain as above that G' is finite (and in this case G is abelian). Thus it can be assumed that B is locally cyclic. Let L be the largest soluble normal subgroup of G. Then $C_L(B) = B$, so that L/B is abelian, and L' is contained in B. As G' is polycyclic-by-finite, it follows that L' is cyclic. Moreover, $C_L(L')$ is a nilpotent normal subgroup of G, and so it is contained in B. Thus L/B has order at most 2, and in particular G/B is finite. It follows that $G/C_G(B)$ has order at most 2. On the other hand, the subgroup $C_G(B)$ is central-by-finite, and hence even abelian, as G has no non-trivial periodic normal subgroups. Therefore $C_G(B) = B$, and G/B has order at most 2. Suppose that G is not abelian. Then G' is infinite cyclic, and G contains an element y acting on B as the inversion. Since y acts trivially on B/G', it follows that B/G' has order 2, and G is a finite extension of an infinite cyclic subgroup.

Lemma 2.3. *Let G be a group satisfying the minimal condition on subgroups having infinite index in their normal closure, and let x be an element of infinite order of G. Then the index $|\langle x \rangle^G : \langle x \rangle|$ is finite, and in particular x has finitely many conjugates in G.*

Proof. Assume that the lemma is false, and let $\langle y \rangle$ be an infinite cyclic subgroup of G which is minimal with respect to the condition that the index $|\langle y \rangle^G : \langle y \rangle|$ is infinite. If p and q are distinct prime numbers, the subgroups $\langle y^p \rangle$ and $\langle y^q \rangle$ both have finite index in their normal closure, so that also $\langle y \rangle = \langle y^p, y^q \rangle$ has finite index in $\langle y \rangle^G$. This contradiction proves the lemma.

Lemma 2.4. *Let G be an FC-group satisfying the minimal condition on subgroups having infinite index in their normal closure. Then the commutator subgroup G' of G is finite.*

Proof. Assume that the set $\mathcal{L}$ of all subgroups of G having infinite index in their normal closure is not empty, and let H be a minimal element of $\mathcal{L}$. Then every proper subgroup K of H has finite index in K^G, and hence also in K^H, so that the commutator subgroup H' of H is finite (see [13] Part 1, p.127). The normal closure N of H' in G is finite, and so the index $|H^G : HN|$ is infinite. Since HN/N contains a subgroup which is minimal with respect to the condition of having infinite index in its normal closure in G/N, it can be assumed without loss of generality that H is abelian. The subgroup H cannot be generated by two proper subgroups, so that it is a group of type p^∞ for some prime p. As G is an FC-group, it follows that H is contained in the centre of G, and this contradiction proves that every subgroup of G has finite index in its normal closure. Thus the commutator subgroup G' of G is finite.

The above lemmas have the following easy consequence.

Corollary 2.5. *Let G be a group satisfying the minimal condition on subgroups having infinite index in their normal closure. If G is generated by elements of infinite order, then the commutator subgroup G' of G is finite.*

Proof. It follows from Lemma 2.3 and Lemma 2.1 that G is an FC-group, so that G' is finite by Lemma 2.4.

Lemma 2.6. *Let G be a locally nilpotent group satisfying the minimal condition on subgroups having infinite index in their normal closure. Then either G is a Chernikov group or every subgroup of G is subnormal.*

Proof. Let H be a non-subnormal subgroup of G. Then H is not subnormal in the locally nilpotent group H^G, and so the index $|H^G : H|$ is infinite. It follows that G satisfies the minimal condition on non-subnormal subgroups, and hence either G is a Chernikov group or all its subgroups are subnormal (see [6]).

Lemma 2.7. *Let G be a nilpotent group of finite exponent satisfying the minimal condition on subgroups having infinite index in their normal closure. Then the commutator subgroup G' of G is finite.*

Proof. Assume that the set $\mathcal{L}$ of all infinite subgroups of G having infinite index in their normal closure is not empty, and let H be a minimal element of $\mathcal{L}$. Then H cannot be generated by two infinite proper subgroups, and so the abelian group of finite exponent H/H' is finite. It follows that H is finite, and this contradiction shows that every infinite subgroup of G has finite index in its normal closure. As G has finite exponent, Theorem 2.2 yields that G' is finite.

Lemma 2.8. *Let G be a locally nilpotent group satisfying the minimal condition on subgroups having infinite index in their normal closure. Then either G is a Chernikov group or its commutator subgroup G' is finite.*

Proof. Suppose that G is not a Chernikov group. Then every subgroup of G is subnormal by Lemma 2.6, and in particular G is soluble (see [10]). Assume now that the lemma is false, and let G be a counterexample with minimal derived length. The group G is periodic by Lemma 2.5, and contains an element x with infinitely many conjugates by Lemma 2.4. Let K be the smallest non-trivial term of the derived series of G. Then the factor group G/K either is Chernikov or has finite commutator subgroup. Since G is a Baer group, it follows that G/K has finite commutator subgroup, and so $\langle x \rangle^G K/K$ is finite. Thus $L = \langle x \rangle^G \cap K$ is an infinite abelian normal subgroup of G. Moreover, the Baer group $\langle x \rangle^G$ is abelian-by-finite, so that it is nilpotent of finite exponent, and Lemma 2.7 yields that the commutator subgroup $(\langle x \rangle^G)'$ is finite. Then $C = C_L(x)$ has finite index in L, and so is an infinite abelian group of finite exponent. Clearly $C = U \times V$, where U and V both are infinite. Let

$$U_1 > U_2 > \cdots > U_n > \cdots$$

be a properly descending chain of subgroups of U. If $\langle x, U_i \rangle = \langle x, U_j \rangle$ with $i \leq j$, then

$$U_i = \langle x, U_j \rangle \cap U_i = U_j(\langle x \rangle \cap U_i),$$

and so $|U_i : U_j| \leq |\langle x \rangle|$. It follows that the chain

$$\langle x, U_1 \rangle \geq \langle x, U_2 \rangle \geq \cdots \geq \langle x, U_n \rangle \geq \cdots$$

is infinite, and there exists a positive integer k such that $\langle x, U_k \rangle$ has finite index in its normal closure. It can be shown similarly that V contains a subgroup V_0 such that $\langle x, V_0 \rangle$ has finite index in its normal closure. Since $U \cap V = 1$, the subgroup $\langle x, U_k \rangle \cap \langle x, V_0 \rangle$ is finite, so that $\langle x, U_k \rangle^G \cap \langle x, V_0 \rangle^G$ is also finite. In particular $\langle x \rangle^G$ is finite, and this contradiction proves the lemma.

Lemma 2.9. *Let G be a locally finite group satisfying the minimal condition on subgroups having infinite index in their normal closure. Then G is hyperabelian-by-finite.*

Proof. Clearly it is enough to prove that, if G is infinite, then it contains an abelian non-trivial normal subgroup. To show this it can be assumed that G is not a Chernikov group, so that it contains an abelian subgroup of the form $A = \underset{n \in \mathbb{N}}{\mathrm{Dr}}\, A_n$, where every A_n has prime order (see [15]). Then there exists a positive integer k such that the subgroup $B = \underset{n \geq k}{\mathrm{Dr}}\, A_n$ has finite index in its normal closure. The core X of B in B^G has also finite index in B^G, so that its normal closure X^G is an infinite nilpotent subgroup of G. Then $Z(X^G)$ is an abelian non-trivial normal subgroup of G.

It is now possible to characterize locally finite groups satisfying the minimal condition on subgroups having infinite index in their normal closure.

Theorem 2.10. *Let G be a locally finite group satisfying the minimal condition on subgroups having infinite index in their normal closure. Then either G is a Chernikov group or its commutator subgroup G' is finite.*

Proof. The group G is hyperabelian-by-finite by Lemma 2.9. Suppose that G is not a Chernikov group, and let H and R be the Hirsch–Plotkin radical and the last term of the upper Hirsch–Plotkin series of G, respectively. As G/R is finite and $C_R(H) \leq H$, also H is not a Chernikov group, and hence its commutator subgroup H' is finite by Lemma 2.8. In particular H is nilpotent, G/H' is not a Chernikov group and H/H' is the Hirsch–Plotkin radical of G/H'. Thus without loss of generality it can be assumed that H is abelian, so that it contains a subgroup of the form $A = U \times V$, where $U = \underset{n \in \mathbb{N}}{\mathrm{Dr}}\, U_n$ and $V = \underset{n \in \mathbb{N}}{\mathrm{Dr}}\, V_n$, and the subgroups U_n and V_n all have prime order. For every positive integer m put $U_m^* = \underset{n \geq m}{\mathrm{Dr}}\, U_n$ and $V_m^* = \underset{n \geq m}{\mathrm{Dr}}\, V_n$. Let x be an element of H. If $\langle x, U_i^* \rangle = \langle x, U_j^* \rangle$ with $i \leq j$, then

$$U_i^* = \langle x, U_j^* \rangle \cap U_i^* = U_j^* (\langle x \rangle \cap U_i^*),$$

and so $|U_i^* : U_j^*| \leq |\langle x \rangle|$. It follows that the chain

$$\langle x, U_1^* \rangle \geq \langle x, U_2^* \rangle \geq \cdots \geq \langle x, U_n^* \rangle \geq \cdots$$

is infinite, and there exists a positive integer h such that $\langle x, U_h^* \rangle$ has finite index in $\langle x, U_h^* \rangle^G$. Similarly there exists a positive integer k such that $\langle x, V_k^* \rangle$ has finite index in its normal closure. Since $U \cap V = 1$, the subgroup $\langle x, U_h^* \rangle \cap \langle x, V_k^* \rangle$ is finite, so that also $\langle x, U_h^* \rangle^G \cap \langle x, V_k^* \rangle^G$ is finite, and x has finitely many conjugates in G. Thus H is contained in the FC-centre of G. Assume that the set $\mathcal{L}$ of all subgroups X of H having infinite index in X^G is not empty, and let K be a minimal element of $\mathcal{L}$. Clearly K is infinite and cannot be generated by two proper subgroups, so that it is a group of type p^∞ for some prime p. Then $A \cap K$ is finite, and without loss of generality it can be assumed that $A \cap K = 1$, so that the chains

$$KU_1^* > KU_2^* > \cdots > KU_n^* > \cdots$$

and

$$KV_1^* > KV_2^* > \cdots > KV_n^* > \cdots$$

are properly descending, and there exist positive integers r and s such that KU_r^* and KV_s^* both have finite index in their normal closure. It follows that K has finite index in K^G, and this contradiction proves that the index $|X^G : X|$ is finite for every subgroup X of H. In particular every subgroup of the reduced abelian group A^G has finite index in its normal closure in G, and hence there exists a finite G-invariant subgroup E of A^G such that all subgroups of A^G/E are normal in G/E (see [4], Theorem 2.11). Clearly G/E is not a Chernikov group, and without loss of generality we may suppose that every subgroup of A^G is normal in G. Let g be an element of G. It can be assumed that $A \cap \langle g \rangle = 1$, and as above there exist positive integers i and j such that $\langle g, U_i^* \rangle$ and $\langle g, V_j^* \rangle$ both have finite index in their normal closure. It follows that $\langle g \rangle = \langle g, U_i^* \rangle \cap \langle g, V_j^* \rangle$ has finite index in $\langle g \rangle^G$, so that $\langle g \rangle^G$ is finite, and G is an FC-group. Therefore the commutator subgroup G' of G is finite by Lemma 2.4.

We consider now the case of non-periodic groups with the prescribed property. Our next lemma shows in particular that such groups are soluble-by-finite.

Lemma 2.11. *Let G be a non-periodic group satisfying the minimal condition on subgroups having infinite index in their normal closure. Then G is finite-by-abelian-by-finite.*

Proof. Let a be an element of infinite order of G. The centralizer $C_G(a)$ has finite index in G by Lemma 2.3. If x is any element of $C_G(a)$, the abelian subgroup $\langle a, x \rangle$ is generated by elements of infinite order, and hence is contained in the FC-centre of G. In particular $C_G(a)$ is an FC-group, and so it is finite-by-abelian by Lemma 2.4. Thus G is finite-by-abelian-by-finite.

Lemma 2.12. *Let G be a group, and let A be an abelian normal subgroup of G of infinite total rank. If every subgroup of A has finitely many conjugates in G, then there exists a sequence $(a_n)_{n \in \mathbb{N}}$ of non-trivial elements of A such that $\langle \langle a_n \rangle^G \mid n \in \mathbb{N} \rangle = \underset{n \in \mathbb{N}}{\mathrm{Dr}} \langle a_n \rangle^G$.*

Proof. Clearly it can be assumed that A is the normal closure of a subgroup B of G which either is the direct product of infinitely many subgroups of prime order or is free abelian of infinite rank. Suppose first that B is periodic, so that $A = B^G$ is also a periodic group whose primary components have prime exponent. Let $a_1 \neq 1$ be an element of A. Then $\langle a_1 \rangle^G$ is finite, and there exists a subgroup K such that $A = \langle a_1 \rangle^G \times K$. Since K has finite index in A and has finitely many conjugates in G, also its core K_1 has finite index in A. In particular K_1 is infinite. Let $a_2 \neq 1$ be an element of K_1. Then $\langle a_1 \rangle^G \langle a_2 \rangle^G = \langle a_1 \rangle^G \times \langle a_2 \rangle^G$ is finite, and as above K_1 contains an infinite G-invariant subgroup K_2 such that $\langle a_1 \rangle^G \langle a_2 \rangle^G \cap K_2 = 1$. This argument allows to construct a sequence $(a_n)_{n \in \mathbb{N}}$ of non-trivial elements of A^G such that $\langle \langle a_n \rangle^G \mid n \in \mathbb{N} \rangle = \underset{n \in \mathbb{N}}{\mathrm{Dr}} \langle a_n \rangle^G$.

Suppose now that B is free abelian. Let a_1 be an element of infinite order of A, and let K be a subgroup of A which is maximal with respect to the condition $\langle a_1 \rangle^G \cap K = 1$. Then $A/\langle a_1 \rangle^G K$ is periodic and A/K has finite torsion-free rank as $\langle a_1 \rangle^G$ is finitely generated. Since K has finitely many conjugates in G, also the factor group A/K_1 has finite rank, where K_1 is the core of K in G. In particular K_1 is not periodic. Let $a_2 \neq 1$ be an element of infinite order of K_1. Then $\langle a_1 \rangle^G \langle a_2 \rangle^G = \langle a_1 \rangle^G \times \langle a_2 \rangle^G$ is finitely generated, and as above K_1 contains a non-periodic G-invariant subgroup K_2 such that $\langle a_1 \rangle^G \langle a_2 \rangle^G \cap K_2 = 1$. This argument allows to construct also in this case a sequence $(a_n)_{n \in \mathbb{N}}$ of non-trivial elements of A such that $\langle \langle a_n \rangle^G \mid n \in \mathbb{N} \rangle = \underset{n \in \mathbb{N}}{\mathrm{Dr}} \langle a_n \rangle^G$.

Recall that a soluble-by-finite group G is said to be an $\mathcal{S}_1$-*group* if it has finite Prüfer rank and the set $\pi(G)$ of prime divisors of orders of elements of G is finite.

Lemma 2.13. *Let G be a non-periodic group satisfying the minimal condition on subgroups having infinite index in their normal closure. Then either G is an $\mathcal{S}_1$-group or its commutator subgroup G' is finite.*

Proof. The group G is finite-by-abelian-by-finite by Lemma 2.11, and without loss of generality it can be assumed that G is abelian-by-finite. Suppose that G is not an S_1-group, and let A be an abelian normal subgroup of finite index of G. Then A has infinite total rank, and by Lemma 2.12 there exists a sequence $(a_n)_{n \in \mathbb{N}}$ of non-trivial elements of A such that

$$\langle \langle a_n \rangle^G \mid n \in \mathbb{N} \rangle = \operatorname*{Dr}_{n \in \mathbb{N}} \langle a_n \rangle^G.$$

If x is an element of G, there exists a positive integer m such that

$$\langle x \rangle \cap \operatorname*{Dr}_{n \geq m} \langle a_n \rangle^G = 1,$$

and it can also be assumed that

$$\langle x \rangle \cap \operatorname*{Dr}_{n \in \mathbb{N}} \langle a_n \rangle^G = 1.$$

Put $U = \operatorname*{Dr}_{n \in \mathbb{N}} \langle a_{2n} \rangle^G$ and $V = \operatorname*{Dr}_{n \in \mathbb{N}} \langle a_{2n-1} \rangle^G$, so that

$$U \cap V = \langle x \rangle \cap \langle U, V \rangle = 1.$$

Write also $U_m = \operatorname*{Dr}_{n \geq m} \langle a_{2n} \rangle^G$ and $V_m = \operatorname*{Dr}_{n \geq m} \langle a_{2n-1} \rangle^G$ for every positive integer m. The chains of subgroups

$$\langle x, U_1 \rangle > \langle x, U_2 \rangle > \cdots > \langle x, U_n \rangle > \cdots$$

and

$$\langle x, V_1 \rangle > \langle x, V_2 \rangle > \cdots > \langle x, V_n \rangle > \cdots$$

are properly descending, and there exist positive integers h and k such that $\langle x, U_h \rangle$ and $\langle x, V_k \rangle$ both have finite index in their normal closure. Then $\langle x \rangle = \langle x, U_h \rangle \cap \langle x, V_k \rangle$ has finite index in $\langle x \rangle^G$, and x has finitely many conjugates in G. Therefore G is an FC-group, and G' is finite by Lemma 2.4.

The consideration of the infinite dihedral group shows that a non-periodic group satisfying the minimal condition on subgroups having infinite index in their normal closure need not have finite commutator subgroup. Our next theorem proves in particular that the infinite dihedral group is always involved in the structure of such groups.

Theorem 2.14. *Let G be a non-periodic group. Then G satisfies the minimal condition on subgroups having infinite index in their normal closure if and only if either the commutator subgroup G' is finite or G contains a finite normal subgroup E such that $G/E = J/E \times D/E$, where J/E is a radicable abelian group satisfying the minimal condition on subgroups and D/E is infinite dihedral.*

Proof. Suppose that G satisfies the minimal condition on subgroups having infinite index in their normal closure, but the commutator subgroup G' is infinite. Then G is an S_1-group by Lemma 2.13, and it is finite-by-abelian-by-finite by Lemma 2.11.

Without loss of generality it can be assumed that G is abelian-by-finite. Let A be an abelian normal subgroup of finite index of G, and let J be the subgroup generated by all Prüfer subgroups of G. Clearly J is contained in A, and there exists a subgroup K such that $A = J \times K$. Assume that the set $\mathcal{L}$ of all infinite subgroups of $\bar{G} = G/J$ having infinite index in their normal closure is not empty, and let $\bar{H}$ be a minimal element of $\mathcal{L}$. Clearly $\bar{H}$ cannot be generated by two infinite proper subgroups, so that in particular $\bar{H}/\bar{H}'$ is periodic. Since every periodic subgroup of $\bar{G}$ is finite, it follows that $\bar{H}'$ is infinite. On the other hand, every infinite proper subgroup of $\bar{H}$ has finite index in its normal closure, and Theorem 2.2 yields that $\bar{H}$ is a finite extension of a cyclic group. Since $\bar{H}$ has a unique maximal subgroup, we obtain that $\bar{H}$ is finite, and this contradiction proves that every infinite subgroup of $\bar{G}$ has finite index in its normal closure. If the commutator subgroup $\bar{G}'$ of $\bar{G}$ is finite, then G' is periodic and the set of all elements of finite order of G is a subgroup, so that G' is finite by Corollary 2.5. This contradiction shows that $\bar{G}'$ is infinite, so that $\bar{G}$ is cyclic-by-finite by Theorem 2.2. Let L be a finitely generated subgroup of G such that $G = JL$. Then $J \cap L$ is a finite normal subgroup of G, and so L is an infinite cyclic-by-finite group. In particular every subgroup of finite index of L is the join of two maximal subgroups, and it follows from the hypothesis that L has finite index in its normal closure. Therefore also $D = L^G$ is a finite extension of an infinite cyclic group, and its commutator subgroup D' is infinite, so that there exists a finite characteristic subgroup E of D containing $J \cap D$ and such that D/E is infinite dihedral. Thus $G/E = JE/E \times D/E$ has the prescribed structure.

Conversely, suppose that G contains a finite normal subgroup E such that $G/E = J/E \times D/E$, where J/E is a radicable abelian group satisfying the minimal condition on subgroups and D/E is infinite dihedral. Let X be a non-periodic subgroup of G. Then $X \cap D$ has finite index in D, and so X has finite index in XD. Since $X^G \leq (XE)^G = (XE)^D \leq XD$, it follows that X has finite index in X^G. On the other hand, all periodic subgroups of G satisfy the minimal condition on subgroups, and hence G satisfies the minimal condition on subgroups having infinite index in their normal closure.

Corollary 2.15. *Let G be a non-periodic group without elements of order* 2. *Then G satisfies the minimal condition on subgroups having infinite index in their normal closure if and only if its commutator subgroup G' is finite.*

3. Conjugacy Classes of Subgroups

Groups satisfying the minimal condition on subgroups with a given property χ are naturally involved in the investigations concerning groups with a finite number of conjugacy classes of χ-subgroups. This is a consequence of the following result.

Lemma 3.1 (see [1], Lemma 4.6.3). *Let G be a group locally satisfying the maximal condition on subgroups. If H is a subgroup of G such that $H^x \leq H$ for some element x of G, then $H^x = H$.*

A group G is called **locally graded** if every finitely generated non-trivial subgroup of G has a proper subgroup of finite index. It is clear that the class of locally graded groups contains that of locally (soluble-by-finite) groups. Moreover, every free group is locally graded, and hence homomorphic images of locally graded groups need not be locally graded. However, if N is a soluble normal subgroup of a locally graded group G, it is easy to show that also the factor group G/N is locally graded (see [9]).

Theorem 3.2. *A locally graded group G has finitely many conjugacy classes of subgroups having infinite index in their normal closure if and only if either the commutator subgroup G' of G is finite or G is a finite extension of an infinite cyclic group.*

Proof. Suppose that G has finitely many conjugacy classes of subgroups having infinite index in their normal closure. It is well-known that G contains a largest locally (nilpotent-by-finite) normal subgroup H (see for instance [13] Part 1, Theorem 2.31), and obviously H contains the FC-centre of G. If X/H is a cyclic non-trivial subgroup of G/H, then $X = \langle x, H \rangle$ for some $x \in G \setminus H$, so that x has infinitely many conjugates in G and in particular the index $|\langle x \rangle^G : \langle x \rangle|$ is infinite. It follows that G/H has finitely many conjugacy classes of cyclic subgroups, and so also finitely many normal subgroups. Let

$$H = H_0 < H_1 < \cdots < H_t = K$$

be the normal series of G defined by choosing H_{i+1}/H_i as the largest locally (nilpotent-by-finite) normal subgroup of G/H_i. Then G/K has no non-trivial locally (nilpotent-by-finite) normal subgroups. Let A/K be an abelian subgroup of finite rank of G/K. Clearly A^G/K is infinite and A/K does not contain non-trivial normal subgroups of G/K, so that the index $|A^G : A|$ is infinite. It follows that G/K has finitely many conjugacy classes of abelian subgroups of finite rank, and so it has finite exponent (see [14], Lemma 1). Assume that G/H is not periodic, and let $i < t$ be the largest non-negative integer such that H_{i+1}/H_i contains an element of infinite order yH_i. Clearly the subgroup $\langle y^n \rangle$ has infinite index in its normal closure for each positive integer n, and so there exists a subgroup L of $\langle y \rangle$ and an element g of G such that L properly contains L^g. Since G/H_{i+1} is periodic, the group $\langle g, H_{i+1} \rangle/H_i$ locally satisfies the maximal condition on subgroups, so that $LH_i = L^g H_i$ by Lemma 3.1 and hence $L = L^g$. This contradiction shows that G/H is periodic. In particular, for every element g of G the group $\langle K, g \rangle$ is an extension of a locally (nilpotent-by-finite) group by a locally finite group, and so it locally satisfies the maximal condition on subgroups. It follows from Lemma 3.1 that every chain of subgroups of K having infinite index in their normal closure is finite. Since all periodic subgroups of K are locally finite, we obtain by Theorem 2.10 and Lemma 2.11 that K is soluble-by-finite. Then the factor group G/K is also locally graded, and hence it is locally finite (see [14], Lemma 3). It follows that G locally satisfies the maximal condition on subgroups, and by Lemma 3.1 it satisfies the minimal condition on subgroups having infinite index in their normal closure. Let m be a positive integer such that every finite subgroup of G of order greater than m has finite index in its normal closure. Suppose first that G contains a finite normal subgroup E such that $|E| > m$. If Y is any finite normal subgroup of G, it follows that YE has finite index in its normal closure, so that also the subgroup Y^G is finite. Thus every element of finite order of G has finitely many conjugates, and hence the commutator subgroup G' of G is finite by Theorem 2.10 and Theorem 2.14. Assume now that G' is infinite, but the orders of finite normal subgroups of G are bounded. Using again Theorem 2.10 and Theorem 2.14, we obtain that G is a finite extension of an infinite cyclic subgroup.

Conversely, let G be a finite extension of an infinite cyclic subgroup. Then every infinite subgroup of G has finite index. Moreover, it is well-known that the finite subgroups of G lie in finitely many conjugacy classes (see for instance [2]), and hence G has finitely many conjugacy classes of subgroups having infinite index in their normal closure.

References

[1] B. Amberg, S. Franciosi and F. de Giovanni, Products of Groups, Oxford Mathematical Monographs, Clarendon Press, Oxford, 1992.

[2] J. F. Bowers and S. E. Stonehewer, A theorem of Mal'cev on periodic subgroups of soluble groups, Bull. London Math. Soc. 5 (1973), 323–324.

[3] R. Brandl, S. Franciosi and F. de Giovanni, Groups with finitely many conjugacy classes of non-normal subgroups, Proc. Roy. Irish Acad., to appear.

[4] C. Casolo, Groups with finite conjugacy classes of subnormal subgroups, Rend. Sem. Mat. Univ. Padova 81 (1989), 107–149.

[5] S. N. Chernikov, Investigation of groups with given properties of the subgroups, Ukrain. Math. J. 21 (1969), 160–172.

[6] S. Franciosi and F. de Giovanni, Groups satisfying the minimal condition on non-subnormal subgroups, Proceedings of the Conference on Infinite Groups (Ravello, Italy 1994), de Gruyter, Berlin, to appear.

[7] F. de Giovanni and G. Vincenzi, Groups satisfying the minimal condition on non-pronormal subgroups, Boll. Un. Mat. Ital., to appear.

[8] L. A. Kurdachenko and V. V. Pylaev, Groups rich with almost normal subgroups, Ukrain. Math. J. 40 (1988), 278–281.

[9] P. Longobardi, M. Maj and H. Smith, A note on locally graded groups, to appear.

[10] W. Möhres, Auflösbarkeit von Gruppen deren Untergruppen alle subnormal sind, Arch. Math. (Basel) 54 (1990), 232–235.

[11] B. H. Neumann, Groups with finite classes of conjugate subgroups, Math. Z. 63 (1955), 76–96.

[12] R. E. Phillips and J. S. Wilson, On certain minimal conditions for infinite groups, J. Algebra 51 (1978), 41–68.

[13] D. J. S. Robinson, Finiteness Conditions and Generalized Soluble Groups, Springer, Berlin, 1972.

[14] H. Smith, Groups with finitely many conjugacy classes of subgroups with large subnormal defect, Glasgow Math. J., to appear.

[15] V. P. Shunkov, On the minimality problem for locally finite groups, Algebra and Logic 9 (1970), 137–151.

Cockcroft Complexes and the Plus Construction

N. D. Gilbert

Abstract. Quillen's plus construction, familiar from algebraic K-theory, may be applied to any topological space relative to a perfect normal subgroup L of the fundamental group G. We discuss the plus construction on a 2-complex for which the Hurewicz map in dimension two is trivial, and consider when the plus construction inherits this property. Applications to group presentations are discussed.

1991 Mathematics Subject Classification: 57M20

Introduction

Let X be a connected 2-complex with fundamental group G, let L be a subgroup of G and let X_L denote the covering space of X with fundamental group L. We say that X is **L-Cockcroft** if the Hurewicz map $\pi_2 X_L \to H_2 X_L$ is zero. This property first arose in Cockcroft's work on subcomplexes of aspherical 2-complexes [9]. Recently, Cockcroft properties have been studied extensively in their own right, for example in [5, 12, 13, 14, 15, 20]. The aim of the present paper is to consider the effect of the plus construction on an L-Cockcroft 2-complex X, where L is a perfect normal subgroup of $\pi_1 X$ and the plus construction is formed relative to L. This work was prompted by an unpublished paper of J.-C. Hausmann, and a recent paper of M. N. Dyer.

Hausmann [17] used the plus construction to study subcomplexes of aspherical 2-complexes. If Y is an aspherical 2-complex obtained from a 2-complex X by attaching 2-cells, and if P is the maximal perfect subgroup of $K := \ker(\pi_1 X \to \pi_1 Y)$ then Hausmann shows that the plus construction X^{+P} is an aspherical 3-complex. Further, he showed that the same conclusion holds with P replaced by the Adams subgroup of K.

The main result of Dyer's paper [11] is as follows. Let X be a 2-complex with fundamental group G, and suppose that P is a non-trivial, finite, proper, normal, superperfect subgroup of G. Then X is not P-Cockcroft. (We remark here that Dyer includes the hypothesis that G/P has cohomological dimension at most 2. However, if the cohomological dimension of G/P exceeds 2 then X is not P-Cockcroft, by [7] Proposition 2.4.) Combined with a theorem of Adams [1], Dyer's result shows that if X is a subcomplex of an aspherical 2-complex, then X is aspherical if G is finitely presented and has no infinite perfect subgroups.

We shall concentrate upon Cockcroft properties for the plus construction. Our main result (Theorem 1.3) shows that the survival or destruction of Cockcroft properties on the formation of the plus construction X^{+L} depend upon Cockcroft properties of X and on the relationship between the second integral homology of L and that of G. We discuss a number of examples to illustrate the types of behaviour that are found.

I am grateful to J. Harlander and to J. Howie for conversations and other exchanges of information that assisted the progress of this work. I am pleased to acknowledge the financial support of the Royal Society and of the University of Durham Special Staff Travel Fund.

1. Cockcroft Properties of the Plus Construction

Let X be a connected space with fundamental group G and let L be a perfect subgroup of G. Quillen's plus construction associates to the pair (X, L) a space X^{+L} obtained from X by attaching cells in dimensions 2 and 3 such that the inclusion $X \hookrightarrow X^{+L}$ is acyclic and such that $L = \ker(\pi_1 X \to \pi_1 X^{+L})$ (see [3]). Our principal aim is to understand Cockcroft properties of 3-complexes obtained from the plus construction on a 2-complex. From [6,12] we have the following equivalent characterisations of the Cockcroft property.

Proposition 1.1. *Let K be a subgroup of the fundamental group of a space Y. Then the following are equivalent:*
(1) *Y is K-Cockcroft; that is, the Hurewicz map $\pi_2 Y \to H_2 Y_K$ is zero,*
(2) *the canonical map $H_2 Y_K \to H_2 K$ is an isomorphism,*
(3) *the canonical map $H_3 K \to \pi_2 Y \otimes_K \mathbb{Z}$ is surjective.*

The properties of Proposition 1.1 should be distinguished from that of a 3-complex being Cockcroft in the sense of Howie and Schneebeli [19]. The latter property requires that $\pi_2 Y = 0$ and that the Hurewicz map $\pi_3 Y \to H_3 Y$ is trivial. If Y is obtained as a plus construction X^{+L} on a 2-complex X, then $\pi_3 Y \to H_3 Y$ is trivial, for $H_3 Y = H_3 X = 0$. However, $\pi_2 Y$ may well be non-zero. Contrary to the case for 2-complexes, Cockcroft conditions need not be inherited by subcomplexes if there are cells in dimension 3. For example, we might begin with a non-Cockcroft space X and add 3-cells to kill $\pi_2 X$. The resulting space is certainly Cockcroft and has a non-Cockcroft subcomplex. On the other hand, if (Y, X) is a 2-connected pair and X is Cockcroft, then Y is also Cockcroft. The following lemma is a standard result on the covering properties of the plus construction (see [2] for example).

Lemma 1.2. *Let X be a space with fundamental group G and let L be a perfect normal subgroup of G. If H is a subgroup of G containing L then the covering space $(X^{+L})_{H/L}$ is homology equivalent to the plus construction $(X_H)^{+L}$.*

Proof. The space X^{+L} is formed by first attaching 2-cells to X so as to kill the subgroup L, obtaining a space Y. We then add 3-cells to Y, attached by the images under the universal covering map $\tilde{Y} \to Y$ of generators for the free $\mathbb{Z}(G/L)$-module $H_2(\tilde{Y}, X_L)$. The covering space $(X^{+L})_{H/L}$ is obtained by attaching 2-cells and 3-cells to X_H, which are translates by coset representatives of H in G of the 2-cells in $Y \setminus X$ and the 3-cells in $X^{+L} \setminus Y$. The cellular chain complex of the pair $((X^{+L})_{H/L}, X_H)$ is zero in all dimensions except 2 and 3: in each of these dimensions we have a free abelian group on the sets of attached cells of the appropriate dimension. The boundary map takes a 3-cell to the 2-cell that determines its attaching map and is therefore an isomorphism, and it follows that $H_i(X^{+L})_{H/L}, X_H) = 0$ for all $i \geq 0$. Hence the covering space X_H is homology equivalent to each of $(X^{+L})_{H/L}$ and $(X_H)^{+L}$.   □

As is well-known, $\pi_2 X^{+L}$ is isomorphic to $H_2 X_L$ via the composition of isomorphisms

$$\pi_2 X^{+L} \to \pi_2 \widetilde{X^{+L}} \to H_2(X_L)^{+L} \to H_2 X_L.$$

Therefore X^{+L} is Cockcroft if and only if the covering map $X_L \to X$ induces the trivial map $H_2 X_L \to H_2 X$. Our main result expands upon this observation.

Theorem 1.3. *Let X be a space with fundamental group G. Let L be a perfect normal subgroup of G and let H be a subgroup of G containing L. Then the following are equivalent:*
(a) *X^{+L} is (H/L)-Cockcroft,*
(b) *the covering map $X_L \to X_H$ induces the trivial map $H_2 X_L \to H_2 X_H$,*
(c) *X is H-Cockcroft and the inclusion $L \hookrightarrow G$ induces the trivial map $H_2 L \to H_2 H$,*
(d) *X is H-Cockcroft and the quotient map $H \to H/L$ induces an isomorphism $H_2 H \to H_2(H/L)$,*
(e) *X is H-Cockcroft and $H_2 L$ is a perfect $\mathbb{Z}H$-module.*

Proof. That (a) implies (b) follows from Lemma 1.2 and a generalisation of the discussion preceding the theorem, since $(X^{+L})_{H/L} \simeq (X_H)^{+L}$ and $H_2(X_H)^{+L} \cong H_2 X_H$. Now (b) implies (c), for the map $\pi_2 X \to H_2 X_H$ factors through $H_2 X_L$, so that X is H-Cockcroft. Then $H_2 X_H \cong H_2 H$ and $H_2 L$ is an image of $H_2 X_L$, whence $H_2 L \to H_2 H$ must be trivial. To see that (c) implies (d), set $Q = H/L$. From the long exact sequence for homology of L obtained from the short exact sequence $0 \to IQ \to \mathbb{Z}Q \to \mathbb{Z} \to 0$ of coefficient modules (where IQ is the augmentation ideal), and using Shapiro's lemma, we deduce that

$$H_2 L \cong H_1(L, IQ) \cong \ker(IQ \otimes_L IQ \to IQ) \cong H_2 Q,$$

since L acts on IQ via $\mathbb{Z}L \to \mathbb{Z}Q$. That (d) implies (e) follows from the Lyndon–Hochschild–Serre spectral sequence of the extension $1 \to L \to H \to H/L \to 1$, which gives a short exact sequence

$$0 \to H_0(H/L, H_2L) \to H_2H \to H_2(H/L) \to 0.$$

It follows that H_2L is a perfect $\mathbb{Z}(H/L)$-module, and so a perfect $\mathbb{Z}H$-module. Finally we see that (e) implies (a), for if X is H-Cockcroft and H_2L is a perfect $\mathbb{Z}H$-module, the composite isomorphism $H_2X_H \to H_2H \to H_2(H/L)$ factors through $H_2(X^{+L})_{H/L}$ via the isomorphism $H_2X_H \to H_2(X^{+L})_{H/L}$, whence $H_2(X^{+L})_{H/L} \to H_2(H/L)$ is an isomorphism, and X^{+L} is (H/L)-Cockcroft.   □

Corollary 1.4. *Let X be a 2-complex and let L be a perfect normal subgroup of $\pi_1 X$. Then the following are equivalent:*
(i)   *the plus construction X^{+L} is aspherical,*
(ii) *the covering space X_L is acyclic,*
(iii) *L is superperfect and X is L-Cockcroft.*

Proof. Suppose that X^{+L} is aspherical. Then $0 = \pi_2 X^{+L} = II_2 X_L$, whence (i) implies (ii). It is trivial that (ii) implies (iii). Finally, (iii) implies (i), for if L is superperfect and X is L-Cockcroft then X^{+L} is 1-Cockcroft, that is, $\pi_2 X^{+L} = 0$. Now the universal cover $\widetilde{X^{+L}}$ is homology equivalent to $(X_L)^+$ by Lemma 1.2, and so homology equivalent to X_L. Therefore $\widetilde{X^{+L}}$ is a simply connected 3-complex with trivial homology in dimensions 2 and 3 and so is contractible.   □

The equivalence of parts (ii) and (iii) of Corollary 1.4 is given by Bogley in [4], Theorem 6.3 part 2.

Let X be a connected 2-complex contained in an aspherical 2-complex Y. In this setting there exists a superperfect normal subgroup L of $\pi_1 X$ such that X is L-Cockcroft ([7], Theorem 3.6), and hence X^{+L} is aspherical. In [17] Hausmann reached the same conclusion by a direct analysis of the plus construction.

If X is a 2-complex with fundamental group G, we say that a subgroup H of G is a ***threshold*** for X if X is H-Cockcroft, but not K-Cockcroft for any proper subgroup of H. The existence of threshold subgroups for Cockcroft 2-complexes is considered on detail in [13,15]. The question arises of the existence of threshold subgroups for Cockcroft properties of X^{+L}. A subgroup H of $\pi_1 X$ that contains L is a threshold subgroup if X^{+L} is (H/L)-Cockcroft and if K is a subgroup of $\pi_1 X$ containing L and properly contained in H then X^{+L} is *not* (K/L)-Cockcroft. The results of [13,15] do not immediately apply to X^{+L} since X^{+L} is a 3-complex. However, part (b) of Theorem 1.3 enables us to consider only 2-complexes and then the methods of [15] can be applied.

Proposition 1.5. *Let X be a 2-complex, L is a perfect normal subgroup of $\pi_1 X$, and H a subgroup of $\pi_1 X$ containing L such that X^{+L} is (H/L)-Cockcroft. Then H contains a threshold subgroup.*

Proof. Let $H_0 \supseteq H_1 \supseteq \cdots$ be a descending chain of subgroups of $\pi_1 X$. We write X_i for the covering space X_{H_i}. Suppose that, for all i, we have $H_i \supseteq L$ and $H_2 X_L \to H_2 X_i$ is trivial. Set $H_\infty = \bigcap_{i \geq 0} H_i$ and $X_\infty = X_{H_\infty}$. Now the map $H_2 X_L \to \varprojlim H_2 X_i$ is trivial and also factors through $H_2 X_\infty$. By Lemma 1 of [15], the map $H_2 X_\infty \to \varprojlim H_2 X_i$ is injective: hence $H_2 X_L \to H_2 X_\infty$ is trivial. Zorn's Lemma now completes the argument. $\qquad\square$

2. Examples

1. Let $\mathcal{P}$ be a balanced, finite presentation of a finite, superperfect group G. An example of such a $\mathcal{P}$ is the presentation $\langle a, b \mid a^5 = b^3 = (ab)^2 \rangle$ of $SL_2(5)$: a balanced, finite presentation of $SL_2(p)$, for any odd prime p, is given by Campbell and Robertson [8]. Then $\mathcal{P}$ may be realised by a finite 2-complex X with a single 0-cell and with equal numbers of 1-cells and 2-cells, and X^{+G} is aspherical.

2. In contrast to example 1, Dyer [11] considers a finite 2-complex X with fundamental group G, and a non-trivial, *proper*, finite, superperfect, normal subgroup P of G. Then combining the main theorem of [11] with Corollary 1.4 shows that X^{+P} is *not* aspherical.

3. The plus construction may well change thresholds, or even destroy Cockcroft properties entirely. Consider the group T given by the presentation $\mathcal{P} = \langle a, b \mid a^2, b^3, (ab)^7 \rangle$. Then T is perfect, $H_2 T = \mathbb{Z}$ and T is the threshold subgroup for the 2-complex X associated to $\mathcal{P}$, see [13]. (In the terminology of [13], X is *absolutely Cockcroft*.) Let Y be the one-point union $X \vee S^1$, so that $\pi_1 Y$ is the free product $G = T * \mathbb{Z}$. Let L be the normal closure of T in G. Then L is perfect, but $H_2 L \to H_2 G = \mathbb{Z}$ is non-trivial. From Theorem 1.3 we see that Y^{+L} is not Cockcroft: however, Y *is* Cockcroft and L is a threshold for Y, see [13], Theorem 4.4.

4. Let G be a group. There exists a normal subgroup $\mathcal{A}(G)$ of G, the *Adams subgroup* of G, minimal with respect to the property that $G/\mathcal{A}(G)$ be locally indicable. Adams constructed $\mathcal{A}(G)$ in [1] and showed that if X is a non-aspherical subcomplex of an aspherical 2-complex Y, then the Adams subgroup $\mathcal{A}$ of $\ker(\pi_1 X \to \pi_1 Y)$ is non-trivial and perfect. In fact, as observed by Cohen [10], Adams' argument shows that the covering space $X_{\mathcal{A}}$ is acyclic. Therefore the plus construction $X^{+\mathcal{A}}$ is aspherical. See the survey [4] for further information. Now let X be a 2-complex such that the Adams subgroup $\mathcal{A}$ of $\pi_1 X$ is perfect. If $H_2 X = 0$ then, since $\pi_1 X/\mathcal{A}$ is locally indicable, we have $H_2 X_{\mathcal{A}} = 0$, and Corollary 1.4 shows that the plus construction $X^{+\mathcal{A}}$ is aspherical. However, the converse is false. Let X be a 2-complex with a single 2-cell whose attaching map is not a proper power in $\pi_1 X^{(1)}$ but is trivial in $H_1 X^{(1)}$.

Then $\pi_1 X$ is locally indicable [18], $\mathcal{A} = 1$ and $X^{+\mathcal{A}} = X$ is aspherical: however, $H_2 X = \mathbb{Z}$.

5. In [16] Harlander studies the occurrence of perfect normal subgroups in one relator groups. Let $\langle x_1, \ldots, x_n \mid r \rangle$ present a torsion-free one relator group G. Then G is the fundamental group of an aspherical 2-complex X. Let P be the maximal perfect subgroup of G. By Lemma 1 of [16], if G^{ab} is free abelian of rank $n - 1$, then P is superperfect and so, by Corollary 1.4, the plus construction X^{+P} is aspherical. An example of a group G as above with $P \neq 1$, again taken from [16], is given by the presentation $\langle a, b, c \mid aba^{-1} = [b, cbc^{-1}] \rangle$, in which the normal closure of b is perfect.

References

[1] J. F. Adams, A new proof of a theorem of W. H. Cockcroft, J. London Math. Soc. 49 (1955) 482–488.

[2] D. J. Benson, Representations and Cohomology II, Cambridge University Press (1992).

[3] A. J. Berrick, An Approach to Algebraic K-theory, Pitman Research Notes in Mathematics 56, Pitman (1982).

[4] W. A. Bogley, J. H. C. Whitehead's aspherical question, in: Two-dimensional Homotopy and Combinatorial Group Theory (C. Hog-Angeloni et al., eds.), London Math. Soc. Lecture Note Ser. 197, Cambridge University Press (1993).

[5] W. A. Bogley, Unions of Cockcroft 2-complexes, Proc. Edinburgh Math. Soc. 37 (1994) 193–199.

[6] J. Brandenburg and M. N. Dyer, On J. H. C. Whitehead's aspherical question I, Comment. Math. Helv. 56 (1981) 431–446.

[7] J. Brandenberg, M. N. Dyer and R. Strebel, On J. H. C. Whitehead's aspherical question II, Contemp. Math. 20 (1983) 65–78.

[8] C. M. Campbell and E. F. Robertson, A deficiency zero presentation for $SL(2, p)$, Bull. London Math. Soc. 12 (1980) 17–20.

[9] W. H. Cockcroft, On two-dimensional aspherical complexes, Proc. London Math. Soc 4 (1954) 375–384.

[10] J. M. Cohen, Aspherical 2-complexes, J. Pure Appl. Algebra 12 (1978) 101–110.

[11] M. N. Dyer, Groups with no infinite perfect subgroups and aspherical 2-complexes, Comment. Math. Helv. 68 (1993) 333–339.

[12] M. N. Dyer, Cockcroft 2-complexes, preprint, University of Oregon (1992).

[13] N. D. Gilbert and J. Howie, Threshold subgroups for Cockcroft 2-complexes, Comm. in Algebra, to appear.

[14] N. D. Gilbert and J. Howie, Cockcroft properties of graphs of 2-complexes, Proc. Roy. Soc. Edinburgh Sect. A 124 (1994) 363–369.

[15] J. Harlander, Minimal Cockcroft subgroups, Glasgow Math. J. 36 (1994) 87–90.

[16] J. Harlander, On perfect subgroups of one relator groups, in: Combinatorial and Geometric Group Theory (A. J. Duncan et al., eds.), London Math. Soc. Lecture Note Ser. 204, Cambridge University Press (1994).

[17] J.-C. Hausmann, Acyclic maps and the Whitehead aspherical problem, preprint.

[18] J. Howie, On locally indicable groups, Math. Z. 180 (1982) 445–461.

[19] J. Howie and H. R. Schneebeli, Homological and topological properties of locally indicable groups, Manuscripta Math. 44 (1983) 71–93.

[20] S. J. Pride, An example of a presentation which is minimally Cockcroft in several different ways, J. Pure Appl. Algebra 88 (1993) 199–204.

On the Solution of the
Dimension Subgroup Problem

Narain Gupta

Abstract. In this note we give an outline of the main steps used in my recent solution of the general dimension subgroup problem while highlighting the central features which allow non-trivial solutions to occur. As it turns out, the dimension quotients all have exponents dividing 2. Constructions in [GK] point towards the possibility that the above bound remains best possible in every level of solubility.

Let $\mathbb{Z}G$ denote the integral group ring of a group and let $\Delta(G) = \text{ideal}\{g - 1 \mid g \in G\}$ be its augmentation ideal. For any $n \geq 1$, the **n-th dimension subgroup** $D_n(G) = G \cap (1 + \Delta(G)^n)$ of G is easily seen to contain the n-th term $\gamma_n(G)$ of the lower central series of G. It had long been conjectured that $D_n(G) = \gamma_n(G)$ for all n and all groups G. However, it is now known that for each $n \geq 4$, there exist groups G_n such that their dimension quotients $D_n(G_n)/\gamma_n(G_n)$ are non-trivial (the dimension quotients being trivial for $n \leq 3$). The **dimension subgroup problem** amounts to finding the structure of the **n-th dimension quotient** $D_n(G)/\gamma_n(G)$ of a given group G. Without loss of generality, the group G may be assumed to be finitely generated and nilpotent. We proceed with the main steps used to solve the dimension subgroup problem.

1. Translating to Free Group Rings

The first necessary step is to translate the problem into the transparent language of *free group rings*. Identifying G as the quotient group F/R of a free group F of finite rank, it follows that the problem of identifying $D_n(G)$ reduces to identifying the subgroup

$$D(n, \mathfrak{r}) \equiv F \cap (1 + \mathfrak{r} + \mathfrak{f}^n) \mod R\gamma_n(F)$$

which amounts to solving a general congruence of the form

$$w - 1 \equiv 0 \mod \mathfrak{r} + \mathfrak{f}^n,$$

where $\mathfrak{f} = \Delta(F)$, $\mathfrak{r} = ZF\Delta(R)$ are the augmentation ideals of $\mathbb{Z}F$ (R being a normal in F).

2. Pre-Solvable Presentation of $G = F/R$

Let G be a finitely generated nilpotent group of arbitrary but fixed nilpotency class precisely $n - 1 \geq 2$. Then G is solvable of derived length precisely, say, ℓ which is clearly bounded by $2^\ell \leq n$. So G is a homomorphic image of the relatively free group $F/\delta_\ell(F)\gamma_n(F)$, where $F = \langle x_1, \ldots, x_m \mid \emptyset \rangle$ is free of rank m ($\delta_0(F) = F$ and $\delta_k(F) = [\delta_{k-1}(F), \delta_{k-1}(F)]$ are the terms of the derived series of F). Since F/RF' is finitely generated abelian, it follows (see, for instance, [MKS] p. 149) that G admits a *pre-abelian* presentation of the form

$$G = \langle x_{0,1}, \ldots, x_{0,m(0)} \mid x_{0,1}{}^{e(0,1)}\xi_{0,1}, \ldots, x_{0,m(0)}{}^{e(0,m(0))}\xi_{0,m(0)}, R_1 \rangle,$$

where $F_0 = \langle x_{0,1}, \ldots, x_{k,m(0)} \mid \emptyset \rangle$ ($m = m(0)$, $F_0 = F$), $\xi_{0,i} \in [F_0, F_0]$, $R_1 = R \cap [F_0, F_0]$, $e(0, p) \geq 0$ and for each pair $(e(0, i), e(0, j)) \neq (0, 0)$, either $e(0, i)|e(0, j)$ or $e(0, j)|e(0, i)$.

Repeated application of the above shows the existence of a presentation of G such that, for each $0 \leq k \leq \ell - 1$, the k-th derived subgroup $\delta_k(G)$ admits a *pre-abelian* presentation of the form:

$$\delta_k(G) = \langle x_{k,1}, \ldots, x_{k,m(k)} \mid x_{k,1}{}^{e(k,1)}\xi_{k,1}, \ldots, x_{k,m(k)}{}^{e(k,m(k))}\xi_{k,m(k)}, R_{k+1} \rangle,$$

where
(i) $F_k = \langle x_{k,1}, \ldots, x_{k,m(k)} \mid \emptyset \rangle < [F_{k-1}, F_{k-1}](< \delta_k(F))$ is a free subgroup of F;
(ii) $e(k, p) \geq 0$ and, for each pair $(e(k, i), e(k, j)) \neq (0, 0)$, either $e(k, i)|e(k, j)$ or $e(k, j)|e(k, i)$;
(iii) $\xi_{k,i} \in [F_k, F_k] < \delta_{k+1}(F)$ and $R_{k+1} = R_k \cap [F_k, F_k](\geq [R_k, f_k])$;
(iv) if $x_{k,i} \in \gamma_{n(k,i)}(F)\backslash\gamma_{n(k,i)+1}(F)$ and $x_{k,j} \in \gamma_{n(k,j)}(F)\backslash\gamma_{n(k,j)+1}(F)$ such that $n(k, i) < n(k, j)$, then $x_{k,i} < x_{k,j}$.
      [We shall refer to $n(k, p)$ as the *outer-weight* of $x_{k,p}$.]

3. Reduction to Partial Dimension
Congruences of Metabelian Type

Define $\mathfrak{f}_0 = \mathfrak{f}$ $(= \mathbb{Z}F(F - 1))$, $\mathfrak{r}_0 = \mathfrak{r}$ $(= \mathbb{Z}F(R - 1))$, $\mathfrak{r}(1) = \mathbb{Z}F([R, F] - 1)$, $\mathfrak{r}_1 = \mathbb{Z}F'(R \cap F' - 1)$; and, for each $k \geq 1$, define the ideals (in the *free group ring* $\mathbb{Z}F_k \subseteq \mathbb{Z}F$) $\mathfrak{f}_k = \mathbb{Z}F_k(F_k - 1)$; $\mathfrak{r}(k + 1) = \mathbb{Z}F_k([R_k, F_k] - 1)$; $\mathfrak{r}_k = \mathbb{Z}F_k(R_{k-1} \cap F_k - 1)$; and, in addition, define $\mathfrak{r}_k^* = (R_{k-1} \cap F_k - 1) = \{r_k - 1; r_k \in R_k\}$.

Recall from Section 1 that $D(n, \mathfrak{r}) = F \cap (1 + \mathfrak{r} + \mathfrak{f}^n)$, and define

$$D(n, \mathfrak{r}_0) = D(n, \mathfrak{r}) = F_0 \cap (1 + \mathfrak{r}_0 + \mathfrak{f}_0^2 \cap \mathfrak{f}^n),$$

and, for $k \geq 1$,

$$D(n, \mathfrak{r}_k) = F_k \cap (1 + \mathfrak{r}_k + \mathfrak{f}_k^2 \cap \mathfrak{f}^n).$$

If u is an element of the ideal $\mathfrak{r}_k$, then we may view u as

$$u \equiv \sum_i n_i (r_i - 1) \equiv (\Pi_i r_i^{n_i} - 1) = r - 1 \quad (\text{mod } \mathfrak{f}_k \mathfrak{r}_k + \mathfrak{r}_k \mathfrak{f}_k + \mathfrak{r}_{k+1}),$$

for some $r \in R_k$. Thus

$$D(n, \mathfrak{r}_k) = F_k \cap (1 + \mathfrak{r}_k^* + \mathfrak{f}_k \mathfrak{r}_k + \mathfrak{r}_k \mathfrak{f}_k + \mathfrak{f}_k^2 \cap \mathfrak{f}^n + \mathfrak{r}_{k+1})$$
$$= R_k([F_k, F_k] \cap (1 + \mathfrak{f}_k \mathfrak{r}_k + \mathfrak{r}_k \mathfrak{f}_k + \mathfrak{f}_k^2 \cap \mathfrak{f}^n + \mathfrak{r}_{k+1})).$$

Defining $G(n, \mathfrak{r}_k)(\leq [F_k, F_k])$, for each $k \geq 0$, by

$$F_{k+1} \cap (1 + \mathfrak{f}_k \mathfrak{r}_k + \mathfrak{r}_k \mathfrak{f}_k + \mathfrak{f}_k^2 \cap \mathfrak{f}^n + \mathfrak{r}_{k+1}) = G(n, \mathfrak{r}_k) D(n, \mathfrak{r}_{k+1})$$

proves that

$$D(n, \mathfrak{r}_k) = R_k G(n, \mathfrak{r}_k) D(n, \mathfrak{r}_{k+1}) \text{ for each } k \geq 0.$$

Repeated application of the above gives

$$D(n, \mathfrak{r}) = R_0 G(n, \mathfrak{r}_0) R_1 G(n, \mathfrak{r}_1) D(n, \mathfrak{r}_2) \cdots R_\ell G(n, \mathfrak{r}_\ell)$$
$$= R_0 R_1 \cdots R_{\ell-1} G(n, \mathfrak{r}_0) G(n, \mathfrak{r}_1) \cdots G(n, \mathfrak{r}_\ell)$$

thus solving completely the dimension subgroup problem in terms of the **partial dimension** subgroups $G(n, \mathfrak{r}_k)$, $k \geq 0$, $2^k \leq n$, which are characterized by the solutions of the congruences

$$w - 1 \equiv 0 \quad \text{mod } (\mathfrak{f}_k \mathfrak{r}_k + \mathfrak{r}_k \mathfrak{f}_k + \mathfrak{f}_k^2 \cap \mathfrak{f}^n + \mathfrak{r}_{k+1}).$$

A uniform solution of the above general congruence is all that is required to completely solve the dimension subgroup problem. Because of the ideal $\mathfrak{r}_{k+1}$, the above congruence has the framework of a free metabelian group ring for each k.

4. Normal Forms for Elements of
$\mathfrak{f}_k \mathfrak{r}_k + \mathfrak{r}_k \mathfrak{f}_k$ and $\mathfrak{f}_k^2 \cap \mathfrak{f}^n$

Define $\mathfrak{a}_{k+1} = \mathbb{Z} F_k([F_k, F_k] - 1) \ (\subseteq \mathfrak{f}_k \mathfrak{f}_k)$ to be the ideal of $\mathbb{Z} F_k$ generated by all elements $(h - 1)$, $h \in [F_k, F_k]$, and $\mathfrak{r}(k + 1) = \mathbb{Z} F_k([R_k, F_k] - 1) \ (\subseteq \mathbb{Z} F_k(R_k \cap [F_k, F_k] - 1) \subseteq \mathfrak{f}_k \mathfrak{r}_{k+1} + \mathfrak{r}_{k+1})$. It is easily seen that, for $k \geq 0$, modulo $\mathfrak{r}_k \mathfrak{a}_{k+1} +$

$\mathfrak{f}_k\mathfrak{r}(k+1) + \mathfrak{f}_k^2 \cap \mathfrak{f}^n$, the additive subgroup $(\mathfrak{f}_k\mathfrak{r}_k + \mathfrak{r}_k\mathfrak{f}_k; +)$ of $(\mathbb{Z}F_k; +)$ is generated by all distinct ordered products

$$(x_{k,1} - 1)^{c(1)} \cdots (x_{k,p} - 1)^{c(p)-1}(\overline{x}_{k,p} - 1) \cdots (x_{k,m(k)} - 1)^{c(m(k))},$$

where $c(i) \geq 0$, $c(p) \geq 1$, $c(1) + \cdots + c(m(k)) = q \geq 2$ and $\overline{x}_{k,p} = x_{k,p}{}^{e(k,p)}\xi_{k,p}$.
Next, for the normal form of elements of $\mathfrak{f}_k^2 \cap \mathfrak{f}^n$, first note that if $\mathfrak{n}(k, i)$, $\mathfrak{n}(k, j)$ are the outer-weights of $x_{k,i}$, $x_{k,j}$ then

$$(x_{k,i} - 1)(x_{k,j} - 1) \in \mathfrak{f}_k\mathfrak{f}_k \cap \mathfrak{f}^{\mathfrak{n}(k,i)+\mathfrak{n}(k,j)}.$$

An arbitrary element of $\mathfrak{f}_k^2 \cap \mathfrak{f}^n$ is a linear sum of products of the form

$$\mu = (x_{k,i(1)} - 1) \cdots (x_{k,i(t)} - 1) \in \mathfrak{f}_k^2 \cap \mathfrak{f}^n$$

with $\mathfrak{n}(k, i(1)) + \cdots + \mathfrak{n}(k, i(t)) \geq n$ and $t \geq 2$. It is easily seen that

$$\mu \equiv (x_{k,1} - 1)^{f_1} \cdots (x_{k,m(k)} - 1)^{f_{m(k)}}$$

modulo $\mathfrak{f}_k a_{k+1} \cap \mathfrak{f}^n + ([F_k, F_k] \cap \gamma_n(F) - 1)$, where

$$f_1\mathfrak{n}(k, 1) + \cdots + f_{m(k)}\mathfrak{n}(k, m(k)) = \mathfrak{n}(k, i(1)) + \cdots + \mathfrak{n}(k, i(t)) \geq n.$$

Let $x(k)$ be the ideal in $\mathbb{Z}F_k$ defined by

$$x(k) = \mathfrak{r}_k a_{k+1} + \mathfrak{f}_k\mathfrak{r}(k+1) + \mathfrak{f}_k a_{k+1} \cap \mathfrak{f}^n + ([F_k, F_k] \cap \gamma_n(F) - 1).$$

Then combining the two normal forms above it follows that, modulo $x(k)$, an arbitrary element of $\mathfrak{f}_k\mathfrak{r}_k + \mathfrak{r}_k\mathfrak{f}_k + \mathfrak{f}_k^2 \cap \mathfrak{f}^n$ is of the form $u + v$ where u is a $\mathbb{Z}$-linear sum of the ordered products of the form

$$(x_{k,1} - 1)^{c(1)} \cdots (x_{k,p} - 1)^{c(p)-1}(\overline{x}_{k,p} - 1) \cdots (x_{k,m(k)} - 1)^{c(m(k))},$$

with $c(i) \geq 0, c(p) \geq 1, c(1) + \cdots + c(m(k)) = q \geq 2$ and $\overline{x}_{k,p} = x_{k,p}{}^{e(k,p)}\xi_{k,p}$,
and v is in the ideal generated by the ordered products of the form

$$(x_{k,1} - 1)^{f_1} \cdots (x_{k,m(k)} - 1)^{f_{m(k)}}$$

with $f_i \geq 0$ with $f_1\mathfrak{n}(k, 1) + \cdots + f_{m(k)}\mathfrak{n}(k, m(k)) \geq n$.

5. Solution of the k-th Partial Dimension Congruence

Let $0 \leq k \leq \ell - 1$ be arbitrary but fixed. In the free group ring $\mathbb{Z}F_k$, consider the k-th partial dimension congruence

$$w - 1 \equiv 0 \quad \mod (\mathfrak{f}_k\mathfrak{r}_k + \mathfrak{r}_k\mathfrak{f}_k + \mathfrak{f}_k^2 \cap \mathfrak{f}^n + \mathfrak{r}_{k+1}).$$

where $w \in [F_k, F_k]$ and $\mathfrak{r}_k \neq 0$.

Using the analysis in Section 4, it follows that there exist $r \in [R_k, F_k]$, $h \in [F_k, F_k] \cap \gamma_n(F)$ such that

$$wrh - 1 \equiv u + v \quad \mod x(k), \tag{5.1}$$

where

$$x(k) = \tau_k a_{k+1} + f_k \tau(k+1) + f_k a_{k+1} \cap f^n + \tau_{k+1},$$

u is a $\mathbb{Z}$-linear of products of the form

$$(x_{k,1} - 1)^{c(1)} \cdots (x_{k,p} - 1)^{c(p)-1}(x_{k,p}{}^{e(k,p)} \xi_{k,p} - 1) \cdots (x_{k,m(k)} - 1)^{c(m(k))},$$

and v is in the ideal generated by ordered products of the form

$$(x_{k,1} - 1)^{f_1} \cdots (x_{k,m(k)} - 1)^{f_{m(k)}}.$$

We write

$$u = u_1 + u_2,$$

where u_1 ia a linear sum of products of the form

$$(x_{k,1} - 1)^{c(1)} \cdots (x_{k,p} - 1)^{c(p)-1} x_{k,p}{}^{e(k,p)} (\xi_{k,p} - 1) \cdots (x_{k,m(k)} - 1)^{c(m(k))},$$

and u_2 is a linear sum of products of the form

$$(x_{k,1} - 1)^{c(1)} \cdots (x_{k,p} - 1)^{c(p)-1}(x_{k,p}^{e(k,p)} - 1) \cdots (x_{k,m(k)} - 1)^{c(m(k))},$$

Since $wrh - 1 \in a_{k+1}$, $u_1 \in a_{k+1}$ and $x(k) \subseteq a_{k+1}$, it follows from (5.1) that

$$u_2 + v = 0 \quad \text{(absolutely)}. \tag{5.2}$$

Thus, in particular,

$$wrh - 1 \equiv u_1 \quad \mod (x(k)). \tag{5.3}$$

where $u_1 = u - u_2$ is a linear sum products of the form

$$(x_{k,1} - 1)^{c(1)} \cdots (x_{k,p} - 1)^{c(p)-1} x_{k,p}{}^{e(k,p)} (\xi_{k,p} - 1) \cdots (x_{k,m(k)} - 1)^{c(m(k))}.$$

A detailed analysis of (5.2) and (5.3) yields the following:

Firstly, each $\xi_{k,p}$ lies in the second centre of F_k. This reduces u_1 in (5.3) to a linear sum of products

$$(x_{k,i} - 1)(\xi_{k,p} - 1)(i \leq p) \quad \text{and} \quad (\xi_{k,p} - 1)(x_{k,j} - 1)(p \leq j). \tag{5.4}$$

Further, using (5.4),

$$u_2 = (g - 1) + \sum_{1 \leq p \leq m} (x_{k,p} - 1)(y_{k,p} - 1) \tag{5.5}$$

where

$$g \equiv \Pi_{1 \le i < j \le m}[\xi_{k,i}, x_{k,j}]^{a_{k,ij}} \quad (a_{k,ij} \ge 0)$$
$$\equiv \Pi_{1 \le i < j \le m}[x_{k,i}{}^{e(k,i)}, x_{k,j}]^{a_{k,ij}} \mod [R_k, F_k]$$

and

$$y_{k,p} = \Pi_{i<p}\xi_{k,i}^{-a_{k,ip}} \Pi_{p<j}\xi_{k,j}^{b_{k,pj}} \in \gamma_{n-n(k,p)}(F) \text{ for each } p,$$

implying each $\sum_{1 \le p \le m}(x_{k,p} - 1)(y_{k,p} - 1) \in \mathfrak{f}_k \mathfrak{a}_{k+1} \cap \mathfrak{f}^n \subseteq x(k)$, and the congruence(5.1) reduces to the congruence

$$wrh - 1 \equiv (g - 1) \mod x(k),$$

or equivalently,

$$wrhg^{-1} - 1 \equiv 0 \mod x(k), \tag{5.6}$$

where

$$x(k) = \mathfrak{r}_k \mathfrak{a}_{k+1} + \mathfrak{f}_k \mathfrak{r}(k+1) + \mathfrak{f}_k \mathfrak{a}_{k+1} \cap \mathfrak{f}^n + \mathfrak{r}_{k+1}.$$

Since (5.6) implies

$$wrhg^{-1} \in F_{k+1} \cap (1 + \mathfrak{f}_k \mathfrak{a}_{k+1}),$$

by the well-known Magnus embedding $F \cap (1 + \mathfrak{f}\mathfrak{a}) = F''$ (see, [G] p. 5]) of the free metabelian groups adopted to F_k it follows that

$$wrhg^{-1} \in [F_{k+1}, F_{k+1}],$$

which in turn implies that in fact

$$wrhg^{-1} - 1 \in \mathfrak{a}_{k+1}\mathfrak{a}_{k+1} \cap x(k)$$
$$\subseteq \mathfrak{r}_{k+1}\mathfrak{f}_{k+1} + \mathfrak{f}_{k+1}\mathfrak{r}(k+1) + \mathfrak{f}_{k+1}^2 \cap \mathfrak{f}^n + \mathfrak{r}_{k+2}.$$

It follows that, modulo $\mathfrak{r}_{k+1}\mathfrak{f}_{k+1} + \mathfrak{f}_{k+1}\mathfrak{r}(k+1) + \mathfrak{f}_{k+1}^2 \cap \mathfrak{f}^n + \mathfrak{r}_{k+2}$,

$$w - 1 \equiv gr^{-1}h^{-1} - 1, \tag{5.7}$$

which solves the k-th partial dimension congruence modulo the $(k+1)$-th partial dimension congruence.

6. Exponents of the n-th Dimension Quotients

Recall that the n-th dimension subgroup $D(n, \mathfrak{r})$ of F relative to the normal subgroup R is given by

$$D(n, \mathfrak{r}) = R_0 G(n, \mathfrak{r}_0) R_1 G(n, \mathfrak{r}_1) \cdots R_{\ell-1} G(n, \mathfrak{r}_{\ell-1})$$

where, for each $k \geq 0$, $G(n, \tau_k)$ has been characterized by the solutions (see 5.7)

$$w - 1 \equiv gr^{-1}h^{-1} - 1 \quad \mathrm{mod}\ \tau_{k+1}\mathfrak{f}_{k+1} + \mathfrak{f}_{k+1}\tau(k+1) + \mathfrak{f}_{k+1}^2 \cap \mathfrak{f}^n + \tau_{k+2}.$$

Thus, $w \in D(n, \tau)$ implies

$$w \equiv (g_1 r_1^{-1} h_1^{-1})(g_2 r_2^{-1} h_2^{-1}) \cdots (g_\ell - 1 r_{\ell-1}^{-1} h_{\ell-1}^{-1}) \quad (\mathrm{mod}\ \gamma_n(F)\delta_1(F))$$

with $r_k^{-1} h_k^{-1} \in (R_{k-1} \cap F_k([F_k, F_k] \cap \gamma_n(F))$, so that

$$w \equiv g_1 g_2 \cdots g_{\ell-1} \quad (\mathrm{mod}\ R\gamma_n(F)). \tag{6.1}$$

It follows from our proof that if w is of the form (6.1) above then $w - 1 \in \tau + \mathfrak{f}^n$. Thus, (6.1) yields the complete solution of the dimension subgroup problem.
    Further,

$$(g_1 g_2 \cdots g_{\ell-1})(g_1 g_2 \cdots g_{\ell-1}) \equiv g_1 g_1^{(g_2 \cdots g_n)^{-1}} (g_2 \cdots g_{\ell-1})(g_2 \cdots g_{\ell-1})$$

$$\equiv g_1 g_1 (g_2 \cdots g_{\ell-1})(g_2 \cdots g_{\ell-1})$$

and, by induction,

$$w^2 \equiv (g_1 \cdots g_{\ell-1})(g_1 \cdots g_{\ell-1}) \equiv g_1 g_1 g_2 g_2 \cdots g_{\ell-1} g_{\ell-1} = g_1^2 \cdots g_{\ell-1}^2.$$

Thus, to prove that $w^2 \equiv 1$ $(\mathrm{mod}\ R\gamma_n(F))$ it suffices to prove that

$$g_{k+1}^2 \equiv 1 \quad (\mathrm{mod}\ (R \cap F_{k+1})\gamma_n(F)) \text{ for each } k \geq 0.$$

This, indeed, is shown to be true. The proof requires further analysis of (5.2) and is rather technical. The reader is referred to [G] for details. Finally, we remark that an embedding theorem proved in [GK] points towards the existence of non-trivial solutions in each of the higher solvability levels.

References

[G]   Narain Gupta, Free Group Rings, Contemp. Math. 66 (1987), American Mathematical Society, Providence, Rhode Island.

[GK]   Narain Gupta and Yuri Kuz'min, Groups with non-central dimension quotients, J. Pure Appl. Algebra (to appear).

[MKS] Wilhelm Magnus, Abraham Karrass and Donald Solitar, Combinatorial Group Theory, Interscience, New York (1966).

Triangle Groups and Their Generalisations

James Howie, Vasileios Metaftsis and Richard M. Thomas

Abstract. We describe the classification and structure of the finite generalised triangle groups.

1991 Mathematics Subject Classification: 20F05

1. Triangle Groups

The triangle group $T = T(p, q, r)$ defined by the presentation $\langle a, b \mid a^p = b^q = (ab)^r = 1 \rangle$ arises geometrically as a discrete group of isometries of the elliptic, euclidean or hyperbolic plane, in which the elements a, b and ab represent rotations through $\frac{2\pi}{p}$, $\frac{2\pi}{q}$ and $\frac{2\pi}{r}$ about the vertices of a triangle with angles $\frac{\pi}{p}$, $\frac{\pi}{q}$ and $\frac{\pi}{r}$ respectively. These groups were first studied intensively in [FK]. Here we just note some well-known properties.

1. The elliptic/euclidean/hyperbolic trichotomy. If $\frac{1}{p} + \frac{1}{q} + \frac{1}{r} > 1$ then T is a finite subgroup of $SO(3)$, acting orthogonally on the sphere S^2.

If $\frac{1}{p} + \frac{1}{q} + \frac{1}{r} = 1$, then T is abelian-by-finite, and acts by isometries on the euclidean plane E^2.

If $\frac{1}{p} + \frac{1}{q} + \frac{1}{r} < 1$, then T is a Fuchsian group, that is a discrete, cocompact group of isometries of the hyperbolic plane H^2. In particular, it contains a nonabelian free subgroup.

2. Faithful representation. The groups $SO(3)$, $Isom^+(E^2)$ and $Isom^+(H^2) \cong PSL(2, \mathbb{R})$ of orientation-preserving isometries of E^2 and H^2 respectively, all embed into $PSL(2, \mathbb{C}) \cong Isom^+(H^3)$, the group of orientation-preserving isometries of the hyperbolic 3-space. It follows that T always has a faithful representation $T \to PSL(2, \mathbb{C})$. The embeddings can be realised geometrically by extending the actions of $SO(3)$, $Isom^+(E^2)$ and $Isom^+(H^2)$ on a sphere, horosphere and hyperplane respectively, to the whole of H^3. Algebraically,

$$SO(3) \cong PSU(2) = \left\{ \begin{pmatrix} a & b \\ -\bar{b} & \bar{a} \end{pmatrix} \;\middle|\; |a|^2 + |b|^2 = 1 \right\} / \{\mp I\},$$

and

$$Isom^+(E^2) \cong \left\{ \begin{pmatrix} a & b \\ 0 & \bar{a} \end{pmatrix} \mid |a| = 1 \right\} / \{\mp I\};$$

while the embedding

$$PSL(2, \mathbb{R}) \hookrightarrow PSL(2, \mathbb{C})$$

is the natural one.

3. The Tits alternative. A theorem of Tits [T] asserts that any subgroup G of a matrix group, over any field of characteristic zero, satisfies the following property, known as the Tits alternative: either G has a soluble subgroup of finite index, or G contains a non-abelian free subgroup. By property 2 above, this Tits alternative holds for all triangle groups. This can also be seen directly, using the trichotomy 1: a triangle group T has an abelian subgroup of finite index (if T is elliptic or euclidean) or a free subgroup of rank 2 (if T is hyperbolic).

2. Generalised Triangle Groups

Definition. A *generalised triangle group* is a group defined by a presentation of the form

$$G = \langle a, b \mid a^p = b^q = W(a, b)^r = 1 \rangle$$

where p, q, r are integers greater than 1, and W is a word of the form

$$a^{\alpha_1} b^{\beta_1} \dots a^{\alpha_k} b^{\beta_k},$$

$k \geq 1, 0 < \alpha_i < p, 0 < \beta_i < q$ for all i, which is not a proper power.

These groups were first investigated by Baumslag, Morgan and Shalen in [BMS], who proved a number of facts about them. Intensive investigations have been carried out by Rosenberger and others [FHR, FHR2, FLR, FLR2, FR, FR2, LR, R2]. One motivation for their study is a result of Gordon and Luecke [GL]: if the 3-manifold $M(\alpha, k)$ obtained by Dehn surgery on the knot $k \subset S^3$ according to the slope-parameter $\alpha \in \mathbb{Q} \cup \{\infty\}$ is a connected sum, then $\alpha \in \mathbb{Z}$. A key step in the proof of this uses the fact that generalised triangle groups are non-trivial.

It would be of interest to find appropriate generalisations of properties 1–3 of triangle groups above. For example:

Conjecture (Rosenberger [R2]). There is a Tits alternative for generalised triangle groups. That is, if G is a generalised triangle group then either G is soluble-by-finite or G contains a free subgroup of finite index.

A significant amount of work has been done in this direction [FLR, FR, LR, R2] and the conjecture has been proved in the case $r \geq 3$, as well as in many cases where $r = 2$. It remains open in general, however.

There is not such a clean trichotomy into finite/abelian-by-finite/huge as there is for triangle groups. In one direction the result is the same.

Lemma. *If $\frac{1}{p} + \frac{1}{q} + \frac{1}{r} \leq 1$ then G is infinite.*

Proof. See [BMS].   □

The existence (almost always) of free subgroups is part of the Tits alternative question, which we will not address here. In this paper we consider the generalised triangle groups which are finite. By the above Lemma, only the case $\frac{1}{p} + \frac{1}{q} + \frac{1}{r} > 1$ occurs. The main tool uses representations, which brings us to the second property of triangle groups mentioned above. Clearly it is too much to expect generalised triangle groups to have faithful representations into $PSL(2, \mathbb{C})$ (or any other fixed Lie group) in general. In this context, the following is as good as we may expect.

Definition. A representation ρ of G in a group H is called **special** (or **essential**) if $\rho(a)$, $\rho(b)$ and $\rho(W)$ have orders p, q and r respectively.

Theorem. *Every generalised triangle group admits a special representation to $PSL(2, \mathbb{C})$.*

Proof. See [BMS] or [FHR]. The proof is based on the idea of the trace polynomial. An element $\{\mp A\}$ of $PSL(2, \mathbb{C})$ has order $m \geq 2$ if and only if $\mathrm{tr}(A) = \mp(\alpha + \bar{\alpha})$ for some primitive m-th root α of -1. If we fix suitable values for the traces of $A = \rho(a)$ and $B = \rho(b)$ so that they have orders p and q respectively, then $\mathrm{tr}(W(A, B))$ is a polynomial of degree exactly k in $\lambda = \mathrm{tr}(AB)$. The Fundamental Theorem of Algebra allows us to fix any value we want for this trace also, so $W(A, B)$ can be chosen to have order r. (For more details see [FHR].)   □

If G is a finite generalised triangle group, and ρ is a special representation then $\rho(G)$ is a finite subgroup of $PSL(2, \mathbb{C})$, generated by two elements of order p and q. Such subgroups are completely understood and we have a finite list of possible values for $\mathrm{tr}(\rho(ab))$, i.e. of possible roots for the trace polynomial.

It can also been shown that, if the trace polynomial has a multiple root, then G has an element of infinite order (see [HMT], Theorem 2.5). Hence we have a bound on the degree of the trace polynomial, i.e. on the length of W.

This leaves a finite number of possible presentations to check. Many of these have been shown infinite by Rosenberger and others (see [Co, FLR, FR, LR, R2]). Using a mixture of representation-theoretic, geometric and computational techniques we proved in [HMT] that most of the others were infinite. Apart from eleven finite groups, this left only two undecided cases, which were finally solved by Lévai, Rosenberger and Souvignier [LRS]. Combining the results of [HMT] and [LRS], we have the following.

Theorem. *Let G be a finite generalised triangle group, and not a triangle group. Then G is one of:*

(1) $\langle a, b \mid a^2, b^3, (abab^2)^2 \rangle \cong A_4 \times C_2$, *of order 24;*

(2) $\langle a, b \mid a^2, b^3, (ababab^2)^2 \rangle$, *of order 48;*

(3) $\langle a, b \mid a^2, b^5, (abab^2)^2 \rangle \cong \langle a, b \mid a^2, b^3, (abababab^2)^2 \rangle \cong A_5 \times C_2$, *of order 120;*

(4) $\langle a, b \mid a^3, b^3, (abab^2)^2 \rangle \cong A_5 \times C_3$, *of order 180;*

(5) $\langle a, b \mid a^2, b^4, (ababab^3)^2 \rangle$, *of order 192;*

(6) $\langle a, b \mid a^3, b^3, (aba^2b^2)^2 \rangle$, *of order 288;*

(7) $\langle a, b \mid a^2, b^3, (ababab^2ab^2)^2 \rangle$, *of order 576;*

(8) $\langle a, b \mid a^2, b^3, (ababab^2abab^2)^2 \rangle$, *of order 720;*

(9) $\langle a, b \mid a^2, b^5, (ababab^4)^2 \rangle \cong \langle a, b \mid a^2, b^5, (abab^2ab^4)^2 \rangle$, *of order 1, 200;*

(10) $\langle a, b \mid a^2, b^3, (ababab^2)^3 \rangle$, *of order 1, 440;*

(11) $\langle a, b \mid a^2, b^3, (ababababab^2ab^2)^2 \rangle$, *of order 2, 880;*

(12) $\langle a, b \mid a^2, b^3, (ababababab^2ab^2abab^2ab^2)^2 \rangle$ *of order* 424, 673, 280 $= 2^{20} \cdot 3^4 \cdot 5$.

Indeed, the presentation can be shown to be equivalent to one of the 14 presentations shown under the equivalence relation generated by automorphisms of the cyclic groups generated by a and b, inversion and cyclic permutation of the third relator, and interchange of a and b in the case where $p = q$.

The order of the last group (12) was found by Lévai, Rosenberger and Souvignier [LRS], using the computer system GAP [S]. We call it the **LRS-monster**.

3. Structure of the Finite Generalised Triangle Groups

It would be interesting to have some sort of understanding of the structure of generalised triangle groups—or at least of the twelve finite ones listed above. It turns out that the representations we have found go a long way towards yielding such an understanding.

Suppose $\rho : G \to PSL(2, \mathbb{C})$ is a special representation. Form the pullback square

$$
\begin{array}{ccc}
\tilde{G} & \to & SL(2, \mathbb{C}) \\
\pi \downarrow & & \downarrow \\
G & \overset{\rho}{\to} & PSL(2, \mathbb{C})
\end{array}
$$

so that $\tilde{G}$ is a central extension of G by $\mathbb{Z}_2$.

Lemma. *Let G be a finite generalised triangle group. Then the central extension $\tilde{G}$ is independent of the choice of special representation ρ.*

Proof. Let A, B, C be the cyclic subgroups of G (of orders p, q, r respectively) generated by a, b, W, and let $\tilde{A} = \pi^{-1}(A)$, $\tilde{B} = \pi^{-1}(B)$, $\tilde{C} = \pi^{-1}(C)$. Since $\rho : A \to PSL(2, \mathbb{C})$ is injective, $\tilde{A}$ is a subset of $SL(2, \mathbb{C})$, and is necessarily cyclic (of order $2p$) since $SL(2, \mathbb{C})$ has only one element of order 2. Hence the central extension $\tilde{A} \to A$ is independent of ρ. The same holds for the central extensions $\tilde{B} \to B$ and $\tilde{C} \to C$. To complete the proof, it suffices to show that

$$res : H^2(G; \mathbb{Z}_2) \to H^2(A; \mathbb{Z}_2) \times H^2(B; \mathbb{Z}_2) \times H^2(C; \mathbb{Z}_2)$$

is injective. To see this, construct a topological space X by gluing a $K(C, 1)$ space to $K(A, 1) \vee K(B, 1)$ along a path representing W. Then $\pi_1(X) \cong G$, so X can be made into a $K(G, 1)$-space by attaching cells in dimensions greater or equal to three. In particular the restriction map above factors through an injection

$$H^2(G; \mathbb{Z}_2) \to H^2(X; \mathbb{Z}_2).$$

The other part of the restriction map fits into a Mayer–Vietoris sequence

$$H^1(A; \mathbb{Z}_2) \times H^1(B; \mathbb{Z}_2) \times H^1(C; \mathbb{Z}_2) \xrightarrow{j} H^1(S^1; \mathbb{Z}_2) \to H^2(X; \mathbb{Z}_2) \to$$

$$H^2(A; \mathbb{Z}_2) \times H^2(B; \mathbb{Z}_2) \times H^2(C; \mathbb{Z}_2)$$

so the result follows unless the map j in this sequence is zero. In particular, it follows if r is even, for then

$$H^1(C; \mathbb{Z}_2) \to H^1(S^1; \mathbb{Z}_2)$$

is an isomorphism. But the only finite generalised triangle group for which r is odd is the group (10) defined by the presentation

$$G = \langle a, b \mid a^2 = b^3 = (ababab^2)^3 = 1 \rangle$$

and in this case it is clear that $H^1(A; \mathbb{Z}_2) \to H^1(S^1; \mathbb{Z}_2)$ is an isomorphism (since a occurs an odd number of times in W). $\qquad\square$

Remark. The above argument also works for many infinite generalised triangle groups, but not for all of them. For example, if $p = q = r = 3$ then the space X is aspherical [DH], and so we have $H^2(G; \mathbb{Z}_2) \cong H^2(X; \mathbb{Z}_2) \cong H^1(S^1; \mathbb{Z}_2) \cong \mathbb{Z}_2$, while $H^2(A; \mathbb{Z}_2) = H^2(B; \mathbb{Z}_2) = H^2(C; \mathbb{Z}_2) = 0$. Indeed, if G is the triangle group $\langle a, b \mid a^3 = b^3 = (ab)^3 = 1 \rangle$, then the faithful representation $\rho : G \to PSL(2, \mathbb{C})$ gives rise to two distinct lifts $\tilde{G} \to SL(2, \mathbb{C})$, depending on whether an even or odd number of $\rho(a), \rho(b), \rho(ab)$ are deemed to have trace $+1$, rather than -1. In the first case the central extension $\tilde{G} \to G$ splits, in the second it does not.

Corollary. *Let G be a finite generalised triangle group admitting n inequivalent special representations* $\rho_1, \ldots, \rho_n$ *into* $PSL(2, \mathbb{C})$. *Then there is a special representation* $\sigma : G \to PSL(2n, \mathbb{C})$ *such that* $\mathrm{Ker}\,\sigma$ *is contained in* $\mathrm{Ker}\,\rho_1 \cap \cdots \cap \mathrm{Ker}\,\rho_n$ *as a subgroup of index dividing* 2^{n-1}.

Proof. Let $\tilde{\rho}_i : \tilde{G} \to SL(2, \mathbb{C})$ $(i = 1, \ldots, n)$ be the pullbacks of $\rho_i : G \to PSL(2, \mathbb{C})$. By the above Lemma the group $\tilde{G}$ does not depend on ρ_i. Moreover, if z is the non-trivial element of the kernel of $\tilde{G} \to G$, then $\rho_i(z) = -I$ for all i. Let

$$\tilde{\rho} = \tilde{\rho}_1 \times \cdots \times \tilde{\rho}_n : \tilde{G} \to SL(2, \mathbb{C})^n,$$

the n-th direct power. Composing $\tilde{\rho}$ with the diagonal embedding of $SL(2, \mathbb{C})^n$ into $SL(2n, \mathbb{C})$, we obtain a representation $\tilde{\sigma} : \tilde{G} \to SL(2n, \mathbb{C})$ such that $\tilde{\sigma}(z) = -I$, and indeed, $\tilde{\sigma}^{-1}(Z(SL(2n, \mathbb{C}))) = \{1, z\}$. Projectivising, we get a representation $\sigma : G \to PSL(2n, \mathbb{C})$. The last statement concerning $\mathrm{Ker}\,\sigma$ is immediate from the construction. $\qquad\square$

In practice, many of our finite generalised triangle groups have faithful representations into $PSL(4, \mathbb{C})$ arising this way (where there are precisely two inequivalent special representations).

For example, the group

$$(6) : \langle a, b \mid a^3, b^3, (aba^2b^2)^2 \rangle$$

admits two special representations up to conjugacy, each with image isomorphic to A_4 (they are distinguished by whether a, b are mapped to conjugate 3-cycles in A_4 or not). The resulting representation $\sigma : G \to PSL(4, \mathbb{C})$ is faithful, with image isomorphic to $\tilde{A}_4 \vee \tilde{A}_4$. Here $\tilde{A}_4 \cong SL(2, 3) \subset SL(2, \mathbb{C})$ denotes the double cover of $A_4 \subset PSL(2, \mathbb{C})$, and $\vee$ denotes a **central product**, that is, the direct product with a central subgroup (of order 2, in all cases considered here) identified.

In the same way, the groups

$$(7) : \langle a, b \mid a^2, b^3, (ababab^2ab^2)^2 \rangle,$$
$$(8) : \langle a, b \mid a^2, b^3, (ababab^2abab^2)^2 \rangle,$$
$$(9) : \langle a, b \mid a^2, b^5, (ababab^4)^2 \rangle,$$
$$(10) : \langle a, b \mid a^2, b^3, (ababab^2)^3 \rangle, \text{ and}$$
$$(11) : \langle a, b \mid a^2, b^3, (ababababab^2ab^2)^2 \rangle$$

each admit two special representations to $PSL(2, \mathbb{C})$, and the resulting representation into $PSL(4, \mathbb{C})$ is faithful, with image isomorphic to $\tilde{A}_4 \vee \tilde{S}_4$, $\tilde{S}_3 \vee \tilde{A}_5$, $\tilde{D}_{10} \vee \tilde{A}_5$, $\tilde{A}_4 \vee \tilde{A}_5$ and $\tilde{S}_4 \vee \tilde{A}_5$, respectively.

The group

$$(2) : \langle a, b \mid a^2, b^3, (ababab^2)^2 \rangle$$

also admits two special representations, onto S_3 and S_4 respectively, but the S_3 representation factors through the S_4 representation, and hence the resulting representation to $\tilde{S}_3 \vee \tilde{S}_4 \subset PSL(4, \mathbb{C})$ maps onto a proper subgroup (of index 6) of $\tilde{S}_3 \vee \tilde{S}_4$. The representation is faithful, and the image is isomorphic to $\tilde{S}_4$.

Something similar happens with the group

$$(5): \langle a, b \mid a^2, b^4, (ababab^3)^2 \rangle.$$

The two special representations have images D_8 and S_4 respectively. The abelianisation of $\tilde{D}_8 \vee \tilde{S}_4$ is elementary abelian of order 8, so the group cannot be generated by two elements, and hence the representation $\sigma : G \to \tilde{D}_8 \vee \tilde{S}_4$ cannot be surjective. In fact σ is faithful, and its image has index 2 in $\tilde{D}_8 \vee \tilde{S}_4$.

This completes the description of the structure of those finite generalised triangle groups that admit precisely two inequivalent special representations to $PSL(2, \mathbb{C})$. Three of the groups,

$$(1): \langle a, b \mid a^2, b^3, (abab^2)^2 \rangle \cong A_4 \times C_2,$$

$$(3): \langle a, b \mid a^2, b^5, (abab^2)^2 \rangle \cong \langle a, b \mid a^2, b^3, (ababab^2)^2 \rangle \cong A_5 \times C_2, \text{ and}$$

$$(4): \langle a, b \mid a^3, b^3, (abab^2)^2 \rangle \cong A_5 \times C_3,$$

each admit only one special representation, with image A_4, A_5, A_5 respectively, and in each case there is an isomorphism $G \cong \rho(G) \times C_k$ for some finite cyclic group C_k ($k \in \mathbb{N}$). Note the differences between the two presentations for the group (3): in the first case ab is mapped to a 3-cycle in A_5, so the trace polynomial has two roots ± 1, which is compatible with the length of the third relator; in the second presentation, ab maps to a 5-cycle in A_5, so the trace polynomial has four roots $\frac{\pm 1 \pm \sqrt{5}}{2}$, again compatible with the length of the third relator.

Finally, we have the LRS-monster

$$(12): \langle a, b \mid a^2, b^3, (ababababab^2 ab^2 abab^2 ab^2)^2 \rangle.$$

This admits three inequivalent special representations, with images A_4, S_4, A_5. We can thus produce a representation $\sigma : G \to PSL(6, \mathbb{C})$. The image of this representation is isomorphic to $(\tilde{A}_4 \times \tilde{S}_4 \times \tilde{A}_5)/\{\pm I\}$, which has order $2^9 \cdot 3^3 \cdot 5$, and so its kernel has order $2^{11} \cdot 3$. We do not know the smallest degree of a faithful representation of this group, although we do make some further comments about this in the next section.

4. Subgroups of the LRS-Monster

In this section, we make some comments about the Sylow p-subgroups of the LRS-monster G of order $2^{20} \cdot 3^4 \cdot 5$.

As we noted above, G has three inequivalent special representations with homomorphic images A_4, S_4 and A_5, and hence there are homomorphisms of G onto $A_4 \times S_4 \times A_5$ and $A_4 \times S_4$. Let K and L respectively be the kernels of these homomorphisms, so that K has order $2^{13} \cdot 3$ and L has order $2^{15} \cdot 3^2 \cdot 5$.

Using Quotpic [HR], we discovered that L has a normal subgroup M of index 3. Applying the nilpotent quotient algorithm to M yielded a maximal nilpotent 2-quotient of order 2^{12}. So we have a normal subgroup N of M of order $2^3 \cdot 3 \cdot 5$ such that N involves A_5 and N/N' has odd order; we see that N must be isomorphic to $SL(2, 5)$. Since $Z(N)$ is characteristic in N which is normal in G, $Z(N)$ is normal in G.

Let $\overline{G} = G/K$, and let A be the full pre-image of $A_4 \times S_4$ in G. We see that $A \cap N = Z(N)$, so that G is a central product $A \vee N$ (as was pointed out in [LRS]). Now an irreducible complex $(A \times N)$-module is a tensor product of an irreducible A-module and an irreducible N-module, and every such tensor product yields an irreducible $(A \times N)$-module. Since $G = A \vee N$ is a quotient of $A \times N$ by a cyclic subgroup T of order 2, and since the irreducible complex representations of $N = SL(2, 5)$ are well known, the problem of determining the minimal degree of a faithful complex representation of G reduces to determining the minimal degree of a suitable faithful complex representation of A.

Let us now investigate the structure of a Sylow 2-subgroup of G. Since $G = A \vee N$, a Sylow 2-subgroup is of the form $P \vee Q_8$, where P is a Sylow 2-subgroup of A. To investigate the structure of P, we used Quotpic (as above) and Cayley [Ca].

Let U be the subgroup $\langle a, b^{-1}ab, bab^{-1}ab^{-1}ab^{-1}ab^{-1}, bab^{-1}abab^{-1}abab^{-1}\rangle$ of G. We found that U has index 9 and presentation

$$\langle c, d, e, f \mid c^2, d^2, e^2, (cef)^2, (cdf^{-2})^2, (cef^{-2})^2, (df^{-1}ef^{-1})^2, (cdcdf^{-1})^2,$$
$$(cdcef^{-1})^2, (cdef^{-1}d)^2, (df^2ef^{-1})^2, (ef^2ef^{-1})^2\rangle.$$

Let H be the subgroup $\langle c, de, df, fcd, f^2d, fdf^{-1}, fef^{-1}\rangle$ of U. Then H has index 3 in U, and we may derive a presentation for H on 7 generators and 30 relators.

Now H maps onto A_5, and we may use Quotpic to determine the kernel R of this map. Since H has index 27 in G and H/R is isomorphic to A_5, R is isomorphic to a Sylow 2-subgroup of A, and hence has order 2^{18}. The nilpotent quotient algorithm gives factors of orders 2^7, 2^5, 2^5 and 2 in R.

We now turn our attention to a Sylow 3-subgroup of G. Since $G = A \vee N$ with $A \cap N$ of order 2 and N isomorphic to $SL(2, 5)$, we see that a Sylow 3-subgroup of G is of the form $Q \times C_3$, where Q is a Sylow 3-subgroup of A. We see that Q is isomorphic to a Sylow 3-subgroup of G/M, and Quotpic gives that G/M has presentation

$$\langle a, b \mid a^2, b^3, (ababababab^2ab^2abab^2ab^2)^2, (ab)^{12}, (abababab^2)^3\rangle.$$

Cayley shows that Q is the non-abelian group of order 27 and exponent 3.

5. Orbifolds

It turns out that many generalized triangle groups can be realised as the groups of three-dimensional orbifolds, with underlying space S^3 and singular set consisting of a knotted graph. The first examples of this were produced by Helling, Mennicke and Vinberg [HMV], who showed that the groups defined by the presentations $\langle a, b \mid a^p = b^q = (abab^{-1}a^{-1}b^{-1})^r = 1 \rangle$ are orbifold groups where the singular set is a trefoil knot with a tunnel added, provided $\frac{1}{p} + \frac{1}{q} + \frac{1}{r} > 1$. Hagelberg [H] proved that the group defined by the presentation $\langle a, b \mid a^p = b^q = (aba^{-1}b^{-1})^r = 1 \rangle$ is an orbifold group under similar restrictions, with singular set the Hopf link plus a tunnel. Further generalised triangle groups were realised as 3-dimensional orbifold groups by Hagelberg and Vesnin [HV].

In the most interesting cases, the orbifolds that occur turn out to have hyperbolic structures, so that the groups involved admit faithful representations to $PSL(2, \mathbb{C})$. The questions that arise concern which groups admit such faithful representations, and when is the image of such a representation discrete, or of finite covolume. Some further work on such questions has been done by Hagelberg, Maclachlan and Rosenberger [HMR].

For finite groups there is no question of a hyperbolic structure, but nevertheless the results of [HMV], [H] and [HV] show that many of the finite generalised triangle groups do indeed arise as orbifold groups. Specifically, the groups numbered 1,2,5,6,7,8,9 and 10 in our Theorem are the groups of 3-dimensional orbifolds with underlying space S^3. Moreover it follows from work of Dunbar [D] that each of these orbifolds has a spherical structure. It follows that each of the groups is isomorphic to a subgroup of $SO(4)$, the isometry group of S^3. We are grateful to Marcel Hegelberg and Andrei Vesnin for conversations on these matters.

Consider the remaining groups in our list. Since A_5 is a subgroup of $SO(3) \subset SO(4)$ that does not contain $-I_4$, we can obtain group number 3, namely $A_5 \times C_2$, as a subgroup $A_5 \times \{\pm I_4\}$ of $SO(4)$, and hence as a 3-dimensional spherical orbifold group. The underlying space is S^3, with singular set an unknotted complete graph on 4 vertices, in other words the 1-skeleton of an embedded 3-simplex.

We do not know of an analogous construction for the group $A_5 \times C_3$, which is number 4 in our list. As far as we are aware, it is unknown whether this group is the group of a 3-dimensional orbifold or not. The same applies to group number 11 in our list.

On the other hand, we are able to rule out the LRS monster, number 12 in our list, as the group of a 3-dimensional orbifold with a geometric structure. Since the only geometry in question is the spherical one, it suffices to show that the group cannot be embedded in $SO(4)$, and this follows from our analysis of its representations in the previous section. Since the group A does not have a faithful 2-dimensional representation (over $\mathbb{C}$), and $N \cong SL(2, 5)$ does not have a faithful 1-dimensional representation, the central product $A \vee N$ cannot have any faithful 4-dimensional complex representation. Alternatively, one can argue using the structure of the Sylow

3-subgroups, since $SO(4)$ does not contain an elementary abelian subgroup of order 27.

6. Deficiency Zero Analogues

A finite group is said to have **deficiency zero** if it has a presentation with an equal number of generators and relators. Study of such groups has been of interest for a number of years, particularly with regards to the question as to their possible structures; see [JR] for a nice survey of these issues.

If a finite group G is defined by a presentation of the form $\langle a, b \mid a^p, b^q, V(a, b) \rangle$, we may define a group H by the presentation $\langle a, b, z \mid a^p b^{-q}, a^p z^{-1}, V(a, b) \rangle$. The generator z here is clearly redundant, but is included for convenience. It is clear that $z = a^p = b^q$ is central in H.

Suppose that H/H' is a finite group; in particular, $z^i \in H'$ for some $i > 0$. Let Z be the subgroup $\langle z \rangle$ of H, so that H/Z is isomorphic to G. By [S2], $H' \cap Z$ is a homomorphic image of the Schur multiplier $M(G)$ of G, and hence is finite. So $\langle z^i \rangle$ is finite, so that $\langle z \rangle$ is finite, and hence H is a finite group of deficiency zero. Of course, there can be several groups H corresponding to the same group G, since (for example) replacing a subword $ab^i a$ of $V(a, b)$ by $ab^{q-i} a$ makes no difference to G but could well change the (isomorphism class of the) corresponding group H.

There have been quite a few interesting examples of finite groups of deficiency zero with presentations like this. For example, Kenne [K] constructed examples of soluble groups of deficiency zero with derived length six by considering the presentations

$$\langle a, b \mid a^2 b^{-6}, (ab^{-1})^3 ab^{-2} ab^k a^{-1} b \rangle,$$

where $k \equiv 3$ (mod 6). Further similar examples may be found in [CHRT], [DJ], [NO] and [R] (among others). For the case of deficiency zero groups linked to the (ordinary) triangle groups, see [C], [CM], [CCS] and [N] for example.

We may clearly perform a similar sort of construction on the finite generalized triangle groups. For example, if G is the generalized triangle group (5) of order 192 with presentation $\langle a, b \mid a^2, b^4, (ababab^3)^2 \rangle$, then we get a finite group H_1 of deficiency zero with presentation

$$\langle a, b \mid a^2 b^{-4}, (ababab^3)^2 \rangle,$$

which has order 4224 and is soluble of derived length 4. The derived factors of H_1 have orders 44, 12, 4 and 2 respectively. On the other hand, the group H_2 with presentation

$$\langle a, b \mid a^2 b^4, (ababab^3)^2 \rangle,$$

while also being soluble of derived length 4, has derived factors of orders 4, 12, 4 and 2 respectively.

For another example, consider the generalized triangle group (7) of order 576 with presentation $\langle a, b \mid a^2, b^3, (ababab^2ab^2)^2 \rangle$. This time the deficiency zero analogue with presentation

$$\langle a, b \mid a^2b^{-3}, (ababab^2ab^2)^2 \rangle$$

has order 9216 and is (again) soluble of derived length 4, this time with derived factors of orders 48, 12, 8 and 2. There is clearly an abundance of examples of finite groups of deficiency zero obtained in this way, including insoluble examples involving A_5.

References

[BMS]   Baumslag, G., Morgan, J. W. and Shalen, P. B., Generalized triangle groups, Math. Proc. Camb. Phil. Soc. 102 (1987), 25–31.

[CHRT]   Campbell, C. M., Heggie, P. M., Robertson, E. F. and Thomas, R. M., Finite one-relator products of two cyclic groups with the relator of arbitrary length, J. Austral. Math. Soc. 53A (1992), 351–367.

[Ca]   Cannon, J. J., An introduction to the group theory language Cayley, in: Computational Group Theory (M. D. Atkinson, ed.), 145–183, Academic Press, 1984.

[Co]   Conder, M. D. E., Three-relator quotients of the modular group, Quart. J. Math. 38 (1987), 427 – 447.

[CCS]   Conway, J. H., Coxeter, H. S. M. and Shephard, G. H., The centre of a finitely generated group, Tensor 25 (1972), 405–418.

[C]   Coxeter, H. S. M., The binary polyhedral groups and other generalizations of the quaternion group, Duke Math. J. 7 (1940), 367–379.

[CM]   Coxeter, H. S. M. and Moser, W. O. J., Generators and Relations for Discrete Groups, Ergeb. Math. Grenzgeb. 14, Springer-Verlag, 1984.

[DJ]   Doostie, H. and Jamali, A. R., A class of deficiency zero soluble groups of derived length 4, Proc. Royal Soc. Edinburgh (A) 121 (1992), 163–168.

[D]   Dunbar, W. D., Geometric orbifolds, Revista Math. 1 (1988), 67–99.

[DH]   Duncan, A. J. and Howie, J., One relator products with high-powered relators, in: Geometric Group Theory, Volume I (G. A. Niblo and M. A. Roller, eds.), 48–74, London Math. Soc. Lecture Note Ser. 181, Cambridge University Press, 1993.

[FHR]   Fine, B., Howie, J. and Rosenberger, G., One-relator quotients and free products of cyclics, Proc. Amer. Math. Soc. (2) 102 (1988), 249–254.

[FHR2]   Fine, B., Howie, J. and Rosenberger, G., Ree-Mendelsohn pairs in generalized triangle groups, Comm. Algebra 17 (1989), 251–258.

[FLR]   Fine, B., Levin, F. and Rosenberger, G., Free subgroups and decompositions of one-relator products of cyclics. Part 1: The Tits alternative, Arch. Math. 50 (1988), 97–109.

[FLR2]   Fine, B., Levin F. and Rosenberger, G., Free subgroups and decompositions of one-relator products of cyclics. Part 2: Normal torsion-free subgroups and FPA decompositions, J. Indian Math. Soc. 49 (1985), 237–247.

[FR]   Fine, B. and Rosenberger, G., A note on generalized triangle groups, Abh. Math. Sem. Univ. Hamburg. 56 (1986), 233–244.

[FR2]   Fine, B. and Rosenberger, G., Complex representations and one-relator products of cyclics, Contemp. Math. 74 (1988), 131–147.

[FK]   Fricke, R. and Klein, F., Vorlesungen über die Theorie der automorphen Funktionen, Volume 1, Teubner, Leipzig, 1897.

[GL]   Gordon, C. McA. and Luecke, J., Only integral Dehn surgeries can yield reducible manifolds, Math. Proc. Cambridge Philos. Soc. 102 (1987), 97–101.

[H]   Hagelberg, M., Generalized triangle groups and 3-dimensional orbifolds, Preprint, Bielefeld, 1992.

[HMR]   Hagelberg, M., Maclachlan, C. and Rosenberger, G., On discrete generalised triangle groups, Proc. Edinburgh Math. Soc, to appear.

[HV]   Hagelberg, M. and Vesnin, A. Yu., On geometric structures of dihedral orbifolds, Preprint, 1994.

[HMV]   Helling, H., Mennicke, J. and Vinberg, E. B., On some generalized triangle groups and 3-dimensional orbifolds, Preprint, Bielefeld, 1991.

[HR]   Holt, D. F. and Rees, S., A graphics system for displaying finite quotients of finitely presented groups, in: Groups and Computation (L. Finkelstein and W. M. Kantor, eds.), 113–126, DIMACS Ser. Discrete Math. Theoret. Comput. Sci. 11, American Mathematical Society, 1993.

[HMT]   Howie, J., Metaftsis, V. and Thomas, R. M., Finite generalised triangle groups, Trans. Amer. Math. Soc., to appear.

[JR]   Johnson, D. L. and Robertson, E. F., Finite groups of deficiency zero, in: Homological Group Theory (C. T. C. Wall, ed.), 275 – 289, London Math. Soc. Lecture Note Ser. 36, Cambridge University Press, 1979.

[K]   Kenne, P., Some new efficient soluble groups, Comm. Alg. 18 (1990) 2747–2753.

[LRS]   Lévai, L., Rosenberger, G. and Souvignier, B., All finite generalized triangle groups, Trans. Amer. Math. Soc., to appear.

[LR]   Levin, F. and Rosenberger, G., On free subgroups of generalized triangle groups, Part II, in: Proceedings of the Ohio State meeting in 1992 in honour of H. Zassenhaus (S. Seghal et al., eds.), to appear.

[N]   Neumann, B. H., Some finite groups with few defining relations, J. Austral. Math. Soc. 38 (1985), 230–240.

[NO]   Newman, M. F. and O'Brien, E. A., A computer-aided analysis of some finitely presented groups, J. Austral. Math. Soc. Ser. A 53 (1992), 369–376.

[R]   Robertson, E. F., A comment on finite nilpotent groups of deficiency zero, Canad. Math. Bull. 23 (1980), 313–316.

[R2]   Rosenberger, G., On free subgroups of generalized triangle groups, Algebra i Logika, 28 (1989), 227–240.

[S]   Schönert, M. et al., GAP - Groups, Algorithms and Programming, Lehrstuhl D für Mathematik, RWTH Aachen, 1992.

[S2]   Schur, I., Über die Darstellung der endlichen Gruppen durch gebrochene lineare Substitutionen, J. Reine Angew. Math. 127 (1904), 20–50.

[T]   Tits, J., Free subgroups in linear groups, J. Algebra 20 (1972), 250–270.

Some Results on Hadamard Groups

Noboru Ito

1. Introduction and General Remarks

Let G be a group of order $2n$ such that G contains a central involution e^*. Let D be a transversal of G with respect to a subgroup N generated by e^*. Then there exist exactly 2^n transversals. For every transversal D we have that $D = De$, where e denotes the identity element of G, $D \cap De^* = \emptyset$ and $G = D \cup De^*$. If for some transversal D we have that $|D \cap Da| = n/2$ for every element a of G other than e and e^*, where $|X|$ denotes the number of elements in a finite set X, namely if D is a $(n, n/2, N)$ relative difference set, then we call D and G an Hadamard subset and an Hadamard group respectively. Hadamard groups are introduced so that we may scrutinize the structure of automorphism groups of Hadamard matrices (See [2]).

Let D be any transversal. Then it holds obviously that

$$|D \cap Da| = |D \cap Da^{-1}| \quad \text{for any element } a \text{ of } G \tag{1}$$

and that

$$|D \cap Da| + |De^* \cap Da| = n \quad \text{for any element } a \text{ of } G. \tag{2}$$

From (1) and (2) follows the following lemma.

Lemma 1. *Let D be any transversal. If a is an element of order 4 such that $a^2 = e^*$, then we have that $|D \cap Da| = n/2$.*

Proof. It holds that $|D \cap Da| = |Da \cap Da^2| = |Da \cap De^*|$. So by (2) we get the assertion. $\square$

Lemma 2. *Let H be a non-trivial subgroup of G. If a transversal D is a union of left cosets of H, then D is not an Hadamard subset.*

Proof. If D is a union of left cosets of H, then we have that $Dh = D$ for any non-identity element h of H. $\square$

2. Examples and Propositions

Example 1. Let C_m and Q denote the cyclic group of order m and the quaternion group of order 8 respectively. Then C_4 and Q are Hadamard groups with the property that every transversal is an Hadamard subset. The converse is true (we disregard the trivial case of C_2).

Proposition 1. *Let G be an Hadamard group with the property that every transversal is an Hadamard subset. Then G is isomorphic to either C_4 or Q.*

Proof. If G contains a non-trivial subgroup H not containing e^*, then there exists a transversal D which is a union of left cosets of H. So by Lemma 2 we may assume that every non-trivial subgroup of G contains e^*. Thus G is a 2-group with a unique involution e^*. Then it is well known that G is either cyclic or generalized quaternion ([8], p. 148). By Proposition 7 of (2) cyclic 2-groups whose orders are greater than 4 are not Hadamard groups. So we may assume that G is a generalized quaternion group of order greater than 8. Then G is generated by two elements a and b such that $a^{2^{m+1}} = b^4 = e$ with $m \geq 2$, $a^{2^m} = b^2$, and $b^{-1}ab = a^{-1}$. We have that $a^{2^m} = e^*$. So we may choose D so that D consists of elements $e, a, \ldots, a^{2^m-1}, b, ba, \ldots$, and ba^{2^m-1}. Then we have that $|D \cap Da| = 2(2^m - 1) = 2^m$. The last equality holds, since, by assumption, D is an Hadamard subset. But this implies that $m = 1$, which is a contradiction.   □

Example 2. $G = C_4 \times C_2$ is an Hadamard group. Let a and b be generators for C_4 and C_2 respectively. Since the subgroup generated by a^2b is also a direct factor of G, there are essentially two ways to choose e^*.

    (i) $e^* = b$. This case corresponds to a circulant Hadamard matrix of order 4.

    By Proposition 2 of [2] the order of an Hadamard group is a multiple of 8, if it is greater than 4. Now in our terminology the conjecture of Ryser concerning circulant Hadamard matrices ([6], p. 134) may be stated as follows: Let $G = C_{4m} \times C_2$ and b the generator for C_2. If G is an Hadamard group with $e^* = b$, then $m = 1$ (See [3], Proposition 1).

    (ii) $e^* = a^2$. We may choose D so that D consists of e, a, be^* and ab.

    Moreover let $G = C_8 \times C_2$ and let a and b be generators for C_8 and C_2 respectively. If we put $e^* = a^4$, then G is an Hadamard group. (By proposition 5 of [2] the subgroup generated by e^* is not a direct factor of G). In fact, we may choose D so that D consists of elements $e, a, a^2, a^3, b, abe^*, a^2b$ and a^3be^*.

    However, we have the following proposition.

Proposition 2. *Let G be an Hadamard group of order 8n and S a Sylow 2-subgroup of G. Assume that $S = C_{2^{m+1}} \times C_2$ and that a and b are generators for $C_{2^{m+1}}$ and C_2 respectively. If $e^* = a^{2^m}$, then $m = 1$ or 2.*

Proof. Assume that $m \geq 3$. First we show that G is 2-nilpotent. Let N be the normalizer of S in G. Notice that the subgroup K generated by e^* and b and the subgroup L generated by a^2 are characteristic subgroups of S. Since e^* is central in G, both K and L are central in N, and hence S is central in N. Therefore by a splitting theorem of Burnside G is 2-nilpotent ([1], p. 419). Let T be the normal Sylow 2-complement of G.

For any natural number i, let $e(i)$ denote a primitive 2^i-th root of unity. Put $\mathbf{K} = \mathbb{Q}(e(m+1))$, where $\mathbb{Q}$ denotes the field of rational numbers. Now let D be an Hadamard subset corresponding to e^* and λ a linear representation of G whose kernel equals $T C_2$. Then we have that $\lambda(e^*) = -1$. Hence by Proposition 4 of [2] we have that $\lambda(D^{-1})\lambda(D) = 4n = 2^{m+1}t$, where $t = |T|$. Since $\lambda(D^{-1})$ is the complex conjugate of $\lambda(D)$, we have that $\lambda(D) = (4n)^{1/2}E$, where E is a unit. Since this is true for any algebraic conjugate of λ, by a theorem of Kronecker E is a root of unity ([7], p. 122). Further only 2 ramifies in $\mathbf{K}$ ([7], p. 112). Hence t is a square. Since $m \geq 3$, $2^{1/2}$ belongs to $\mathbf{K}$. Thus E belongs to $\mathbf{K}$.

Now let us consider the coset decomposition of G with respect to T, and put $|Ta^i \cap D| = w_i$, $|Ta^i e^* \cap D| = x_i$, $|Tba^i \cap D| = y_i$ and $|Tba^i e^* \cap D| = z_i$ for $0 \leq i \leq 2^m - 1$. Since $w_i + x_i = y_i + z_i = t$, which is odd, we have that $w_i \neq x_i$ and $y_i \neq z_i$ for every i. Put $\lambda(a) = e(m+1)$ and $r = 2^m - 1$. Then we have that

$$\lambda(D) = (w_0 - x_0 + y_0 - z_0) + (w_1 - x_1 + y_1 - z_1)\lambda(a) + \cdots + (w_r - x_r + y_r - z_r)\lambda(a^r).$$

Since $X^{r+1} + 1 = 0$ is a defining equation for $e(m+1)$, every number in $\mathbf{K}$ is uniquely expressible as a polynomial (over $\mathbb{Q}$) of $e(m+1)$ of degree at most r. So if the number d of $i's$ such that $w_i - x_i + y_i - z_i \neq 0$ exceeds 2, then since $2^{1/2} = e(3)(1 + \mathbf{i})$, where $\mathbf{i}$ is the imaginary unit, we get a contradiction against $\lambda(D) = (4n)^{1/2}E$. Otherwise, we consider a linear representation μ of G such that the kernel of equals T and $\mu(e^*) = \mu(b) = -1$. Then the number corresponding to d becomes $2^m - d \geq 3$. So considering $\mu(D)$ instead of $\lambda(D)$ we get a similar contradiction. □

Remark. We conjecture that if $m \geq 3$, the G is not an Hadamard group regardless of the choice of e^*.

Example 3. $G = C_4 \times C_4$ is an Hadamard group. Let a and b be generators for two C_4's. Then we may choose e^* and D as follows: $e^* = a^2$. D consists of elements $e^*, be^*, b^2e^*, b^3, a, ab, ab^2$ and ab^3e^*.

Does there exist an Hadamard group of order greater than 16 which is the direct product of two isomorphic subgroups?

3. Remarks on the Strong Hadamard Conjecture and a Conjecture of Ryser

First of all we state the strong Hadamard conjecture.

Strong Hadamard conjecture: For any multiple of 8, say $8n$, there exists an Hadamard group of order $8n$.

If this has an affirmative answer, then an Hadamard matrix of order $4n$ exists for any positive integer n. So from a group theoretical viewpoint an affirmative answer to this conjecture provides an optimal solution to the existence problem of Hadamard matrices.

By Proposition 3 of [2] we may assume that n is odd. Let $G(n)$ be a group generated by two elements a and b such that $a^{4n} = e$, $b^2 = a^{2n}$ and $b^{-1}ab = a^{-1}$. Then $G(n)$ has order $8n$ and contains a unique involution, namely a^{2n}. So we put $e^* = a^{2n}$ and N denotes a subgroup generated by e^*. Let D be any transversal of $G(n)$ with respect to N. Then, since $(ba^i)^2 = (a^n)^2 = e^*$, by Lemma 1 we have that $|D \cap Dba^i| = |D \cap Da^n| = 2n$ for $0 \le i \le 4n - 1$. Further the following lemma holds.

Lemma 3. *Let D be any transversal of $G(n)$ with respect to N. Then we have that* $|D \cap Da^{n-i}| + |D \cap Da^{n+i}| = 4n$, *where* $0 \le i \le n$.

Proof. It holds that $|D \cap Da^{n-i}| = |Da^i \cap Da^n|$ and $|D \cap Da^{n+i}| = |Da^n \cap Da^i e^*|$. Hence by (2) we have the lemma. □

Thus in order to show that $G(n)$ is an Hadamard group, it is sufficient to show the existence of a transversal D such that $|D \cap Da^i| = 2n$ for $1 \le i \le n-1$. Furthermore, if either $4n = q+1$ or $2n = q+1$, where q is a prime power, then $G(n)$ is an Hadamard group of either quadratic residue type or Paley type (see [2] and [3]). Thus for any n, $G(n)$ may be said to be a reasonable candidate for an Hadamard group. At any rate it is easy to find an appropriate transversal D when n is small enough.

To proceed further we label elements of a transversal D as follows: $d_0, d_1, \ldots,$ $d_{2n-1}, e_0, e_1, \ldots, e_{2n-1}$. Here d_i equals a^i or $a^i e^*$ and e_i equals ba^i or $ba^i e^*$, where $0 \le i \le 2n - 1$. Next a sign function s is defined on $G(n)$ as follows: Let g be an element of $G(n)$. Then $s(g) = -1$ or 1, and $s(g) = 1$ if and only if $g = a^i$ or ba^i, where $0 \le i \le 2n - 1$. Lastly we define two polynomials $a(X)$ and $b(X)$ in $\mathbb{Q}[X]/(X^{2n} + 1)$, where $(*)$ denotes the ideal generated by $*$, as follows:

$$a(X) = s(d_0) + s(d_1)X + \cdots + s(d_{2n-1})X^{2n-1}$$

and

$$b(X) = s(e_0) + s(e_1)X + \cdots + s(e_{2n-1})X^{2n-1}.$$

Now we have the following lemma.

Lemma 4. *Let D be a transversal of $G(n)$ with respect to N. Then D is an Hadamard subset if and only if the following equation holds :*

$$a(X^{-1})a(X) + b(X^{-1})b(X) = 4n. \tag{3}$$

Proof. The coefficients of X^i in $a(X^{-1})a(X)$ and $b(X^{-1})b(X)$ are equal to

$$s(d_0)s(d_i) + \cdots + s(d_{2n-1-i})s(d_{2n-1}) - s(d_{2n-i})s(d_0) - \cdots - s(d_{2n-1})s(d_{i-1}) \tag{4}$$

and

$$s(e_0)s(e_i) + \cdots + s(e_{2n-1-i})s(e_{2n-1}) - s(e_{2n-i})s(e_0) - \cdots - s(e_{2n-1})s(e_{i-1}) \tag{5}$$

respectively.

Suppose that d_j belongs to Da^i, where $0 \le i,\ j \le 2n - 1$. If $i + j \le 2n - 1$, then we have that $d_j a^i = d_{j+i}$ and $s(d_j)s(d_{j+i}) = 1$. If $i + j \ge 2n$, then we have that $d_j a^i = d_{j+i-2n}e^*$ and $s(d_j)s(d_{j+i-2n}) = -1$. The same holds for e_j instead of d_j. Therefore we have that $|D \cap Da^i| = 2n$ if and only if the sum of the coefficients (4) and (5) equals 0.

We say that $a(X)$ and $b(X)$ are complementary if they satisfy (3). By Lemma 4 the strong Hadamard conjecture has an affirmative answer if there exists a pair of complementary polynomials $a(X)$ and $b(X)$. In (3) X may take a complex value α, provided that $\alpha^{2n} + 1 = 0$. In particular, if we take $\alpha = i$, then we obtain that

$$\begin{aligned}
&(s(d_0) - s(d_2) + \cdots - s(d_{2n-4}) + s(d_{2n-2}))^2 \\
&+ (s(d_1) - s(d_3) + \cdots - s(d_{2n-3}) + s(d_{2n-1}))^2 \\
&+ (s(e_0) - s(e_2) + \cdots - s(e_{2n-4}) + s(e_{2n-2}))^2 \\
&+ (s(e_1) - s(e_3) + \cdots - s(e_{2n-3}) + s(e_{2n-1}))^2 = 4n.
\end{aligned} \tag{6}$$

(6) can be regarded as a necessary condition for $a(X)$ and $b(X)$ to be complementary.

Further we add the following remark. Let $v(a(X)) = (s(d_0), s(d_1), \ldots, s(d_{2n-1}))$ and $v(b(X)) = (s(e_0), s(e_1), \ldots, s(e_{2n-1}))$ be the coefficient vectors of $a(X)$ and $b(X)$ respectively. Since we are considering modulo $X^{2n} + 1$, and since the coefficient vectors can be replaced by their cyclic shifts, we may assume that $s(d_0) = s(d_{2n-1})$ and $s(e_0) = s(e_{2n-1})$. Then we divide $v(a(X))$ and $v(b(X))$ into subvectors so that the signs are equal within the same subvector and signs in every consecutive subvectors are opposite. Now let l_i and m_i be the number of subvectors of size i in $v(a(X))$ and $v(b(X))$ respectively. Then we have that

$$l_1 + 2l_2 + \cdots + rl_r = 2n \tag{7}$$

and

$$m_1 + 2m_2 + \cdots + sm_s = 2n \tag{8}$$

where r and s denote the largest of sizes of subvectors in $v(a(X))$ and $v(b(X))$ respectively. It is easy to see that the equality $|D \cap Da| = 2n$ is equivalent to the equality

$$l_2 + \cdots + (r-1)l_r + m_2 + \cdots + (s-1)m_s = 2n. \tag{9}$$

Thus, if $|D \cap Da| = 2n$, from (7), (8) and (9) we get that

$$l_1 + l_2 + \cdots + l_r + m_1 + m_2 + \cdots + m_s = 2n. \tag{10}$$

The left hand side of (10) equals the sum of the numbers of subvectors in $v(a(X))$ and $v(b(X))$. (10) also can be regarded as a necessary condition for $a(X)$ and $b(X)$ to be complementary. Now we consider Ryser's conjecture concerning circulant Hadamard matrices which is stated in Example 2 of Section 2. For a situation of this conjecture see [5].

Put $H(n) = C_{4n} \times C_2$. Let c and e^* be generators for C_{4n} and C_2 respectively. Let D be a transversal of $H(n)$ with respect to C_2. Since e^* is not a square in $H(n)$, the equalities $|D \cap Dc^i| = 2n$ for $0 \le i \le 2n$ are likely to give a strong restriction to D supporting the affirmative answer to the conjecture.

We label elements of D as follows: $c_0, c_1, \ldots, c_{4n-1}$. Here c_i equals c^i or $c^i e^*$, where $0 \le i \le 4n - 1$. A sign function s on $H(n)$ is defined as in the case of $G(n)$, and a polynomial $c(X)$ in $\mathbb{Q}[X]/(X^{4n} - 1)$ is defined as follows:

$$c(X) = s(c_0) + s(c_1)X + \cdots + s(c_{4n-1})X^{4n-1} .$$

Then, as in the case of $G(n)$, we see that the equation

$$c(X^{-1})c(X) = 4n \tag{11}$$

is a necessary and sufficient condition for D to be an Hadamard subset. We call $c(X)$ an Hadamard polynomial if it satisfies (11). In (11) X may take any complex value β, provided that $\beta^{4n} = 1$. The expression $c(\beta^{-1})c(\beta)$ is rather complicated in general. However, if we take $\beta = 1$, $\beta = -1$ and $\beta = i$, then we obtain that

$$(s(c_0) + s(c_1) + \cdots + s(c_{4n-2}) + s(c_{4n-1}))^2 = 4n \tag{12}$$

$$(s(c_0) - s(c_1) + \cdots + s(c_{4n-2}) - s(c_{4n-1}))^2 = 4n \tag{13}$$

and

$$(s(c_0) - s(c_2) + \cdots + s(c_{4n-4}) - s(c_{4n-2}))^2$$
$$+ (s(c_1) - s(c_3) + \cdots + s(c_{4n-3}) - s(c_{4n-1}))^2 = 4n . \tag{14}$$

These are regarded as necessary conditions for $c(X)$ to be an Hadamard polynomial. The coefficient vector $v(c(X))$ is defined as above: $v(c(X)) = (s(c_0), s(c_1), \ldots, s(c_{4n-1}))$. Since we are considering modulo $X^{4n} - 1$, we may assume that $s(c_0)s(c_{4n-1}) = -1$. Then we decompose $v(c(X))$ into subvectors as in the case of $a(X)$ and $b(X)$. Let n_i be the number of subvectors of size i in $v(c(X))$. Then we have that

$$n_1 + 2n_2 + \cdots + tn_t = 4n \tag{15}$$

where t denotes the largest of sizes of subvectors of $v(c(X))$. As before, we have that $|D \cap Da| = 2n$ if and only if

$$n_2 + 2n_3 + \cdots + (t-1)n_t = 2n. \tag{16}$$

Similarly we have that $|D \cap Da^2| = 2n$ if and only if

$$2(n_2 + n_3 + \cdots + n_t) = 2n. \tag{17}$$

Therefore if $|D \cap Da| = |D \cap Da^2| = 2n$, then from (15), (16) and (17) we obtain that

$$\begin{aligned} &n_1 = n, \ n_2 = n_4 + 2n_5 + \cdots + (t-3)n_t \quad \text{and} \\ &n_3 + 2n_4 + \cdots + (t-2)n_t = n. \end{aligned} \tag{18}$$

(18) can be regarded as a necessary condition for c(X) to be an Hadamard polynomial.

As a final remark we mention that it is easy to find transversals D both in $G(n)$ and in $H(n)$, where $|D \cap Da^{2i+1}| = 2n$ and $|D \cap Dc^{2j+1}| = 2n$ hold for every i and j such that $1 \leq 2i + 1 \leq 2n - 1$ and $1 \leq 2j + 1 \leq 4n - 1$.

References

[1]   B. Huppert, Endliche Gruppen I, Springer-Verlag 1967.

[2]   N. Ito, On Hadamard groups, J. Algebra 168 (1994).

[3]   N. Ito, On Hadamard groups II, J. Algebra 169 (1994).

[4]   N. Ito, Note on Hadamard groups of quadratic residue type, Hokkaido Math. J. 22 (1993), 373–378.

[5]   C. Lin and W. D. Wallis, On the circulant Hadamard matrix conjecture, in: Coding theory, Design theory, Group theory, Proc. Marshall Hall Conference, Wiley 1993, 213–217.

[6]   H. Ryser, Combinatorial Mathematics, MAA, 1963.

[7]   T. Takagi, Theory of algebraic integers, Iwanami 1948 (in Japanese).

[8]   H. Zassenhaus, The theory of groups, Second Edition, Chelsea 1958.

The Growth of the Trefoil Group

D. L. Johnson, A. C. Kim and H. J. Song

Abstract. Explicit computations are made of the growth (in the sense of Milnor) of the trefoil group, presented as a free product with amalgamation on two generators. The method uses carefully chosen rewrite rules to find a minimal normal form, and then the numbers of words of each length are expressed in terms of generating functions.

1991 Mathematics Subject Classification: 20E06 (primary); 20F38, 57M25 (secondary)

0. Introduction

Here we mean growth in the sense of Milnor [9]. Thus, for each element g in a group G generated by a finite set X, let $l(g)$ denote the minimal number of letters in a word in $X^{\pm}$ representing g. Then the series

$$G_X(t) := \sum_{g \in G} t^{l(g)} = \sum_{n \geq 0} a_n t^n \in \mathbb{Z}[[t]] \tag{1}$$

is called the **growth series,** or **Poincaré series,** of G with respect to X, and $\{a_n \mid n \geq 0\}$ the sequence of **growth coefficient.** When the set X of generators is a natural one, or is otherwise clear from the context, we suppress it and write $G(t)$ for the growth series. Given a subset $S \subseteq G$, we likewise write $S(t) = \sum_{g \in S} t^{l(g)}$ for the contribution S makes to $G(t)$.

Growth series have been computed explicitly for a number of groups and classes of groups. Those with a geometrical flavour include Coxeter groups [3, Example 26, p. 45], closed surface groups [4], and certain 3-manifold groups [7, 8] of which the former includes the growth of the trefoil group on the Wirtinger generators. We extend this list by describing a method for the groups

$$G_{p,q} = \langle x, y \mid x^p = y^q \rangle, \tag{2}$$

The authors would like to thank the British Council, the Korean Science and Engineering Foundation and the Korean Research Foundation 1994 for their support, without which this collaboration would not have been possible.

which we illustrate in the case $(p, q) = (2, 3)$ of the trefoil group. The easier case $p = q$ is covered in [5].

The method is an extension of that used in [2] for free products, which goes back to [6]. The idea is to express each element g of G in **minimal normal form** (§2), that is, a normal form of length $l(g)$, by applying certain rewriting rules (§1), and then count the number of such forms of each length (§3).

1. Rewriting Rules

Our rewriting rules for the group $G_{2,3}$ in (2) are applied to elements of the free group $F = \langle x, y \mid \ \rangle$ in four stages, which may be stated informally as follows.

S1. Collect any power of the central element $x^2 = y^3$ to the left and write it as a power of x^2.

S2. Use any power of x^2 (or x^{-2}) to eliminate occurrences of x^{-1} and y^{-1} (or x and y).

S3. Preclude the occurrence of y^2 (or y^{-2}) with y^{-2} or x^{-1} (or y^2 and x).

S4. If any of the pairs y^2 and y^{-1}, y and y^{-2}, x^{-1} and x occur, ensure that all occurrences of the first-named are to the left.

These rules correspond to the following system of 17 equations:

S1: $ay^3b = ax^2b, \ ay^{-3}b = ax^{-2}b, \ ax^2b = x^2ab, \ ax^{-2}b = x^{-2}ab;$

S2: $x^2ax^{-1}b = axb, \ x^2axb = ax^{-1}b, \ x^2ay^{-1}b = ay^2b, \ x^{-2}ayb = ay^{-2}b;$

S3: $axby^{-2}c = ax^{-1}byc, \ ay^{-2}bxc = aybx^{-1}c, \ ay^{-2}by^2c = ayby^{-1}c,$
$\quad\ ax^{-1}by^2c = axby^{-1}c, \ ay^2bx^{-1}c = ay^{-1}bxc, \ ay^2by^{-2}c = ay^{-1}byc;$

S4: $ay^{-1}by^2c = ay^2by^{-1}c, \ ay^{-2}byc = ayby^{-2}c, \ axbx^{-1}c = ax^{-1}bxc.$

Here, a, b, c are arbitrary elements of the free group F on $\{x, y\}$ such that the left-hand sides are all reduced, and the application of the rule consists in replacing the left-hand side by the right-hand side in each case. Note that this system

(i)   is non-increasing: $\forall$ rule $w = w'$, $w' < w$, where " $<$ " is some l-respecting well-ordering of F;

(ii) respects the natural map $v : F \to G_{2,3}$: $\forall$ rule $w = w'$, $wv = w'v$.

Provided we can show in addition that the resulting irreducibles

$$T = \{w \in F \mid w \text{ is not the left-hand side of any rule}\}$$

satisfy the condition of

(iii) transversality, i.e., $w, w' \in T, w \neq w' \Rightarrow wv \neq w'v$,
then it is clear that T is a minimal normal form for $G_{2,3}$.

2. The Minimal Normal Form

To describe the irreducibles in this case, we adopt the following notation. Put $X = \langle x \rangle \backslash \{e\}$, $Y = \langle y \rangle \backslash \{e\}$, and for $A \subseteq X, B \subseteq Y$, let $\pi(A; B)$ denote a typical product of elements taken alternately from A and B. Thus, $\{\pi(X; Y)\} = F$, for example. Further, put a bar under (or over) an argument A or B to denote that the product begins (or ends) with an element of that set, the empty product being allowed in all cases. Then the irreducibles T are as follows.

$$A : (x^2)^k \pi(x; y, y^2), \quad k > 0,$$
$$B : \pi_1(\overline{x}; y, y^2)y^2\pi_2(\underline{x}; y, y^{-1}),$$
$$C : \pi(x; y, y^{-1}),$$
$$D : \pi_1(x^{-1}; \overline{y, y^{-1}})x^{-1}\pi_2(x; \underline{y, y^{-1}}),$$

together with the inverses of the sets A and B.

The verification of the transversality condition (iii) above is as follows. First note that v is one-to-one on each of B, B^{-1}, C, D and that their images are pairwise disjoint subsets of $G_{2,3}$. Letting

$$\rho : G_{2,3} \rightarrow \langle x \mid x^2 \rangle * \langle y \mid y^3 \rangle$$

denote the natural map, we check case-by-case that if $u, v \in T$ with $u \neq v$, $u\rho = v\rho$ and $u \in A^{\pm 1}$, then u and v have distinct images under the homomorphism

$$\sigma : G_{2,3} \rightarrow \langle a \mid \rangle,$$
$$x \mapsto a^3$$
$$y \mapsto a^2.$$

Finally, since Ker $\rho = \langle x^2 \rangle$, it follows that Ker $\rho \cap$ Ker $\sigma = \{1\}$. Thus, (iii) holds and $T = A^{\pm 1} \cup B^{\pm 1} \cup C \cup D$ is a minimal normal form.

3. Counting

It remains to count the elements of T; we treat the sets A, C, B, D in turn. For A, we first compute the growth series of the set of $\pi(x; y, y^2)$, which we write $[x; y, y^2]$. A typical such word has the form

$$x_0 y_1 x \ldots y_k x y_{k+1},$$

where

$$k \geq 0, \quad x_0 \in \{1, x\}, \quad y_{k+1} \in \{1, y, y^2\},$$
$$y_1 \in \{y, y^2\}, \quad 1 \leq i \leq k.$$

Hence,

$$[x; y, y^2] = (1 + t)\left(\sum_{k \geq 0}((t + t^2)t\right)^k (1 + t + t^2)$$

$$= \frac{(1 + t)(1 + t + t^2)}{1 - t^2 - t^3}.$$

It follows that

$$A^{\pm 1}(t) = 2\frac{t^2}{1 - t^2}\frac{(1 + t)(1 + t + t^2)}{1 - t^2 - t^3}.$$

Similarly we obtain

$$C(t) = \frac{(1 + t)(1 + 2t)}{1 - 2t^2}.$$

Turning to B the product $\pi_1(\bar{x}; y, y^2)$ cannot end in y or y^2, which means that the factor $(1 + t + t^2)$ in the numerator of the above expression for $[x; y, y^2]$ is replaced by unity. Hence, adapting $C(t)$ similarly to get $\pi(\underline{x}; y, y^{-1})$,

$$B^{\pm 1}(t) = 2\frac{1 + t}{1 - t^2 - t^3}t^2\frac{1 + t}{1 - 2t^2}.$$

In similar fashion, adapting C to the needs of D, we obtain

$$D(t) = \frac{1 + 2t}{1 - 2t^2}t\frac{1 + 2t}{1 - 2t^2}.$$

A simple calculation now yields the growth series of $G_{2,3}$ as the sum of those for $A^{\pm 1}, C, B^{\pm 1}, D$.

Proposition. $G_{2,3}(t) = \dfrac{(1 + t)(1 + 2t + t^2 - 7t^3 - 10t^4 + 6t^5 + 6t^6 + 4t^7)}{(1 - t)(1 - t^2 - t^3)(1 - 2t^2)^2}.$

4. Remarks

(i) The proposition confirms a conjecture, based on machine calculations, of N. J. Smythe [10].

(ii) Further confirmation is provided by computations of D. F. Holt, who has constructed a finite-state automation for this group.

(iii) Calculation by one of us (Song), with $G_{2,5}$ and $G_{6,15}$ for example, suggest that the growth function of $G_{p,q}$ is always a rational function.

(iv) A more general theory for amalgams, taking more favourable (larger) generating sets to obtain 'admissible inclusions', has been developed by J. M. Alonso [1].

(v) Although the rule [1]

$$\frac{1}{G(t)} = \frac{1}{G_1(t)} + \frac{1}{G_2(t)} - \frac{1}{A(t)}$$

for $G = G_1 *_A G_2$ fails here, a weaker version (for the Euler characteristic) holds: $\chi(G_{2,3})$ and $\frac{1}{G_{2,3}(1)}$ are both equal to zero.

References

[1]   J. M. Alonso, Growth functions of amalgams, in: Arboreal group theory (R. C. Alperin, ed.), Math. Sci. Res. Inst. Publ. 19, 1–34, Springer-Verlag, New York 1991.

[2]   N. Billington, The growth of stratified groups, PhD thesis, La Trobe University, Melbourne 1986.

[3]   N. Bourbaki, Groupes et algèbres de Lie, Chapters 4, 5, 6, Hermann, Paris 1968.

[4]   J. W. Cannon, The growth of the closed surface groups and the compact hyperbolic groups, Preprint, Madison.

[5]   M. Edjvet and D. L. Johnson, The growth of certain amalgamated free products and HNN-extensions, J. Austral. Math. Soc. Ser. A 52 (1992), 285–298.

[6]   V. E. Govorov, On the dimension of graded algebras, Mat. Zametki 14 (1973), 209–216; Math. Notes 14 (1973), 678–682.

[7]   M. A. Grayson, Geometry and growth in three dimensions, PhD thesis, Princeton 1983.

[8]   D. L. Johnson and H.-J. Song, The Poincaré series of the Gieseking group, in: Discrete groups and geometry (W. J. Harvey and C. Maclachlan, eds.), London Math. Soc. Lecture Note Ser. 173, Cambridge University Press 1992.

[9]   J. Milnor, A note on curvature and the fundamental groups, J. Differential Geometry 2 (1968), 1–7.

[10] N. J. Smythe, Growth functions and Euler series, Invent. Math. 77 (1984), 517–531.

Derivations of Group Algebras
of the Infinite Dihedral Group

Naoki Kawamoto

Abstract. We consider the Lie algebraic structure for the derivation algebras of group algebras of the infinite dihedral group over a field of characteristic zero.

1991 Mathematics Subject Classification: 16W25, 17B65, 20C07.

1. Introduction

Let G be a group, and L be the derivation algebra Der kG of the group algebra kG over a field k. Then L is a Lie algebra with the Lie product $[\delta_1, \delta_2] = \delta_1\delta_2 - \delta_2\delta_1$ ($\delta_1, \delta_2 \in L$).

We assume throughout the paper that the ground field k is of characteristic 0. If G is finite and abelian, then it is easy to see that Der $kG = 0$. If $k = \mathbb{C}$ and if G is finite, then the derivations of kG are inner [1, p. 490], and Der kG is a direct sum of special linear Lie algebras $sl(n, \mathbb{C})$ [2]. More interesting cases occur for infinite groups. If G is an infinite cyclic group, then Der kG is the Witt algebra, i.e. the centerless Virasoro algebra, and in general if $G = \mathbb{Z}^n$, then Der kG is an infinite-dimensional simple Lie algebra [3]. These generalized Witt algebras form an interesting class of infinite-dimensional simple Lie algebras [4].

In this paper we consider the case that G is the infinite dihedral group $D_\infty = \langle r, s \mid s^2 = 1, (sr)^2 = 1 \rangle$. Inn kD_∞ is the set $\{\text{ad } x \mid x \in kD_\infty\}$ of inner derivations of kD_∞, where ad $x(y) = [x, y] = xy - yx$ ($y \in kD_\infty$). Our main result is that the set Der kD_∞ of derivations of kD_∞ is the semidirect product of Inn $kD_\infty + B$ and A (Theorem 2), where $A = \langle \delta_{a_i} \mid i \in \mathbb{Z} \rangle$, $\delta_{a_i}(r) = r^{1+i} - r^{1-i}$, $\delta_{a_i}(s) = 0$, and $B = \langle \delta_{b_i} \mid i \in \mathbb{Z} \rangle$, $\delta_{b_i}(r) = sr^i - sr^{-i}$, $\delta_{b_i}(s) = 0$.

2. Derivations of kD_∞

The group algebra kD_∞ is generated by $\{r, s\}$ and has a basis $\{r^i, sr^i \mid i \in \mathbb{Z}\}$. Let δ be a derivation of kD_∞, and let

$$\delta(r) = \sum_i a_i r^i + \sum_i b_i sr^i, \tag{2.1}$$

$$\delta(s) = \sum_i c_i r^i + \sum_i d_i sr^i, \tag{2.2}$$

where a_i, b_i, c_i, d_i ($\in k$) are equal to 0 except for a finite number of i's. Then the derivation δ is determined by the coefficients a_i, b_i, c_i, d_i.

Since $(sr)^2 = 1$, it holds that $rs = sr^{-1}, r^{-1}s = sr$. Hence we have

$$\delta(s^2) = \delta(s)s + s\delta(s)$$
$$= (\sum_i c_i r^i + \sum_i d_i sr^i)s + s(\sum_i c_i r^l + \sum_i d_i sr^i)$$
$$= \sum_i c_i sr^{-i} + \sum_i d_i r^{-i} + \sum_i c_i sr^i + \sum_i d_i r^i$$
$$= \sum_i (d_i + d_{-i})r^i + \sum_i (c_i + c_{-i})sr^i,$$

and

$$\delta(srsr) = \delta(s)s + s\delta(r)sr + sr\delta(s)r + r^{-1}\delta(r)$$
$$= \sum_i c_i sr^{-i} + \sum_i d_i r^{-i} + \sum_i a_i r^{1-i} + \sum_i b_i sr^{1-i}$$
$$\quad + \sum_i c_i sr^{i+2} + \sum_i d_i r^i + \sum_i a_i r^{i-1} + \sum_i b_i sr^{i+1}$$
$$= \sum_i (d_{-i} + d_i + a_{1-i} + a_{1+i})r^i + \sum_i (c_{-i} + c_{i-2} + b_{1-i} + b_{i-1})sr^i.$$

Since $\delta(s^2) = \delta((sr)^2) = \delta(1) = 0$, we obtain that

$$d_i + d_{-i} = 0, \tag{2.3}$$

$$c_i + c_{-i} = 0, \tag{2.4}$$

$$a_{1+i} + a_{1-i} = 0, \tag{2.5}$$

$$b_i + b_{-i} + c_{-1-i} + c_{-1+i} = 0 \tag{2.6}$$

for any $i \in \mathbb{Z}$.

3. Inner Derivations

Let $\delta_{d_i} = \mathrm{ad}\, r^i$ $(i \in \mathbb{Z})$. Then

$$\delta_{d_i}(r) = [r^i, r] = 0,$$
$$\delta_{d_i}(s) = [r^i, s] = sr^{-i} - sr^i,$$

and $\delta_{d_0} = 0$, $\delta_{d_{-i}} = -\delta_{d_i}$ $(i \in \mathbb{Z})$. So (2.3) is the non-trivial relation of the coefficients of δ_{a_i}. Hence if δ has non-zero coefficients d_i in (2.2), then we can choose some $d_i' \in k$ such that

$$(\delta - \sum_i d_i' \delta_{d_i})(s) = \sum_i c_i' r^i.$$

We denote $\delta - \sum_i d_i' \delta_{d_i}$ by the same symbol δ, and we may assume that $d_i = 0$ $(i \in \mathbb{Z})$ in (2.2).

Let $\delta_{c_i} = \mathrm{ad}\, sr^i$ $(i \in \mathbb{Z})$. Then

$$\delta_{c_i}(r) = [sr^i, r] = sr^{i+1} - sr^{i-1},$$
$$\delta_{c_i}(s) = [sr^i, s] = r^{-i} - r^i.$$

If there are non-zero coefficients c_i, then there exist $c_i' \in k$ $(i \in \mathbb{Z})$ such that

$$(\delta - \sum_i c_i' \delta_{c_i})(s) = 0.$$

Now denote $\delta - \sum_i c_i' \delta_{c_i}$ by δ, and we may assume that $c_i = d_i = 0$ $(i \in \mathbb{Z})$ in (2.2), that is, $\delta(s) = 0$.

We note that $\{\delta_{d_i}, \delta_{c_j} \mid i, j \in \mathbb{Z}, \; i > 0\}$ is a basis of $\mathrm{Inn}\, k D_\infty$.

4. Outer Derivations

A derivation δ of $k D_\infty$ is called an outer derivation if $\delta \notin \mathrm{Inn}\, k D_\infty$. We define two kinds of outer derivations of $k D_\infty$.

First, for any $i \in \mathbb{Z}$ let

$$\delta_{a_i}(1) = 0,$$
$$\delta_{a_i}(r) = r^{1+i} - r^{1-i},$$
$$\delta_{a_i}(s) = 0,$$

and for $n \in \mathbb{Z}$ define

$$\delta_{a_i}(r^n) = n(r^{n+i} - r^{n-i}),$$
$$\delta_{a_i}(sr^n) = s\delta_{a_i}(r^n).$$

Then we extend δ_{a_i} linearly to a k-endomorphism of $k D_\infty$.

We claim that δ_{a_i} is a derivation of kD_∞. For $n, m \in \mathbb{Z}$ we have

$$\delta_{a_i}(r^n)r^m + r^n\delta_{a_i}(r^m) = n(r^{n+i} - r^{n-i})r^m + mr^n(r^{m+i} - r^{m-i})$$
$$= (n+m)(r^{n+m+i} - r^{n+m-i})$$
$$= \delta_{a_i}(r^n r^m),$$

$$\delta_{a_i}(r^n)sr^m + r^n\delta_{a_i}(sr^m) = n(r^{n+i} - r^{n-i})sr^m + mr^n s(r^{m+i} - r^{m-i})$$
$$= (m-n)s(r^{m-n+i} - r^{m-n-i})$$
$$= \delta_{a_i}(r^n sr^m),$$

$$\delta_{a_i}(sr^n)r^m + sr^n\delta_{a_i}(r^m) = ns(r^{n+i} - r^{n-i})r^m + msr^n(r^{m+i} - r^{m-i})$$
$$= (m+n)s(r^{m+n+i} - r^{m+n-i})$$
$$= \delta_{a_i}(sr^n r^m),$$

$$\delta_{a_i}(sr^n)sr^m + sr^n\delta_{a_i}(sr^m) = ns(r^{n+i} - r^{n-i})sr^m + msr^n s(r^{m+i} - r^{m-i})$$
$$= (m-n)s(r^{m-n+i} - r^{m-n-i})$$
$$= \delta_{a_i}(sr^n sr^m).$$

Thus δ_{a_i} is a derivation of kD_∞.

If there are non-zero coefficients a_i in (2.1), then for some $a_i' \in k$ we have

$$\left(\delta - \sum_i a_i'\delta_{a_i}\right)(s) = \sum_i b_i'sr^i.$$

So substitute δ for $\delta - \sum_i a_i'\delta_{a_i}$ if necessary, and we may assume that $a_i = 0$ $(i \in \mathbb{Z})$ in (2.1). Hence we have

$$\delta(r) = \sum_i b_i sr^i, \tag{4.1}$$

$$\delta(s) = 0, \tag{4.2}$$

where $b_i + b_{-i} = 0$ $(i \in \mathbb{Z})$.

Second, for any $i \in \mathbb{Z}$ let

$$\delta_{b_i}(1) = 0,$$
$$\delta_{b_i}(r) = sr^i - sr^{-i},$$
$$\delta_{b_i}(s) = 0,$$

and for $n \in \mathbb{Z}_{>0}$ let

$$\delta_{b_i}(r^n) = \sum_{h=0}^{n-1}(sr^{i+n-1-2h} - sr^{-(i+n-1-2h)}),$$

$$\delta_{b_i}(sr^n) = s\delta_{b_i}(r^n),$$

$$\delta_{b_i}(r^{-n}) = -\delta_{b_i}(r^n),$$

$$\delta_{b_i}(sr^{-n}) = -s\delta_{b_i}(r^n).$$

Then δ_{b_i} is extended to a k-endomorphism of kD_∞.

We claim that δ_{b_i} is a derivation of kD_∞. For $n, m \in \mathbb{Z}_{>0}$ we have

$$\delta_{b_i}(r^n)r^m + r^n\delta_{b_i}(r^m)$$

$$= \sum_{h=0}^{n-1} s(r^{i+n-1-2h} - r^{-(i+n-1-2h)})r^m$$

$$+ \sum_{h=0}^{m-1} r^n s(r^{i+m-1-2h} - r^{-(i+m-1-2h)})$$

$$= \sum_{h=0}^{n-1} sr^{i+m+n-1-2h} + \sum_{h=0}^{m-1} sr^{i-n+m-1-2h}$$

$$- (\sum_{h=0}^{n-1} sr^{-i+m+n-1-2h} + \sum_{h=0}^{m-1} sr^{-i-n+m-1-2h})$$

$$= \sum_{h=0}^{n+m-1} s(r^{i+n+m-1-2h} - r^{-(i+n+m-1-2h)})$$

$$= \delta_{b_i}(r^n r^m).$$

$$\delta_{b_i}(r^n)r^{-m} + r^n\delta_{b_i}(r^{-m})$$

$$= \delta_{b_i}(r^n)r^{-m} - r^n\delta_{b_i}(r^m)$$

$$= \sum_{h=0}^{n-1} sr^{i-m+n-1-2h} - \sum_{h=0}^{n-1} sr^{-i-m+n-1-2h}$$

$$- \sum_{h=0}^{m-1} sr^{i-n+m-1-2h} + \sum_{h=0}^{m-1} sr^{-i-n+m-1-2h}$$

$$= \sum_{h=0}^{n-1} sr^{i-m+n-1-2h} - \sum_{h=0}^{m-1} sr^{i-n+m-1-2h}$$

$$- (\sum_{h=0}^{n-1} sr^{-i-m+n-1-2h} - \sum_{h=0}^{m-1} sr^{-i-n+m-1-2h})$$

$$
= \begin{cases}
\displaystyle\sum_{h=0}^{n-m-1} s(r^{i+n-m-1-2h} - r^{-(i+n-m-1-2h)}) \\
\qquad\qquad = \delta_{b_i}(r^n r^{-m}) \qquad (n > m), \\[2mm]
0 \qquad (n = m), \\[2mm]
\displaystyle - \sum_{h=0}^{m-n-1} s(r^{i+m-n-1-2h} - r^{-(i+m-n-1-2h)}) \\
\qquad\qquad = -\delta_{b_i}(r^{m-n}) = \delta_{b_i}(r^n r^{-m}) \qquad (n < m).
\end{cases}
$$

$$
\delta_{b_i}(r^{-n})r^m + r^{-n}\delta_{b_i}(r^m)
$$
$$
= -\delta_{b_i}(r^n)r^m + r^{-n}\delta_{b_i}(r^m)
$$
$$
= -\sum_{h=0}^{n-1} sr^{i+m+n-1-2h} + \sum_{h=0}^{n-1} sr^{-i+m+n-1-2h}
$$
$$
+ \sum_{h=0}^{m-1} sr^{i+n+m-1-2h} - \sum_{h=0}^{m-1} sr^{-i+n+m-1-2h}
$$
$$
= -\Big(\sum_{h=0}^{n-1} sr^{i+m+n-1-2h} - \sum_{h=0}^{m-1} sr^{i+n+m-1-2h}\Big)
$$
$$
+ \sum_{h=0}^{n-1} sr^{-i+m+n-1-2h} - \sum_{h=0}^{m-1} sr^{-i+n+m-1-2h}
$$
$$
= \begin{cases}
\displaystyle\sum_{h=0}^{m-n-1} s(r^{i+m-n-1-2h} - r^{-(i+m-n-1-2h)}) \\
\qquad\qquad = \delta_{b_i}(r^{m-n}) = \delta_{b_i}(r^{-n}r^m) \qquad (n < m), \\[2mm]
0 \qquad (n = m), \\[2mm]
\displaystyle - \sum_{h=0}^{n-m-1} s(r^{i+n-m-1-2h} - r^{-(i+n-m-1-2h)}) \\
\qquad\qquad = -\delta_{b_i}(r^{n-m}) = \delta_{b_i}(r^{-n}r^m) \qquad (n > m).
\end{cases}
$$

By definition $\delta_{b_i}(r^{-n}) = -\delta_{b_i}(r^n)$, and $\delta_{b_i}(r^n)r^m = r^{-m}\delta_{b_i}(r^n)$ $(m \in \mathbb{Z})$. Hence we have

$$
\begin{aligned}
\delta_{b_i}(r^{n+m}) &= \delta_{b_i}(r^n)r^m + r^n\delta_{b_i}(r^m) \\
&= r^{-m}\delta_{b_i}(r^n) + \delta_{b_i}(r^m)r^{-n} \\
&= -r^{-m}\delta_{b_i}(r^{-n}) - \delta_{b_i}(r^{-m})r^{-n}.
\end{aligned}
$$

Thus

$$\delta_{b_i}(r^{-m})r^{-n} + r^{-m}\delta_{b_i}(r^{-n}) = -\delta_{b_i}(r^{n+m}) = \delta_{b_i}(r^{-n}r^{-m}).$$

Therefore we have for any $n, m \in \mathbb{Z}$

$$\delta_{b_i}(r^n)r^m + r^n\delta_{b_i}(r^m) = \delta_{b_i}(r^n r^m). \tag{4.3}$$

It is clear that $\delta_{b_i}(r^n)s = -s\delta_{b_i}(r^n)$. Hence from (4.3) for any $n, m \in \mathbb{Z}$

$$\delta_{b_i}(r^n)sr^m + r^n\delta_{b_i}(sr^m) = s(\delta_{b_i}(r^{-n})r^m + r^{-n}\delta_{b_i}(r^m)) = \delta_{b_i}(r^n sr^m),$$
$$\delta_{b_i}(sr^n)r^m + sr^n\delta_{b_i}(r^m) = s(\delta_{b_i}(r^n)r^m + r^n\delta_{b_i}(r^m)) \doteq \delta_{b_i}(sr^n r^m),$$
$$\delta_{b_i}(sr^n)sr^m + sr^n\delta_{b_i}(sr^m) = s(\delta_{b_i}(r^n)sr^m + r^n\delta_{b_i}(sr^m)) = \delta_{b_i}(sr^n sr^m).$$

Thus we can conclude that δ_{b_i} is a derivation of kD_∞.

Now it is clear that the derivation δ satisfying the conditions (4.1) and (4.2) is a linear combination of $\delta_{b_i}(r)$:

$$\delta = \sum_i b_i'\delta_{b_i} \qquad (b_i' \in k).$$

Hence we have the following

Theorem 1. *Any derivation δ of kD_∞ is a linear combination of the inner derivations and derivations $\{\delta_{a_i}, \delta_{b_j} \mid i, j \in \mathbb{Z}\}$.*

5. The Derivation Algebra

Let

$$A = \langle \delta_{a_i} \mid i \in \mathbb{Z}\rangle, \quad B = \langle \delta_{b_i} \mid i \in \mathbb{Z}\rangle.$$

We consider Lie algebraic structure of A and B.

Clearly $\delta_{a_0} = 0$, $\delta_{a_{-i}} = -\delta_{a_i}$ ($i \in \mathbb{Z}$), and the set $\{\delta_{a_i} \mid i \in \mathbb{Z}_{>0}\}$ is linearly independent. We have

$$[\delta_{a_i}, \delta_{a_j}] = (i + j)\delta_{a_{i-j}} - (i - j)\delta_{a_{i+j}}, \tag{5.1}$$

for any $i, j \in \mathbb{Z}$, because

$$[\delta_{a_i}, \delta_{a_j}](r) = \delta_{a_i}(r^{1+j} - r^{1-j}) - \delta_{a_j}(r^{1+i} - r^{1-i})$$
$$= (1 + j)(r^{1+i+j} - r^{1-i+j}) - (1 - j)(r^{1+i-j} - r^{1-i-j})$$
$$\quad - (1 + i)(r^{1+i+j} - r^{1+i-j}) + (1 - i)(r^{1-i+j} - r^{1-i-j})$$
$$= (j - i)(r^{1+i+j} - r^{1-(i+j)}) + (i + j)(r^{1+i-j} - r^{1-(i-j)})$$
$$= ((i + j)\delta_{a_{i-j}} - (i - j)\delta_{a_{i+j}})(r),$$

and

$$[\delta_{a_i}, \delta_{a_j}](s) = ((i+j)\delta_{a_{i-j}} - (i-j)\delta_{a_{i+j}})(s) = 0.$$

The algebra $A = \langle \delta_{a_i} \mid i \in \mathbb{Z} \rangle$ is isomorphic to a subalgebra of the Witt algebra.

It is clear that $\delta_{b_0} = 0$, $\delta_{b_1} = \mathrm{ad}\, s$, $\delta_{b_{-i}} = -\delta_{b_i}$ $(i \in \mathbb{Z})$, and the set $\{\delta_{b_i} \mid i \in \mathbb{Z}, i \geq 2\}$ is linearly independent.

We claim that

$$[\delta_{b_i}, \delta_{b_j}] = 0 \tag{5.2}$$

for any $i, j \in \mathbb{Z}$. (5.2) follows from

$$
\begin{aligned}
&[\delta_{b_i}, \delta_{b_j}](r) \\
&= \delta_{b_i}(sr^j - sr^{-j}) - \delta_{b_j}(sr^i - sr^{-i}) \\
&= 2s(\delta_{b_i}(r^j) - \delta_{b_j}(r^i)) \\
&= 2(\sum_{h=0}^{j-1} r^{i+j-1-2h} + \sum_{h=0}^{i-1} r^{-(j+i-1-2h)} - (\sum_{h=0}^{i-1} r^{j+i-1-2h} + \sum_{h=0}^{j-1} r^{-(i+j-1-2h)})) \\
&= 2(\sum_{h=0}^{i+j-1} r^{i+j-1-2h} - \sum_{h=0}^{i+j-1} r^{i+j-1-2h}) \\
&= 0,
\end{aligned}
$$

and

$$[\delta_{b_i}, \delta_{b_j}](s) = 0.$$

Now we have

$$
[\delta_{a_i}, \delta_{b_j}] =
\begin{cases}
(j-1)\delta_{b_{j+i}} - (j+1)\delta_{b_{j-i}} & (i = 1), \\
(j-1)\delta_{b_{j+i}} - (j+1)\delta_{b_{j-i}} - 2\sum_{h=1}^{i-1} \delta_{b_{i+j-2h}} & (i \geq 2),
\end{cases}
\tag{5.3}
$$

because for $i \geq 2$ we have

$$
\begin{aligned}
&[\delta_{a_i}, \delta_{b_j}](r) \\
&= \delta_{a_i}(sr^j - sr^{-j}) - \delta_{b_j}(r^{1+i} - r^{1-i}) \\
&= s\delta_{a_i}(r^j) - s\delta_{a_i}(r^{-j}) - \delta_{b_j}(r^{1+i}) - \delta_{b_j}(r^{i-1}) \\
&= js(r^{j+i} - r^{j-i}) + js(r^{-j+i} - r^{-j-i}) \\
&\quad - s\sum_{h=0}^{i} (r^{j+i-2h} - r^{-(j+i-2h)}) - s\sum_{h=0}^{i-2} (r^{j+i-2-2h} - r^{-(j+i-2-2h)})
\end{aligned}
$$

$$= js(r^{j+i} - r^{-(j+i)}) - js(r^{j-i} - r^{-(j-i)})$$
$$- s(r^{j+i} - r^{-(j+i)}) - s(r^{j-i} - r^{-(j-i)})$$
$$- 2s \sum_{h=0}^{i-2}(r^{j+i-2-2h} - r^{-(j+i-2-2h)})$$

$$= ((j-1)\delta_{b_i+j} - (j+1)\delta_{b_j-i} - 2\sum_{h=0}^{i-2}\delta_{b_{j+i-2-2h}})(r),$$

and

$$[\delta_{a_i}, \delta_{b_j}](s) = ((j-1)\delta_{b_i+j} - (j+1)\delta_{b_j-i} - 2\sum_{h=1}^{i-1}\delta_{b_{j+i-2h}})(s) = 0.$$

The case $i = 2$ will be shown in a similar manner.

From the relations (5.1), (5.2) and (5.3) we have the following

Theorem 2. Der $kD_\infty = (\text{Inn } kD_\infty + B) \oplus A$ *as vector spaces, where B is an infinite-dimensional abelian Lie algebra and A is the Lie algebra with the basis $\{a_i \mid i \in \mathbb{Z}_{>0}\}$ and the multiplication $[a_i, a_j] = (i+j)a_{i-j} - (i-j)a_{i+j}$.*

References

[1] C. W. Curtis and I. Reiner, Representation Theory of Finite Groups and Associative Algebras, Wiley-Interscience, New York, 1962.

[2] T. Ikeda and N. Kawamoto, On derivation algebras of group algebras, Proceedings of the Third International Conference on Non Associative Algebra and Its Applications (to appear).

[3] N. Kawamoto, Generalizations of Witt algebras over a field of characteristic zero, Hiroshima Math. J., 16 (1986), 417–426.

[4] J. M. Osborn, Examples and conjectures relating to Lie algebras, preprint, 1993.

Cyclic Conjugacy Separability of Groups

G. Kim and C. Y. Tang*

Abstract. Finitely generated nilpotent groups are known to be cyclic conjugacy separable (Dyer [3]). In this paper we prove that conjugacy separable finite extensions of conjugacy separable residually finitely generated torsion-free nilpotent groups are cyclic conjugacy separable. From this it follows that finite extensions of free groups, surface groups and Fuchsian groups are cyclic conjugacy separable.

1991 Mathematics Subject Classification: 20E26, 20E06, 20F10

1. Introduction

Dyer [3] showed that if G is a finitely generated nilpotent group or a free group and if $x, h \in G$ such that $\{x\}^G \cap \langle h \rangle = \emptyset$, then there exists a finite homomorphic image $\overline{G}$ of G in which $\{\overline{x}\}^{\overline{G}} \cap \langle \overline{h} \rangle = \emptyset$. In [9], Tang proved this property for surface groups and formally defined groups with this property to be cyclic conjugacy separable. Dyer [3] made use of cyclic conjugacy separability of finitely generated nilpotent groups and free groups to show that generalized free products of two such groups amalgamating a cyclic subgroup are conjugacy separable. Tang [9] also made use of this property to show the conjugacy separability of generalized free products of two surface groups amalgamating a cyclic subgroup. Thus it seems that it is of interest to study cyclic conjugacy separability on its own. A more general concept is that of subgroup conjugacy separability. Moldavansky [6] showed that free groups are subgroup conjugacy separable. He also showed that supersolvable groups are cyclic conjugacy separable.

In this paper we first show that finite extensions of finitely generated nilpotent groups are cyclic conjugacy separable. Using this we show that conjugacy separable finite extensions of conjugacy separable residually finitely generated torsion-free nilpotent groups are cyclic conjugacy separable. From this it follows that surface groups and finitely generated Fuchsian groups are cyclic conjugacy separable. In a

---
* The second author gratefully acknowledges the partial support by the Natural Science and Engineering Research Council of Canada, Grant No. A-6064.

subsequent paper we shall study the conjugacy separability of generalized free products of two conjugacy separable finite extensions of conjugacy separable residually finitely generated torsion-free nilpotent groups amalgamating a cyclic subgroup.

Our notations and terms are in general standard. In particular we use:

$N \lhd_f G$ to denote N is a normal subgroup of finite index in G.

$\{x\}^G$ denotes the set of conjugates of x in G.

$Z_i(G)$ denotes the i-th center of G.

2. Nilpotent-by-Finite Groups

In [3], Dyer proved that finitely generated nilpotent groups are cyclic conjugacy separable. Her method can not be adapted to finite extensions of finitely generated nilpotent groups. In this section we show that these groups are also cyclic conjugacy separable.

Definition 2.1. A group G is said to be *subgroup conjugacy separable* if for every subgroup H of G and $x \in G$ such that $\{x\}^G \cap H = \emptyset$, there exists $N \lhd_f G$ such that, in $\overline{G} = G/N$, $\{\overline{x}\}^{\overline{G}} \cap \overline{H} = \emptyset$. In particular if G is subgroup conjugacy separable for every cyclic subgroup of G then G is said to be *cyclic conjugacy separable*.

Lemma 2.2. *Let G be a finite extension of a finitely generated nilpotent group, or a free group, B. Let $h \in G$ such that either $|h|$ is finite or $h \in B$. If $x \in G$ such that $\{x\}^G \cap \langle h \rangle = \emptyset$ then there exists $N \lhd_f G$ such that, in $\overline{G} = G/N$, $\{\overline{x}\}^{\overline{G}} \cap \langle \overline{h} \rangle = \emptyset$.*

Proof. Since finite extensions of finitely generated nilpotent groups or free groups are conjugacy separable ([5] and [2]), it is easy to see that the lemma is true if $|h|$ is finite. Thus we need only consider the case of $h \in B$.

Case 1. $x \notin B$. Let $N = B$. The lemma follows immediately.

Case 2. $x \in B$. Let $\{f_1, f_2, \ldots, f_n\}$ be a complete set of coset representatives of B in G. Clearly $f_i^{-1} x f_i \in B$ for $i = 1, \ldots, n$. Moreover, since $\{x\}^G \cap \langle h \rangle = \emptyset$, $\{f_i^{-1} x f_i\}^B \cap \langle h \rangle = \emptyset$ for $i = 1, \ldots, n$. Since B is finitely generated nilpotent or free, by [3], there exists $M_i \lhd_f B$ such that, in $\tilde{B} = B/M_i$, $\{\tilde{f}_i^{-1} \tilde{x} \tilde{f}_i\}^{\tilde{B}} \cap \langle \tilde{h} \rangle = \emptyset$ for each $i = 1, \ldots, n$. Let $M = \cap_{i=1}^n M_i$ and $N = \cap_{i=1}^n f_i^{-1} M f_i$. Then $N \lhd_f G$. Moreover, in $\overline{G} = G/N$, $\{\overline{x}\}^{\overline{G}} \cap \langle \overline{h} \rangle = \emptyset$. This proves the lemma.   □

In fact we can state Lemma 2.2 in a more general form.

Lemma 2.3. *Let G be a finite extension of a group B, where B is cyclic conjugacy separable. Let $h \in G$ such that either $|h|$ is finite or $h \in B$. If $x \in G$ such that $\{x\}^G \cap \langle h \rangle = \emptyset$ and G is residually finite, then there exists $N \lhd_f G$ such that $\{\overline{x}\}^{\overline{G}} \cap \langle \overline{h} \rangle = \emptyset$, where $\overline{G} = G/N$.*

Lemma 2.4. *Let G be a finite extension of a residually finitely generated torsion-free nilpotent group B. Let $h \in G$ be of infinite order. If $h^n \in B$ then, for any integer $\epsilon > 0$, there exists $M_\epsilon \lhd_f G$ such that $M_\epsilon \cap \langle h \rangle = \langle h^{n\epsilon} \rangle$.*

Proof. For convenience, let $h^n = a$. Clearly $a \neq 1$. Since B is residually finitely generated torsion-free nilpotent, there exists $L \lhd B$ such that $a \notin L$ and B/L is a finitely generated torsion-free nilpotent group. Let $\{f_1, \dots, f_m\}$ be a complete set of coset representatives of B in G. Let $M = \cap_{i=1}^{m} f_i^{-1} L f_i$. Then $M \lhd G$ and $\overline{B} = B/M$ can be embedded in the direct product $\times_{i=1}^{m}(B/f_i^{-1} L f_i)$. It follows that $\overline{B}$ is finitely generated torsion-free nilpotent. Moreover $\overline{G}$ is a finite extension of $\overline{B}$. Since $\overline{a} \neq 1$, we can find an integer k such that $\overline{a} \in Z_k(\overline{B}) \backslash Z_{k-1}(\overline{B})$. Let $\tilde{G} = \overline{G}/Z_{k-1}(\overline{B})$. Then $\tilde{a} \in Z(\tilde{B})$. Let $\{\tilde{y}_1, \dots, \tilde{y}_r\}$ be a free basis of $Z(\tilde{B})$ such that $\tilde{a} = \tilde{y}_1^s$ (say). Then, in $\overline{\overline{G}} = \tilde{G}/Z(\tilde{B})^{s\epsilon}$, $|\overline{\overline{a}}| = \epsilon$, whence $|\overline{\overline{h}}| = n\epsilon$. Since $\overline{\overline{G}}$ is residually finite, there exists $\overline{\overline{N}} \lhd_f \overline{\overline{G}}$ such that $\overline{\overline{N}} \cap \langle \overline{\overline{h}} \rangle = 1$. Let M_ϵ be the preimage of $\overline{\overline{N}}$ in G. Then $M_\epsilon \cap \langle h \rangle = \langle h^{n\epsilon} \rangle$ as required. □

Lemma 2.4 simply says that every finite extension of a residually finitely generated torsion-free nilpotent group is weakly potent as defined in [8]. The following is Lemma 2.3 of [8].

Lemma 2.5. *Let $x \in G$, where G is weakly $\langle x \rangle$-potent. If $x^i \sim_G x^j$ then $j = \pm i$.*

Applying Lemma 2.4, we have:

Corollary 2.6. *Let G be a finite extension of a residually finitely generated torsion-free nilpotent group. Let $h \in G$ be of infinite order. If $h^i \sim_G h^j$ then $j = \pm i$.*

To show finite extensions of finitely generated nilpotent groups are cyclic conjugacy separable we first show that finite extensions of finitely generated abelian groups are cyclic conjugacy separable.

Lemma 2.7. *Let G be a finite extension of a finitely generated abelian group B. Then G is cyclic conjugacy separable.*

Proof. Let $x, h \in G$ such that $\{x\}^G \cap \langle h \rangle = \emptyset$. We need to show there exists $N \lhd_f G$ such that, in $\overline{G} = G/N$, $\{\overline{x}\}^{\overline{G}} \cap \langle \overline{h} \rangle = \emptyset$. W.L.O.G. we may assume that B is torsion-free. By Lemma 2.2, we may assume $B \cap \langle h \rangle = \langle h^n \rangle$ for $n > 1$ and $|h| = \infty$. Let $h^n = c^\epsilon$, where $\{c, c_1, \dots, c_s\}$ is a free basis of B. Clearly G/B^ϵ is finite and $B^\epsilon \cap \langle h \rangle = \langle h^n \rangle = \langle c^\epsilon \rangle$. We note that $\langle c^\epsilon \rangle$ is an isolated subgroup of B^ϵ. Thus we may assume that G/B is finite and $B \cap \langle h \rangle = \langle h^n \rangle$ is an isolated subgroup of B.

  Case 1. If $\{x^n\}^G \cap \langle h^n \rangle = \emptyset$, then by Lemma 2.2 there exists $N \lhd_f G$ such that $\{\overline{x}^n\}^{\overline{G}} \cap \langle \overline{h}^n \rangle = \emptyset$, where $\overline{G} = G/N$. Then clearly $\{\overline{x}\}^{\overline{G}} \cap \langle \overline{h} \rangle = \emptyset$, as required.

Hence we need only consider $x^n \sim_G h^{kn}$ for some integer k.

Case 2. Suppose $x^n = h^{kn}$. By Corollary 2.6, we have either $f_i^{-1}h^{nk}f_i \notin \langle h \rangle$ or $f_i^{-1}h^{nk}f_i = h^{\pm kn}$, where $\{f_1, \ldots, f_m\}$ is a complete set of coset representatives of B in G. Let $I = \{i : f_i^{-1}h^{nk}f_i \notin \langle h \rangle\}$. Since G is π_c, there exists $N_1 \vartriangleleft_f G$ such that $f_i^{-1}h^{nk}f_i \notin N_1\langle h \rangle$ for each $i \in I$. Since G is conjugacy separable, we can also find $N_2 \vartriangleleft_f G$ such that $N_2 x \nsim_{G/N_2} N_2 h^{\pm k}$. Choose α so that $B^\alpha \subset N_1 \cap N_2$. Then $B^{\alpha n} \vartriangleleft_f G$. Moreover $\{\overline{x}\}^{\overline{G}} \cap \langle \overline{h} \rangle = \emptyset$, where $\overline{G} = G/B^{\alpha n}$. For if $\overline{x} \sim_{\overline{G}} \overline{h}^\delta$, then $\overline{h}^{\delta n} \sim_{\overline{G}} \overline{x}^n = \overline{h}^{kn}$. Hence $\overline{f_i}^{-1}\overline{h}^{kn}\overline{f_i} = \overline{h}^{\delta n}$ for some f_i, since B is abelian and $h^n \in B$. Now $B^{\alpha n} \subset N_1$. It follows that $i \notin I$ and hence, by Corollary 2.6, $f_i^{-1}h^{kn}f_i = h^{\pm kn}$. Thus $\overline{h}^{\delta n} = \overline{f_i}^{-1}\overline{h^{kn}}\overline{f_i} = \overline{h}^{\pm kn}$. Since $\langle h^n \rangle$ is isolated in B, $\langle h \rangle \cap B^{\alpha n} = \langle h^{n^2\alpha} \rangle$. This implies $\delta n = n^2\alpha s \pm kn$ for some s. Hence $\delta = n\alpha s \pm k$. Since $B^{\alpha n} \subset B^\alpha$, in $\tilde{G} = G/B^\alpha$ we have $\tilde{x} \sim_{\tilde{G}} \tilde{h}^\delta = \tilde{h}^{n\alpha s \pm k} = \tilde{h}^{\pm k}$. This contradicts the choice of N_2, since $B^\alpha \subset N_2$. Hence $\{\overline{x}\}^{\overline{G}} \cap \langle \overline{h} \rangle = \emptyset$, where $\overline{G} = G/B^{\alpha n}$.

Case 3. Suppose $g^{-1}x^n g = h^{kn}$ for some $g \in G$. Let $x' = g^{-1}xg$. Then we have $\{x'\}^G \cap \langle h \rangle = \emptyset$ and $(x')^n = h^{kn}$. Hence, by Case 2, there exists $N \vartriangleleft_f G$ such that $\{\overline{x'}\}^{\overline{G}} \cap \langle \overline{h} \rangle = \emptyset$, where $\overline{G} = G/N$. Since $x' \sim_G x$, it follows that $\{\overline{x}\}^{\overline{G}} \cap \langle \overline{h} \rangle = \emptyset$. $\square$

Theorem 2.8. *Let G be a finite extension of a finitely generated nilpotent group B. Then G is cyclic conjugacy separable.*

Proof. Let $x, h \in G$ such that $\{x\}^G \cap \langle h \rangle = \emptyset$. We shall show that there exists $N \vartriangleleft_f G$ such that, in $\overline{G} = G/N$, $\{\overline{x}\}^{\overline{G}} \cap \langle \overline{h} \rangle = \emptyset$. W.L.O.G. we may assume that B is torsion-free. By Lemma 2.2, we may assume $B \cap \langle h \rangle = \langle h^n \rangle$ for $n > 1$ and $|h| = \infty$. As in Lemma 2.7, it suffices to consider the case of $x^n = h^{kn}$ for some integer k, since the other cases are similar. Hence we suppose $x^n = h^{kn}$.

We proceed by induction on the nilpotent class of B. Let $\tilde{G} = G/Z(B)$. If $\{\tilde{x}\}^{\tilde{G}} \cap \langle \tilde{h} \rangle = \emptyset$ then, by induction hypothesis, the lemma is true. Hence we can assume $\tilde{x} \sim_{\tilde{G}} \tilde{h}^\alpha$ for some α.

Case 1. $|\tilde{h}| = \infty$. Since $x \nsim_G h^{\pm\alpha}$ and G is conjugacy separable, there exists $N_1 \vartriangleleft_f G$ such that $N_1 \subset B$ and $N_1 x \nsim_{G/N_1} N_1 h^{\pm\alpha}$. We note that in $\overline{G} = G/(N_1 \cap Z(B))$ we have $\{\overline{x}\}^{\overline{G}} \cap \langle \overline{h} \rangle = \emptyset$. For if $\overline{x} \sim_{\overline{G}} \overline{h}^\epsilon$ then $\tilde{h}^\alpha \sim \tilde{x} \sim \tilde{h}^\epsilon$, since $N_1 \cap Z(B) \subset Z(B)$. Now $\tilde{G}$ is nilpotent-by-finite and $|\tilde{h}| = \infty$. Therefore, by Lemma 2.5 $\epsilon = \pm\alpha$. This implies $\overline{x} \sim_{\overline{G}} \overline{h}^{\pm\alpha}$ contradicting the choice of N_1, whence $\{\overline{x}\}^{\overline{G}} \cap \langle \overline{h} \rangle = \emptyset$. Now $\overline{G}$ is a finite extension of $\overline{N}_1 = N_1/N_1 \cap Z(B)$. Since $\overline{N}_1 \cong N_1 Z(B)/Z(B) < B/Z(B)$, the nilpotent class of $\overline{N}_1$ is less than the class of B. Hence, by induction, we can find $N \vartriangleleft_f G$ such that $\{Nx\}^{G/N} \cap \langle Nh \rangle = \emptyset$ as required.

Case 2. $|\tilde{h}| < \infty$. Since $Z(B)$ is isolated in B [7, p. 133], we have $\langle h \rangle \cap Z(B) = \langle h^n \rangle = \langle h \rangle \cap B$. Now if $\langle h^n \rangle$ is not isolated in $Z(B)$, then there is a free basis $\{a, a_1, \ldots, a_s\}$ of $Z(B)$, where $h^n = a^\epsilon$ for $\epsilon \neq \pm 1$. Let $\overline{G} = G/Z(B)^\epsilon$. Since

$\overline{Z(B)}$ is finite and $\overline{G}$ is residually finite, there exists $\overline{N} \lhd_f \overline{G}$ such that $\overline{N} \subset \overline{B}$ and $\overline{N} \cap \overline{Z(B)} = 1$. Let $\pi : G \to \overline{G}$ be the natural homomorphism, and let $N = \pi^{-1}(\overline{N}) \lhd_f G$. Then G is a finite extension of a torsion-free nilpotent group $N \cap B$. Moreover $\langle h \rangle \cap Z(N \cap B) = \langle h^n \rangle = \langle h \rangle \cap (N \cap B)$. We show that $\langle h^n \rangle = \langle a^\epsilon \rangle$ is an isolated subgroup of $Z(N \cap B)$. For if $w \in Z(N \cap B)$ such that $w^i \in \langle a^\epsilon \rangle$ then $w \in Z(B)$, since $Z(B)$ is isolated. Since $\{a, a_1, \dots, a_s\}$ is a free basis of $Z(B)$, $w = a^\lambda$ for some λ. Now $w = a^\lambda \in N$ implies $\overline{a}^\lambda \in \overline{N} \cap \overline{Z(B)} = 1$. Hence $a^\lambda \in Z(B)^\epsilon$. This implies $w = a^\lambda \in \langle a^\epsilon \rangle$, showing that $\langle h^n \rangle = \langle a^\epsilon \rangle$ is isolated in $Z(N \cap B)$. Thus by considering $N \cap B$ instead of B if necessary, we may assume that G/B is finite and $B \cap \langle h \rangle = \langle h^n \rangle$ is isolated in $Z(B)$.

Let $\{f_1, \dots, f_m\}$ be a complete set of coset representatives of B in G and $I = \{i : f_i^{-1} h^{nk} f_i \notin \langle h \rangle\}$. As in Lemma 2.7 we can find $N_1 \lhd_f G$ and $N_2 \lhd_f G$ such that $f_i^{-1} h^{nk} f_i \notin N_1 \langle h \rangle$ for each $i \in I$ and $N_2 x \not\sim_{G/N_2} N_2 h^{\pm k}$. Let $Z(B)^\beta \subset N_1 \cap N_2$. Then we claim that $\{\overline{x}\}^{\overline{G}} \cap \langle \overline{h} \rangle = \emptyset$ in $\overline{G} = G/Z(B)^{\beta n}$. For if $\overline{x} \sim_{\overline{G}} \overline{h}^\delta$, then $\overline{h}^{\delta n} \sim_{\overline{G}} \overline{x}^n = \overline{h}^{kn}$. Hence $\overline{f}_i^{-1} \overline{h}^{kn} \overline{f}_i = \overline{h}^{\delta n}$ for some f_i, since $h^n \in Z(B)$. Now $Z(B)^{\beta n} \subset N_1$. It follows that $i \notin I$ and hence $f_i^{-1} h^{kn} f_i = h^{\pm kn}$. Thus $\overline{h}^{\delta n} = \overline{f}_i^{-1} h^{kn} f_i = \overline{h}^{\pm kn}$. Since $\langle h^n \rangle$ is isolated in $Z(B)$, $\langle h \rangle \cap Z(B)^{\beta n} = \langle h^{n^2 \beta} \rangle$. Therefore, $\delta n = n^2 \beta s \pm kn$ for some s. This implies $\delta = n\beta s \pm k$. Since $Z(B)^{\beta n} \subset Z(B)^\beta$, it follows that, in $\tilde{G} = G/Z(B)^\beta$, $\tilde{x} \sim_{\tilde{G}} \tilde{h}^\delta = \tilde{h}^{n\beta s \pm k} = \tilde{h}^{\pm k}$. This contradicts the choice of N_2, since $Z(B)^\beta \subset N_2$. Hence $\{\overline{x}\}^{\overline{G}} \cap \langle \overline{h} \rangle = \emptyset$, where $\overline{G} = G/Z(B)^{\beta n}$. Now $|\overline{h}|$ is finite and $\overline{G}$ is conjugacy separable. It follows that we can easily find $N \lhd_f G$ such that $\{Nx\}^{G/N} \cap \langle Nh \rangle = \emptyset$ as required. $\square$

3. Residually Torsion-Free Nilpotent Groups

We are now ready to show every conjugacy separable finite extension of conjugacy separable residually finitely generated torsion-free nilpotent group is cyclic conjugacy separable. From this it follows that surface groups, free-by-finite groups and finitely generated Fuchsian groups are cyclic conjugacy separable.

Theorem 3.1. *Conjugacy separable residually finitely generated torsion-free nilpotent groups are cyclic conjugacy separable.*

Proof. Let G be conjugacy separable and residually finitely generated torsion-free nilpotent. Clearly G is torsion-free. Let $x, h \in G$ such that $\{x\}^G \cap \langle h \rangle = \emptyset$. Since G is conjugacy separable, we can assume $h \neq 1$. Therefore, there exists $N_1 \lhd G$ such that $h \notin N_1$ and G/N_1 is finitely generated torsion-free nilpotent. Let $\tilde{G} = G/N_1$. If $\{\tilde{x}\}^{\tilde{G}} \cap \langle \tilde{h} \rangle = \emptyset$, then the result follows from Theorem 2.8 (or [3]). Hence we suppose $\{\tilde{x}\}^{\tilde{G}} \cap \langle \tilde{h} \rangle \neq \emptyset$, say, $\tilde{x} \sim_{\tilde{G}} \tilde{h}^\alpha$ for some α. Since G is conjugacy separable and $x \not\sim_G h^{\alpha}$, we can find $N_2 \triangleleft_f G$ such that $N_2 x \not\sim_{G/N_2} N_2 h^{\alpha}$. Let $N = N_1 \cap N_2$. We shall show that $\{\overline{x}\}^{\overline{G}} \cap \langle \overline{h} \rangle = \emptyset$, in $\overline{G} = G/N$. For if $\overline{x} \sim_{\overline{G}} \overline{h}^i$ then $\tilde{h}^{\alpha} \sim \tilde{x} \sim \tilde{h}^i$. Since $\tilde{G}$ is torsion-free nilpotent and $\tilde{h} \neq 1$, it follows that $\alpha = i$. This implies $\overline{x} \sim_{\overline{G}} \overline{h}^{\alpha}$ contradicting the choice of $N \subset N_2$. Hence $\{\overline{x}\}^{\overline{G}} \cap \langle \overline{h} \rangle = \emptyset$. Now $\overline{G}$ is a finite extension of $\overline{N}_2$ and $\overline{N}_2 \cong N_1 N_2/N_1 \leq G/N_1$, whence $\overline{N}_2$ is also finitely generated nilpotent. Therefore, by Theorem 2.8, $\overline{G}$ is cyclic conjugacy separable. Since $\{\overline{x}\}^{\overline{G}} \cap \langle \overline{h} \rangle = \emptyset$, the result follows immediately from theorem 2.8.   □

Since surface groups are generalized free products of free groups amalgamating a cyclic subgroup, they are conjugacy separable [3]. Moreover, by [1], they are residually finitely generated torsion-free nilpotent. Thus we have the following corollary:

Corollary 3.2. *Surface groups are cyclic conjugacy separable.*

Theorem 3.3. *Let G be a conjugacy separable finite extension of a conjugacy separable residually finitely generated torsion-free nilpotent group B. Then G is cyclic conjugacy separable.*

Proof. Let $x, h \in G$ such that $\{x\}^G \cap \langle h \rangle = \emptyset$. By Lemma 2.3 we can assume that $h \notin B$ and $|h| = \infty$. So let $\langle h \rangle \cap B = \langle h^n \rangle$. If $\{x^n\}^G \cap \langle h^n \rangle = \emptyset$ then, by Lemma 2.3, there exists $N \triangleleft_f G$ such that $\{\overline{x}^n\}^{\overline{G}} \cap \langle \overline{h}^n \rangle = \emptyset$, where $\overline{G} = G/N$. It clearly follows that $\{\overline{x}\}^{\overline{G}} \cap \langle \overline{h} \rangle = \emptyset$ as required. So we can assume $x^n \sim_G h^{nk}$ for some integer k.

  Case 1. $x^n = h^{nk}$. Let $\{ f_i : i \in I \}$ be a complete set of coset representatives of B in G, where I is finite. Since B is residually finitely generated torsion-free nilpotent, we have either $\{f_i^{-1} h^{nk} f_i\}^B \cap \langle h^n \rangle = \emptyset$ or $f_i^{-1} h^{nk} f_i \sim_B h^{\pm nk}$ for each $i \in I$. Let $J = \{i \in I : \{f_i^{-1} h^{nk} f_i\}^B \cap \langle h^n \rangle = \emptyset\}$. By Theorem 3.1, B is cyclic conjugacy separable. Thus there exists $M_1 \triangleleft_f B$ such that $\{M_1 f_j^{-1} h^{nk} f_j\}^{B/M_1} \cap \langle M_1 h^n \rangle = \emptyset$ for all $j \in J$. Let $N_1 = \cap_{i \in I} f_i^{-1} M_1 f_i$. Then $N_1 \triangleleft_f G$ and $\{N_1 f_j^{-1} h^{nk} f_j\}^{B/N_1} \cap \langle N_1 h^n \rangle = \emptyset$ for all $j \in J$. Since G is conjugacy separable, there exists $N_2 \triangleleft_f G$ such that $N_2 \subset B$ and $N_2 x \not\sim_{G/N_2} N_2 h^{\pm k}$. Now B is residually finitely generated torsion-free nilpotent and $h^n \neq 1$. Thus there exists $M \triangleleft B$ such that $M \cap \langle h^n \rangle = 1$ and B/M is finitely generated torsion-free nilpotent. Let $N_3 = \cap_{i \in I} f_i^{-1} M f_i$. Then $N_3 \triangleleft G$. Let $N = N_1 \cap N_2 \cap N_3$. Then $\overline{G} = G/N$ is a finite extension of $\overline{N_1 \cap N_2} \cong (N_1 \cap N_2) N_3/N_3 \leq B/N_3$. Since B/N_3 can be embedded in the finitely generated torsion-free nilpotent group $\times_{i \in I} B/(f_i^{-1} M f_i)$, $\overline{N_1 \cap N_2}$ is also finitely generated torsion-free nilpotent. Hence, by Theorem 2.8, $\overline{G}$ is cyclic conjugacy separable. We shall show that $\{\overline{x}\}^{\overline{G}} \cap \langle \overline{h} \rangle = \emptyset$. Suppose $\overline{x} \sim_{\overline{G}} \overline{h}^t$ for some t. Then $\overline{h}^{nk} = \overline{x}^n \sim_{\overline{G}} \overline{h}^{tn}$, whence $\overline{h}^{tn} = \overline{b}^{-1} \overline{f_i}^{-1} \overline{h}^{nk} \overline{f_i} \overline{b}$ for some $b \in B$ and $i \in I$. This implies $h^{tn} \equiv b^{-1} f_i^{-1} h^{nk} f_i b$ mod N. Since $N \subset N_1$, we have $i \notin J$. Therefore $f_i^{-1} h^{nk} f_i \sim_B h^{\pm nk}$. Hence $\overline{h}^{tn} \sim_{\overline{B}} \overline{f_i}^{-1} \overline{h}^{nk} \overline{f_i} \sim_{\overline{B}} \overline{h}^{\pm nk}$. It follows that $\tilde{h}^{tn} \sim_{\tilde{B}} \tilde{h}^{\pm nk}$, where $\tilde{B} = B/N_3$ is finitely generated torsion-free nilpotent. Thus $t = \pm k$. But this implies that $\overline{x} \sim_{\overline{G}} \overline{h}^{\pm k}$ contradicting the choice of N_2. Hence $\{\overline{x}\}^{\overline{G}} \cap \langle \overline{h} \rangle = \emptyset$.

Case 2. $g^{-1}x^n g = h^{nk}$ for some $g \in G$. Let $x' = g^{-1}xg$. Then we have $\{x'\}^G \cap \langle h \rangle = \emptyset$ and $(x')^n = h^{kn}$. Thus, by Case 1, there exists $N \lhd_f G$ such that $\{\overline{x'}\}^{\overline{G}} \cap \langle \overline{h} \rangle = \emptyset$, where $\overline{G} = G/N$. Since $x' \sim_G x$, it is clear that $\{\overline{x}\}^{\overline{G}} \cap \langle \overline{h} \rangle = \emptyset$, as required. $\square$

By [2], finite extensions of free groups are conjugacy separable. Therefore we have:

Corollary 3.4. *Finite extensions of free groups are cyclic conjugacy separable.*

Corollary 3.5. *Finitely generated Fuchsian groups are cyclic conjugacy separable.*

Proof. Finitely generated Fuchsian groups are finite extensions of surface groups. By [1], surface groups are residually finitely generated torsion-free nilpotent groups. Now, by [4], finitely generated Fuchsian groups are conjugacy separable. Hence, by Theorem 3.3, finitely generated Fuchsian groups are cyclic conjugacy separable. $\square$

Remark. Since finitely generated nilpotent groups are cyclic conjugacy separable [3], it is interesting to ask whether polycyclic groups are cyclic conjugacy separable.

References

[1] Baumslag, G., On generalized free products, Math. Z. 78 (1962), 423–438.

[2] Dyer, J. L., Separating conjugates in free-by-finite groups, J. London Math. Soc. (2) 20 (1979), 215–221.

[3] Dyer, J. L., Separating conjugates in amalgamated free products and HNN extensions, J. Austral. Math. Soc. Ser. A 29 (9180), 35–51.

[4] Fine, B., Rosenberger, G., Conjugacy separability of Fuchsian groups and related questions, Contemp. Math. 109 (1990), 11–18.

[5] Formanek, E., Conjugate separability in polycyclic groups, J. Algebra 42 (1976), 1–10.

[6] Moldavansky, D. I., On finite separability of subgroups, Ivanov State University: twenty years (1993), 18–23.

[7] Robinson, D. J. S., A course in the theory of groups, G. T. M. 80. Springer-Verlag, New York–Heidelberg–Berlin 1982.

[8] Tang, C. Y., Conjugacy separability of generalized free products of certain conjugacy separable groups, Canad. Math. Bull., (to appear).

[9] Tang, C. Y., Conjugacy separability of generalized free products of surface groups, preprint.

Some Results and Problems Arising from a Question of Paul Erdös

Pan Soo Kim

1. Introduction

Paul Erdös posed the following question:

Let G be an infinite group. If there is no infinite subset of G whose elements do not mutually commute, is there then a finite bound on the cardinality of each such set of elements?

The affirmative answer to this question was obtained by B. H. Neumann. It is useful to restate his result and some techniques of the proof for the further understanding of subsequent results.

Theorem 1 (B. H. Neumann, see [NE1]). *Let G be an infinite group. If every infinite subset of G contains two elements that commute, then G is centre-by-finite.*

Proof. Ramsey's theorem plays an important role in this theorem, and it can be applied in many similar situations. Suppose there exists an element $g \in G$ having infinitely many conjugates. Then we can find an infinite subset T of G such that $g^s \neq g^t$ whenever $s \neq t$ in T. By Ramsey's theorem, T contains either an infinite subset U whose elements mutually commute or an infinite subset V whose elements mutually do not commute. By the hypothesis, there is no such infinite subset V. By considering U_g, we get a contradiction. So G is an FC-group.

A complete proof can be obtained by the examination of a special subgroup of FC-groups introduced by B. H. Neumann (see Lemma 2 and Lemma 3).

Lemma 2 (B. H. Neumann, see [NE2]). *If G is an FC-group and $|G/Z(G)|$ is infinite where $Z(G)$ is the centre of G, then G contains a subgroup N generated by infinitely many elements x_i, y_i, $i \in I$ (I an infinite index set) where the generators satisfy the* *conditions*

$$[x_i, x_j] = [y_i, y_j] = [x_i, y_j] = 1 \quad \text{if } i \neq j$$
$$[x_i, y_i] = z_i \neq 1.$$

These relations are not to be understood as forming a system of defining relations of N.

With the same notations as above, consider a sequence of elements

$$x_1, \quad y_1, \quad x_1 y_1 x_2, \quad x_1 y_1 y_2, \ldots, \quad x_1 y_1 \cdots x_n y_n x_{n+1}, \quad x_1 y_1 \cdots x_n y_n y_{n+1}, \ldots$$

These elements do not commute with each other. Hence we conclude the following.

Lemma 3. *Let G be an infinite FC-group. If every infinite subset X of G contains a pair of elements that commute, then G is centre-by-finite.*

If for every pair X, Y of infinite subsets of G, there exist x in X, y in Y such that $\langle x, y \rangle$ is abelian (or equivalently $xy = yx$), then G turns out to be abelian. As we have seen above, the centre Z of G is infinite. Consider the sets Zx, Zy for x, y in G. By the hypothesis, $[z_1 x, z_2 y] = 1$ for some $z_1, z_2 \in Z$ and hence $[x, y] = 1$.

Keeping this fact in mind, we generalize in two directions: Let $\mathcal{V}$ be a variety defined by the law $\omega(y_1, y_2, \ldots, y_n) = 1$ where n is the minimal number required to determine $\mathcal{V}$.

Type A. Define the class $\mathcal{V}^*$ as follows: $G \in \mathcal{V}^*$ if G is such that whenever $X_1, X_2, \ldots, X_n$ are infinite subsets of G, then there exist $x_i \in X_i$, $i = 1, 2, \ldots, n$ such that $\langle x_1, x_2, \ldots, x_n \rangle$ is a $\mathcal{V}$-group.

Type B. Similarly define $\mathcal{V}^\sharp$ as follows: $G \in \mathcal{V}^\sharp$ if G is such that whenever $X_1, X_2, \ldots, X_n$ are infinite subsets of G, there exist $x_i \in X_i, i = 1, 2, \ldots, n$ such that $\omega(x_1, x_2, \ldots, x_n) = 1$.

Remark. Let $\mathcal{V}$ be a variety determined by a law $\omega(y_1, y_2, \ldots, y_n) = 1$ and n the minimal number of variables required to define $\mathcal{V}$. If $\mathcal{V} = \mathcal{V}^*$ in the sense that the hypothesis that given infinite subsets $X_1, X_2, \ldots, X_n$ in G there exist $x_i \in X_i$, $i = 1, 2, \ldots, n$ such that $\langle x_1, x_2, \ldots, x_n \rangle \in \mathcal{V}$ implies $G \in \mathcal{V}$, then $\mathcal{V} = \mathcal{V}^*$ for any other law that may determine $\mathcal{V}$. This follows from minimality of n. On the other hand, the class $\mathcal{V}^\sharp$ is dependent upon the specific word $\omega(y_1, y_2, \ldots, y_n)$.

We note that $\mathcal{V} \subset \mathcal{V}^* \subset \mathcal{V}^\sharp$.

The aim of this paper is to collect results and open questions recently developed in connection with type A and B.

Problem. For which variety $\mathcal{V}$ is $\mathcal{V} = \mathcal{V}^*$ or for which law ω is $\mathcal{V} = \mathcal{V}^\sharp$?

2. The Main Results on $\mathcal{V} = \mathcal{V}^*$

Throughout this paper, we assume that G is an infinite group. Notations are standard (see, for example [R]). In the previous section we have seen that $G \in \mathcal{V}^*$ is a $\mathcal{V}$-group for the abelian variety $\mathcal{V}$ defined by $[x, y] = 1$.

To date, we have no example of a variety $\mathcal{V}$ for which an infinite $\mathcal{V}^*$-group is not in $\mathcal{V}$. It is natural to ask for which varieties $\mathcal{V}$, every $\mathcal{V}^*$-group is a $\mathcal{V}$-group. The first observation in this direction, on metabelian varieties, was made by P. Kim, A. Rhemtulla, and H. Smith.

Theorem 4 (see [KRS]). *Let G be an infinite group with the property that, given any four infinite subsets X_1, X_2, X_3, X_4 of G, there exist elements $x_i \in X_i$, $i = 1, 2, 3, 4$, such that $\langle x_1, x_2, x_3, x_4 \rangle$ is metabelian. Then G is metabelian.*

Comments on the proof. The proof of the theorem was broken up into several long steps, by dealing separately with the periodic case, the soluble and locally soluble cases and the case having finite subgroups. Here we give a proof for the non-periodic case only. The reader may refer to [KRS] for the rest of the proof. First we observe the following.

4.1. It can be easily checked that an infinite group G satisfying the hypothesis is metabelian if either G has infinite centre, G is an FC-group, or G is residually finite.

4.2. Suppose G is non-periodic. Let x be an element of infinite order. Then we may assume that G is finitely generated. For any a, b in G, consider sets $\{ax^i \mid i \in \mathbb{Z}\}$, $\{bx^i \mid i \in \mathbb{Z}\}$, $\{x^p \mid p \text{ is prime}\}$, $\{x^q \mid q \text{ is prime}, q \neq p\}$. These sets yield that $\langle a, b, x \rangle$ is metabelian for all $a, b \in G$. Hence $[\langle x \rangle, G]$ is abelian. Let $J = [\langle x \rangle, G]$ and $C = C_G(\langle x \rangle J)$. If J is finite, then $\langle x \rangle J$ is FC-central in G. Hence $|G : C|$ is finite and C is finitely generated. For some nonzero k, $x^k \in C$. By (4.1) C is metabelian and so residually finite. Therefore G is residually finite (see Theorem 9.51 of [R] part II) and so metabelian. We may take J to be infinite. Pick any a, b, c in G and let $S = \{s \in J \mid \langle s, ua, vb, wc \rangle \text{ is metabelian for some } u, v, w \in J\}$. Then by hypothesis $J \setminus S$ is finite and $\langle S \rangle = J$. Since J is a normal abelian subgroup, $[vb, wc] = [b, c]z$ for some $z \in J$ and $[s, ua] = [s, a]$. Hence $[[b, c], [s, a]] = 1$ for all $s \in S$ and $[[b, c], [J, a]] = 1$. Therefore $[G', [J, G]] = 1$. If $[J, G]$ is finite, then for the centralizer D of $[J, G]$ in G, $[J, D, D] = 1$. Hence $[D', J] = 1$ by the Three Subgroup Lemma. Consider infinite subsets aJ, bJ, cJ, dJ where $a, b, c, d \in D$. $[[ua, vb], [wc, zd]] = 1$ for some $u, v, w, z \in J$, which implies $[[a, b], [c, d]] = 1$. So D is metabelian. We conclude as before that G is metabelian. If $[J, G]$ is infinite, then for any $a, b, c, d \in G$ and $u, v, w, z \in [J, G]$, $[[ua, vb], [wc, zd]] = [[a, b], [c, d]]$, which implies that G is metabelian.

The last two authors have extended this result to soluble groups of arbitrary derived length.

Theorem 5 (see [RS1]). *Let G be an infinite group and m a fixed positive integer. If for every n = 2^m infinite subsets $X_1, X_2, \ldots, X_n$ of G, there exist $x_i \in X_i$, $i = 1, 2, \ldots, n$ such that $\langle x_1, x_2, \ldots, x_n \rangle$ is soluble of length at most m, then the group G is soluble of length at most m.*

A further generalization of this to poly-nilpotent groups has been studied by H. Smith.

3. The Main Problems on $\mathcal{V} = \mathcal{V}^{\sharp}$

It is more difficult to cope with $\mathcal{V}^{\sharp}$-groups. So the results dealing with this problem are few in number.

First we observe the following as an easy example.

Lemma 6. *Let $n \geq 2$ be a fixed integer and G an infinite group. If every infinite subset X of G contains an element of order dividing n, then G is of exponent n.*

Proof. Suppose that an element x of G is not of order dividing n. Then $|G : C_G(x)|$ is finite, since the number of those elements is finite. Let D be an infinite subset of $C_G(x)$ in which elements are of order dividing n. Then $(dx)^n = x^n$ for all d in D, which violates the hypothesis.

Therefore a $\mathcal{V}^{\sharp}$-group is a $\mathcal{V}$-group for a variety $\mathcal{V}$ defined by one variable.

A result of similar nature for nilpotent groups was obtained by P. Longobardi, M. Maj and A. Rhemtulla. The proof relies on Ramsey's theorem.

Theorem 7 (see [LR2]). *Let G be an infinite group and $n \geq 2$ be a fixed integer. If for every n infinite subsets $X_1, X_2, \ldots, X_n$ of G, there exist $x_i \in X_i$, $i = 1, 2, \ldots, n$ such that $[x_1, x_2, \ldots, x_n] = 1$, then G is nilpotent of class $n - 1$.*

We do not have such a complete result for any other variety to date. The author obtained a partial answer for the variety of metabelian groups.

Theorem 8 (P. Kim, see [K]). *Let G be an infinite locally soluble group. If for every four infinite subsets X_1, X_2, X_3, X_4, of G, there exist elements $x_i \in X_i$, $i = 1, 2, 3, 4$, such that $[[x_1, x_2], [x_3, x_4]] = 1$, then G is metabelian.*

Comments on the proof. We give a proof for the case of finitely generated soluble groups only. Here we use the special subgroup of FC-groups mentioned in Lemma 2. We prove the result by induction on the solubility length of G. If G has an infinite abelian normal subgroup J, then by an argument similar to (4.2) we can show that G is metabelian. So we may assume that G has no infinite abelian normal subgroup.

By the induction hypothesis, G'' is finite abelian and hence G' is an FC-group. The centralizer H of G'' in G has finite index. In order to apply the argument in (4.2), it is enough to show that H is metabelian, for then G is residually finite and is metabelian. Replacing G by H, we may assume $[G, G''] = 1$. Choose $\{t_1, \ldots, t_n\}$ as a set of generators of G.

Let $t = t_i$ for some i, and let $T = G''[G', \langle t \rangle]$. Modulo G'' we have $T \equiv [G', G'\langle t \rangle]$ and so T is normal in G. Easy calculations show that $[G', \langle t \rangle] \equiv [G', t]$ and so every element of T may be written in the form $[a, t]g''$, where $a \in G'$, $g'' \in G''$. If $T_i = [G', t_i]G''$ is finite for all i, then $[G', G] \le T_1 \cdots T_n$. Hence G is finite-by-nilpotent and hence residually finite.

For the remaining we construct 4 infinite subsets X_i, which would violate the hypothesis. Now we may assume $T = [G', t]G''$ is a non-abelian infinite group, having finite centre. Note that $T \le G'$ is an FC-group and G'' is central. By Lemma 2 T has an N-subgroup generated by x_i, y_i $(i \in I)$ subject to $[x_i, y_i] \ne 1$, but $[x_i, y_j] = [x_i, x_j] = [y_i, y_j] = 1$ for all $i \ne j$. Let $x_i = [a_i, t]g_i''$ $y_j = [b_j, t]f_j''$ for some $a_i, b_j \in G'$, $g_i'', f_j'' \in G''$. Then clearly for each $i \in I$,

$$[x_i, y_i] \ne 1 \quad \text{implies} \quad [[a_i, t], [b_i, t]] \ne 1,$$
$$[x_i, y_j] = 1 \quad \text{implies} \quad [[a_i, t], [b_j, t]] = 1 \quad \text{and so on.}$$

Let $i_0 \in I$ and let I be the disjoint union of $\{i_0\}$, I_1 and J_1 with $|I_1| = |J_1| = \infty$. Set

$$X_1 = \{a_{i_0}b_{i_1} \mid i_1 \in I_1\} \quad X_2 = \{ta_{i_1} \mid i_1 \in I_1\}$$
$$X_3 = \{b_{i_0}a_{j_1} \mid j_1 \in J_1\} \quad X_4 = \{tb_{j_1} \mid j_1 \in J_1\}.$$

We note that $a_i \in G'$ (resp. $b_j \in G'$) are not necessarily distinct. But each X_i is infinite for G'' is finite. For $x_i \in X_i$

$$\begin{aligned}
[[x_1, x_2], [x_3, x_4]] &= [[a_{i_0}b_{i_1}, ta_{i_1}], [b_{i_0}a_{j_1}, tb_{j_1}]] \\
&= [[a_{i_0}, t][b_{i_1}, t]g'', [b_{i_0}, t][a_{j_1}, t]f''] \quad \text{for some } g'', f'' \in G'' \\
&= [[a_{i_0}, t][b_{i_1}, t], [b_{i_0}, t][a_{j_1}, t]] \\
&= [[a_{i_0}, t][b_{i_1}, t]g'', [b_{i_0}, t]]^{g'} \quad \text{for some } g' \in G' \\
&= [[a_{i_0}, t], [b_{i_0}, t]]^{g} \quad \text{for some } g \in G \\
&\ne 1.
\end{aligned}$$

Problem. Can we remove "locally soluble" in the hypothesis of Theorem 8?

O. Puglisi and L. S. Spiezia have studied Engel groups in this direction under restricted conditions.

Theorem 9 (see [PS]). *Let G be an infinite locally soluble or locally finite group. If for any pair X, Y of infinite subsets of G, there exist x in X, y in Y such that* $[x, y, \ldots, y] = 1$, *then G is a k-Engel group.*
$\qquad\underbrace{}_{k}$

Comments on the proof. The result is easily proved for locally soluble groups containing a non-periodic element. If G has an element a of infinite order, then for any x, y in G, $H = \langle a, x, y \rangle$ is soluble. By induction on the solubility length, we may assume that H/N is a soluble k-Engel group and N is a normal abelian subgroup. It is known that a soluble bounded Engel group is a Baer group, and a Baer group is locally nilpotent. So H/N is nilpotent and hence H is residually finite. Therefore H is a k-Engel group by an argument similar to 4.1.

Problem. Let G be an infinite group. Is G a k-Engel group, if for every pair X, Y of infinite subsets of G, there exist x in X, y in Y such that $\langle x, y \rangle$ is a k-Engel subgroup?

4. Some Additional Results

Further questions of a similar nature, with slightly different aspects, have been studied in [LW].

Theorem 10 (J. C. Lennox, J. Wiegold). *A finitely generated soluble group G is polycyclic if and only if every infinite set of elements of G contains a pair generating a polycyclic subgroup.*

The open question left in [LW] was answered by J. R. J. Groves.

Theorem 11 (see [G]). *A finitely generated soluble group such that every infinite subset contains two elements generating a supersoluble group is finite-by-supersoluble.*

Theorems 10 and 11 are concerned with the property that for a group property $\wp$, every infinite subset contains a pair x, y such that $\langle x, y \rangle \in \wp$. And there is good reason to restrict attention to *finitely generated* groups. For instance, Vaughan-Lee and Wiegold (see [VW]) have constructed a group which is infinite, perfect, of prime exponent $p \geq 5$ and locally finite, such that every two-generator subgroup is finite and nilpotent of bounded class.

Also it is known that there exists a 3-generator Engel p-group G all of whose 2-generator subgroups are finite and hence nilpotent, but G is infinite (see [GO]). The latter example leads to study *local finiteness*.

Theorem 12 (A. Rhemtulla, H. Smith, see [RS2]). *If G is a group such that for each positive integer n and for every n infinite subsets $X_1, X_2, \ldots, X_n$ of G there exist $x_i \in X_i$, $i = 1, 2, \ldots, n$ such that $\langle x_1, x_2, \ldots, x_n \rangle$ is finite, then G is locally finite.*

There are interesting results for the nilpotent case. Consider the following assumptions (N_k) and more generally (N):

(N_k) Every infinite subset of G contains a set of $k + 1$ elements which generates a nilpotent subgroup of class at most k.

(N) Every infinite subset of G contains a finite subset X such that $\langle X \rangle$ is nilpotent, and such that class $(\langle X \rangle) < |X|$.

The last result that I want to mention is the following:

Theorem 13 (P. Longobardi, M. Maj, A. Mann, A. Rhemtulla, see [LR1]). *A finitely generated group G satisfies* (N) *if and only if it is finite-by-nilpotent, and in that case it satisfies* (N_k), *for some k.*

References

[G]   J. R. J. Groves, A conjecture of Lennox and Wiegold concerning supersoluble groups, J. Austral. Math. Soc. A 35 (1983), 218–228.

[GO]   E. S. Golod, Some problems of Burnside type, Proc. Internat. Congress Math. (Moscow, 1966), Izdat 'Mir', Moscow, 1968, 284–298; Amer. Math. Soc. Transl. Ser. (2) 84 (1969) 83–88.

[K]   K. P. Kim, A condition for locally soluble groups to be metabelian, Houston J. Math. 20 (1994), 193–199.

[KRS]   P. Kim, A. Rhemtulla, and H. Smith, A Characterization of Infinite Metabelian Groups, Houston J. Math. 17 (1991), 429–437.

[LR1]   P. Longobardi, M. Maj, A. Mann, and A. Rhemtulla, Groups with many nilpotent subgroups, Rend. Sem. Mat. Padova, to appear.

[LR2]   P. Longobardi, M. Maj and A. Rhemtulla, Infinite groups in a given variety and Ramsey's theorem, Comm. Alg. 20 (1992), 127–139.

[LW]   J. C. Lennox and J. Wiegold, Extensions of a problem of Paul Erdös on groups, J. Austral. Math. Soc. A 31 (1981), 459–463.

[NE1]   B. H. Neumann, A problem of Paul Erdös on groups, J. Austral. Math. Soc. A 21 (1976), 467–472.

[NE2]   ——, Groups with finite classes of conjugate subgroups, Math. Z. 63 (1955), 76–96.

[PS]   O. Puglisi and L. S. Spiezia, A combinatorial property of certain infinite groups, Comm. Algebra 22 (1994) 1457–1465.

[R]   D. J. S. Robinson, Finiteness conditions and generalized soluble groups, I, II, Springer-Verlag, Berlin 1972.

[RS1]   A. Rhemtulla and H. Smith, On Infinite Solvable Groups, Infinite Groups and Group
        Rings, Conference Proceedings, World Scientific Publishing (1993) 111–121.

[RS2]   ——, On infinite locally finite groups, Canadian Math. Bull., to appear.

[VW]   M. R. Vaughan-Lee and J. Wiegold, Countable locally nilpotent groups of finite ex-
        ponent without maximal subgroups, Bull.London Math. Soc. 13 (1981), 45–46.

On Locally Graded Groups

Yangkok Kim and Akbar H. Rhemtulla

Abstract. Locally Graded groups are those groups in which every finitely generated non-trivial subgroup has a finite non-trivial quotient. This is a wide class of groups that includes all generalized solvable groups and all residually finite groups. Certain results, which are known to hold for solvable or residually finite groups, are extended to this class of groups.

1. Introduction

In this paper we show that a number of results which are known to hold for residually finite groups or for solvable groups can be extended to the much larger class of locally graded groups. A group G is called *locally graded* if every finitely generated non-trivial subgroup of G has a finite non-trivial quotient. This class of groups has frequently appeared in literature; mainly in the study of groups G that do not have any finitely generated infinite simple group as a section of G, or to avoid the so called "Tarski groups". Some of these papers are [4], [1], [18] and [19]. Typically one studies the structure of locally graded groups under some finiteness condition. Here we have chosen to look at a particular weak form of finiteness condition that characterizes polycyclic-by-finite groups and then deduce, as corollaries, the structure of locally graded groups under several other conditions. The main results of the paper are Theorems A, B, C, D and E.

2. Restrained Groups

We call G a *restrained groups* if $\langle x^{\langle y \rangle} \rangle$ is finitely generated for all x, y in G. If there is some bound $b = b(G) > 0$ such that $\langle x^{\langle y \rangle} \rangle$ can be generated by b elements for all x, y in G, then we shall call G a *uniformly restrained* group. The main result of this section is the following.

Theorem A. *A finitely generated group G is a finite extension of a polycyclic group if and only if it is locally graded and uniformly restrained.*

The proof of this theorem appears in [9], but we include it here to make this paper as self contained as possible. The key observation is the following result which is well known.

Lemma 1. *Let H be a normal subgroup of a finitely generated group G. If G/H is cyclic and $\langle a^{(b)} \rangle$ is finitely generated for all a, b in G then H is finitely generated.*

Proof. For some $g \in G$ we can write $G = H\langle g \rangle$. Since G is finitely generated, there exist $h_1, \ldots, h_r$ in H such that $G = \langle h_1, \ldots, h_r, g \rangle$ and $H = \langle h_1, \ldots, h_r \rangle^G$. For each $i = 1, \ldots, r$, $\langle h_i^{(g)} \rangle$ is finitely generated, say, $\langle h_i^{(g)} \rangle = \langle h_{i1}, \ldots, h_{id(i)} \rangle$. Then $H = \langle h_{i\ell(i)}; \ 1 \leq i \leq r, \ 1 \leq \ell(i) \leq d(i) \rangle$. $\square$

Lemma 2. *Let G be a finitely generated restrained group. Then for every positive integer n, the n-th derived subgroup $G^{(n)}$ is finitely generated. In particular if G is solvable then it is polycyclic.*

Proof. This follows directly from Lemma 1. $\square$

Groups constructed by Grigorchuk in [7] and by Gupta and Sidki in [6] are infinite, finitely generated, periodic and residually finite-p groups. They are restrained since every periodic group is restrained. We can thus see that Lemma 2 can not be extended to larger classes such as residually finite groups; and hence the necessity of replacing "restrained" by "uniformly restrained" in Theorem A.

Proof of Theorem A. Since a finite extension of a polycyclic group is residually finite, it is locally graded. Given a polycyclic-by-finite group G, there is a bound on the number of generators required for every subgroup of G; and hence the group is uniformly restrained. Conversely, let G be a finitely generated uniformly restrained group. Then there is a bound n such that $\langle a^{(b)} \rangle$ requires at most n generators for all a, b in G. Let R be the finite residual of G. Then G/R is residually finite and it has no section isomorphic to a twisted wreath product $A \operatorname{twr}_L B$ where A is a finite abelian p-group, B is a cyclic group and L is a subgroup of B of index larger than n. Thus G/R is a minimax group see [20]. It follows from Lemma 2 that G/R is polycyclic-by-finite. Now use Lemma 1 to deduce that R is finitely generated. Since G is locally graded, either $R = 1$ and we are done, or the finite residual S of R is strictly smaller than R. Once again R/S is polycyclic-by-finite and hence G/S is residually finite. Thus $R \leq S$, which is a contradiction. Thus $R = 1$ and G is polycyclic-by-finite. $\square$

3. Engel Groups and Collapsing Groups

We shall apply Theorem A to the classes of Engel groups and collapsing groups by showing that these classes are restrained. Let $[x, _1y] = [x, y]$ and for $i > 0$, $[x, _{i+1}y] = [x, _iy, y]$. Then G is an *Engel groups* if for each pair (x, y) of elements in G, $[x, _iy] = 1$ for some positive integer i. G is an n-Engel group if $[x, _ny] = 1$ for all x, y in G. The notion of collapsing groups was introduced by Semple and Shalev in [16]. A group G is n-collapsing if for any set S of n elements in G, $|S^n| < n^n$ and G is a collapsing group if it is n-collapsing for some $n > 0$.

Lemma 3. *A group* G *is uniformaly restrained if* (i) *it is an* n-*Engel group for some* $n \geq 1$, *or* (ii) *it is* n-*collapsing for some* $n \geq 1$.

Proof. (i) Let x, y be elements in G. Then the subgroup generated by

$$\{x, [x, y] = [x, _1y], \ldots, [x, _{i+1}y] = [x, _iy, y], \ldots, [x, _ry]\}$$

is precisely the subgroup generated by $\{x, x^y, \ldots, x^{y^r}\}$ as is easily seen by inducting on r. Thus if G is an n-Engel group then $\langle x^{\langle y \rangle} \rangle = \langle x, x^y, \ldots, x^{y^{n-1}} \rangle$.

(ii) Let $S = \{xy^{-1}, xy^{-2}, \ldots, xy^{-n}\}$. Then there exist two distinct functions f, g on the set $\{1, 2, \ldots, n\}$ such that

$$\prod_{i=1,\ldots,n} xy^{f(i)} = \prod_{i=1,\ldots,n} xy^{g(i)}.$$

Let r be the largest integer such that $f(r) \neq g(r)$, let $s(i) = f(1) + \cdots + f(i)$ and $t(i) = g(1) + \cdots + g(i)$. Then we get the equality:

$$xx^{y^{-s(1)}} x^{y^{-s(2)}} \cdots x^{y^{-s(r-1)}} y^{s(r)} = xx^{y^{-t(1)}} x^{y^{-t(2)}} \cdots x^{y^{-t(r-1)}} y^{t(r)}.$$

If $s(r) \neq t(r)$, then $y^k \in \langle x^{\langle y \rangle} \rangle$ for some $k > 0$. Letting m be the least positive integer such that $y^m \in \langle x^{\langle y \rangle} \rangle$, we get $\langle x^{\langle y \rangle} \rangle = \langle y^m, x^{y^i}; 0 \leq i < m \rangle$. Note that $m < n^2$. If $s(r) = t(r)$ and $f(r) \neq g(r)$, then $s(r-1) \neq t(r-1)$, say, $s(r-1) < t(r-1)$. Then $x^{y^{s(r-1)}} \in \langle x, x^y, \ldots, x^{y^{s(r-1)-1}} \rangle$ and $\langle x^{\langle y \rangle} \rangle = \langle x^{y^i}; -s(r-1) < i < s(r-1) \rangle$, requiring fewer than $2n^2$ generators. □

Corollary 4. *If* G *is a locally graded* n-*Engel group then* G *is locally nilpotent.*

Proof. It follows from Lemma 1 and Theorem A that G is locally polycyclic-by-finite. But polycyclic-by-finite Engel groups are nilpotent. Hence the result. □

It is interesting to note that if the group G in Corollary 4 is torsion-free then it is nilpotent of class depending only on n, independent of the number of generators of G, as shown by Zel'manov in [21]. In particular a locally indicable n-Engel group is nilpotent. It also follows from Lemma 3 and Theorem A that if G is a locally graded collapsing group then G is locally polycyclic-by-finite. This provides a major reduction in the work involved towards showing that G is locally nilpotent-by-finite (see the proof of the main result of [16], or see the comment following proof of Theorem B in the next section).

4. Groups with No Free Subsemigroups

In this section we investigate properties of groups G which contain no free subsemi-group on two generators. In other words, for every pair (a, b) of elements of G, $S(a, b)$ has a relation of the form

$$a^{r_1} b^{s_1} \ldots a^{r_j} b^{s_j} = b^{m_1} a^{n_1} \ldots b^{m_k} a^{n_k} \tag{1}$$

where r_i, s_i, m_i, n_i are all non-negative and r_1 and m_1 are positive integers. We shall call G a group without free subsemigroups if it has no free non-abelian subsemigroups; thus taking "free" to mean "free non-abelian." Clearly G has no free subsemigroups if and only if no two generator subgroup of G has free subsemigroup. For this reason there is no loss of generality in assuming that G is finitely generated. We shall give a simple proof to show that a finitely generated solvable group with no free subsemigroup is nilpotent-by-finite. The result was originally proved by Rosenblatt in [15]. We show that groups with no free subsemigroups are restrained, apply Theorem A, and then show that polycyclic groups with no free subsemigroups are nilpotent-by-finite. We shall end this section by proving a corresponding result for locally graded groups with no free subsemigroups.

Lemma 5. *If G has no free subsemigroups then for all a, b in G, $\langle a^{\langle b \rangle} \rangle$ is finitely generated.*

Proof. Consider the semigroup $S(b, b^a)$ generated by b and b^a. By hypothesis,

$$b^{r_1} (b^a)^{s_1} \ldots b^{r_j} (b^a)^{s_j} = (b^a)^{m_1} b^{n_1} \ldots (b^a)^{m_k} b^{n_k},$$

where r_i, s_i, m_i, n_i are non-negative integers and r_1 and m_1 are positive. Hence

$$(a^{-1})^{b^{\lambda_1}} a^{b^{\lambda_2}} (a^{-1})^{b^{\lambda_3}} \ldots a^{b^{\lambda_u}} b^{-\lambda_u} = (a^{-1}) a^{b^{\mu_2}} (a^{-1})^{b^{\mu_3}} \ldots a^{b^{\mu_v - 1}} b^{-\mu_v}$$

where $\lambda_u < \cdots < \lambda_1 < 0$ and $\mu_v < \cdots < \mu_2 < 0$. Let $\lambda = \lambda_u$, $\mu = \mu_v$. If $\lambda \neq \mu$ then $b^{\lambda - \mu} \in \langle a^{\langle b \rangle} \rangle$ which is then finitely generated; and we are done. So assume $\lambda = \mu$. Then $a \in \langle a^{b^{-1}}, \ldots, a^{b^\lambda} \rangle$. By replacing b with b^{-1} we similarly get $a \in \langle a^b, \ldots, a^{b^\nu} \rangle$ for some $\nu > 0$. Thus $\langle a^{b^\nu}, \ldots a^b, a, a^{b^{-1}}, \ldots, a^{b^\lambda} \rangle = \langle a^{\langle b \rangle} \rangle$. $\square$

Lemma 6. *Let $G = A \rtimes T$, the split extension of a finitely generated torsion-free abelian group A by infinite cyclic group $T = \langle t \rangle$. If T acts rationally irreducibly on A and G has no free subsemigroups then G is abelian-by-finite.*

Proof. Let $V = A \otimes_Z \mathbb{Q}$. Then V is an irreducible $\mathbb{Q}T$-module and by Schur's Lemma, $D = \text{End}_{\mathbb{Q}T} V$ is a division ring of finite dimension over $\mathbb{Q}$. Now the image of T in $\text{End}_{\mathbb{Q}} V$ lies in D and generates D. Hence D is an algebraic number field. As a D-space, V is one dimensional. Let α be the image of t in D. Then we can identify V with $\mathbb{Q}(\alpha)$ under addition and the action of t on V being that of multiplication by α. If α is a root of 1, then t^n acts trivially on V and hence the subgroup $\langle A, t^n \rangle$ is abelian of finite index in G. If a is not a root of unity, then D can be embedded in $\mathbb{C}$ so that $|\alpha| < 1$ (see [8], p. 102). By taking a power of α, if necessary, we may assume that $|\alpha| < \frac{1}{4}$. Take any $b \neq e$ in A and consider the semigroup $S(t, t^b)$. By hypothesis there exist positive integers $p, q, r_1, \ldots, r_p, s_1, \ldots, s_{p-1}, u_2, \ldots, u_{q-1}, v_1, \ldots, v_q$ and s_p, u_q non negative such that

$$t^{r_1}(t^b)^{s_1} \ldots t^{r_p}(t^b)^{s_p} = (t^b)^{v_1} t^{u_2} \ldots (t^b)^{v_q} t^{u_q}.$$

Note that $\sum_{i=1}^{p} r_i + \sum_{i=1}^{p} s_i = \sum_{i=2}^{q} u_i + \sum_{i=1}^{q} v_i$ since G/A is infinite. If β corresponds to b in the isomorphism of V and $\mathbb{Q}(\alpha)$, then the above equality translates into

$$\beta \sum_{i=1}^{j} \pm \alpha^{\lambda_i} = \beta \sum_{i=1}^{k} \pm \alpha^{\mu_i}$$

where $0 < \lambda_1 < \cdots < \lambda_j$; $0 = \mu_1 < \mu_2 < \cdots < \mu_k$ and j, k are some positive integers. Since $|\alpha| < \frac{1}{4}$, $|\sum_{i=1}^{j} \pm \alpha^{\lambda_i}| \leq \sum_{i=1}^{j} (\frac{1}{4})^{\lambda_i} < \frac{1}{2}$. On the other side, $|\sum_{i=1}^{k} \pm \alpha^{\mu_i}| \geq 1 - \sum_{i=2}^{k} \frac{1}{4}^{\mu_i} > \frac{1}{2}$, giving a contradiction. Thus G is abelian-by-finite. $\square$

Theorem B. *Let G be a finitely generated solvable group. Then G has no free non-abelian subsemigroups if and only if it is nilpotent by finite.*

Proof. Let G be a finitely generated solvable group with no free subsemigroup. We use induction on the solvability length of G to show that G is nilpotent-by-finite. Clearly there is nothing to prove if G is abelian. Hence, using induction, we may assume that G is abelian-by-nilpotent-by-finite. Taking a subgroup of finite index in G, if necessary, we may assume that G is abelian-by-nilpotent. By Lemma 2, we know that G is polycyclic. Thus, again passing to a subgroup of finite index, if necessary, we may assume that G has a finitely generated torsion-free normal abelian subgroup A and G/A is torsion-free nilpotent. Now, there is a central series $A = A_0, A_1, \ldots, A_s = G$ from A to G with infinite cyclic factors. Say $A_i = A_{i-1}\langle t_i \rangle$, $i = 1, \ldots, s$. It suffices to show that $\langle A, t_i^{n_i} \rangle$ is nilpotent for some $n_i > 0$; for then $\langle A, t_1^{n_1}, \ldots, t_s^{n_s} \rangle$ is nilpotent and of finite index in G, as is required to show.

In order to show that $\langle A, t_i^{n_i} \rangle$ is nilpotent for some $n_i > 0$, consider the series $1 = A_{i0} \triangleleft \cdots \triangleleft A_{im(i)} = A$ where A_{ij} are isolated subgroups of A, normalized by t_i and A_{ij+1}/A_{ij} is of minimal rank. Apply Lemma 6 to $\langle A_{ij+1}, t_i \rangle/A_{ij}$ to get

$[A_{ij+1}, t_i^{n_i}] \leq A_{ij}$ for some $n_i > 0$ and all $j = 0, \ldots, m(i)$. Then $\langle A, t_i^{n_i} \rangle$ is nilpotent of class at most $m(i)$. This completes the proof.   □

It is rather easy to see that collapsing groups have no free subsemigroups Thus it follows from the above theorem that polycyclic collapsing groups are nilpotent-by-finite. It is well-known that $S(a, b)$ is not a free subsemigroup if $\langle a, b \rangle$ is a nilpotent group. In [17] Shalev shows that if $\langle a, b \rangle$ is nilpotent of class c then it satisfies the law $u_c = v_c$ where the words $\{u_i\}, \{v_i\}$ on letters a, b are defined as follows: $u_0 = a$, $v_0 = b$, and for $i \geq 0$, $u_{i+1} = u_i v_i$, $v_{i+1} = v_i u_i$. Thus if $G = \langle x, y \rangle$ is a periodic extension of a locally nilpotent group and a, b are elements in G, then $\langle a^n, b^n \rangle$ is nilpotent for some positive integer n, and hence satisfies the law $u_c = v_c$ for some c and $\langle x, y \rangle$ does not have a free subsemigroup. In a recent paper (to appear) A. Yu. Ol'shanskii and A. Storozhev show that a group with no free subsemigroups need not be periodic extensions of locally solvable groups.

Even under additional conditions on a group G with no free subsemigroups, the structure of G can be quite complicated. Let p be a prime, F the free group of rank two and F/R be isomorphic to the Gupta-Sidki p-group constructed in [6]. Then F/R is infinite, residually finite p-group. Thus $G = F/R'$ is residually torsion-free solvable group. It is also residually finite p-group. For all pairs (a, b) of elements in G, there is a relation of type (1) with $j = k = 1$. But G is not nilpotent-by-finite.

If (a, b) is a pair of elements in G satisfying relation of type (1), then we call the sum $r_1 + \cdots + r_j + n_1 + \cdots + n_k$ the **exponent** of a or $\exp(a)$ of the relation.

Recall that a group G is called **locally indicable** if every finitely generated non-trivial subgroup of G has an infinite cyclic quotient. It was shown in [12] that if G is a locally indicable group and there is a bound N such that for all ordered pairs (a, b) of elements of G there is a relation of the form (1) where the $\exp(a)$ is at most N. Then G is locally nilpotent-by-finite. We do not know if this result can be extended to locally graded groups. Here we show the following.

Theorem C. *Suppose G is a locally graded group and there is a bound N such that for all ordered pairs (a, b) of elements of G there is a relation of the form (1) where the sum of $\exp(a)$ and $\exp(b)$ is at most N. Then G is locally nilpotent-by finite.*

Proof. Since G has no free subsemigroup, it is restrained by Lemma 5. The bound on the sum of $\exp(a)$ and $\exp(b)$ ensures that G is uniformly restrained; and since G is locally graded, it follows from Theorem A that it is polycyclic-by-finite. Now use Theorem B to conclude that G is nilpotent-by-finite.   □

One place where the knowledge that G has no free subsemigroup has immediate application is when G is an orderable (O-) group or a right orderable (RO-) group. We refer the reader to [5], [10] or [2] for basic results and terminology. Recall that G is orderable if there exists a total order relation $\geq$ on G such that for all a, b, h, g in

G, $a \geq b$ implies $hag \geq hbg$. Equivalently, if there exists a normal subset P in G such that $PP = P$, $P \cup P^{-1} = G$ and $P \cap P^{-1} = \{e\}$. G is right orderable if there exists a total order relation $\geq$ on G such that for all a, b, g in G, $a \geq b$ implies $ag \geq bg$. Equivalently, if there exists a subset P in G such that $PP = P$, $P \cup P^{-1} = G$ and $P \cap P^{-1} = \{e\}$. It was shown in [12] that if G is orderable and has no free subsemigroup then all the convex subgroups are normal in G under any order on G. If G is right orderable and has no free subsemigroup then under any right order on G the set of convex subgroups form a series with torsion free abelian factors; and, in particular, G is locally indicable.

5. *CF*-Groups and Other Classes of Restrained Groups

A group G is called a **CF-group** if H/H_G is finite for all subgroups H of G. Here H_G stands for the core of H in G - the largest normal subgroup of G contained in H. G is called a **BCF-group** if there is a bound $N = N(G)$ such that H/H_G is finite of order at most N for all subgroups H of G. CF-groups and BCF-groups were introduced and studied in [3], [19] and [11]. The main result in [19] is that locally graded BCF-groups are abelian-by-finite. We show that if for some integer N, H/H_G is an N-generator finite group for all subgroups H of G, then G is abelian-by-finite. It is not known if every locally graded CF-group is abelian-by-finite.

Theorem D. *Let G be a group and N a positive integer. If H/H_G is an N-generator finite group for all subgroups H of G then G is abelian-by-finite.*

Proof. We first show that G is a restrained group. Consider the subgroup $A = \langle a^{\langle b \rangle} \rangle$, for arbitrary elements a, b in G. If b is of finite order, then A is finitely generated and we are done. So assume that $\langle b \rangle$ is infinite. Then $\langle b^m \rangle$ is normal in G by hypothesis. Thus $a^{-1}b^m a = b^{\pm m}$, and $\langle a^{\langle b \rangle} \rangle = \langle a, a^b, \ldots a^{b^m} \rangle$. Next suppose that G is finitely generated and let R be the finite residual of G so that G/R is residually finite. Suppose it has section W isomorphic to a twisted wreath product $A \operatorname{twr}_L B$ where A is a finite abelian p-group, B is a cyclic group and L is a subgroup of B of index larger than $3N$. Recall the definition of twisted wreath product from Neumann [13]. The base group D of $A \operatorname{twr}_L B$ is a direct product of $|B : L|$ copies of A; B acts faithfully on D permuting transitively the copies of A and L acts faithfully on A. Thus W is the split extension of D by B. It was shown in [20] that if G is a finitely generated residually finite group and N a positive integer, then G is virtually a solvable minimax group if it does not have any section isomorphic to $A \operatorname{twr}_L B$ with $|B : L| > N$. We can assume that $A = \langle a^L \rangle$. Let $B = \langle b \rangle$ and $H = \langle a, a^{b^i}, i = 0, \ldots N \rangle$. Then H can not be generated by fewer than N generators and H_W is 1. Hence the pre image J of H in G requires more than N generators and so does J/J_G. This contradicts the hypothesis. Thus G/R is a minimax group. It follows from Corollary 4.4 of [3] that G/R is abelian-by-finite. By Lemma 2, R is finitely generated so that $R = 1$ as can be seen using the argument similar to that in the proof of Theorem A. Now remove the condition that G is finitely generated. Then it follows that G is locally abelian-by-finite. so that every periodic image of G is locally finite. It follows from Theorem 2 of [19] that G is abelian-by-finite.   □

6. Groups with Finitely Many Intersections of Conjugates

Let G be a finitely generated solvable group. It is known (see [14]) that G is polycyclic if and only if it has the following property.

If $H \leq G$, $(x_1, x_2, \dots)$ is a sequence of elements in G, $H_0 = H$ and

$$H_n = H_{n-1} \cap H^{x_n}, \text{ then the series } H = H_0 \geq H_1 \geq \cdots \quad (2)$$

becomes constant after a finite number of steps.

We prove the following generalization of this result for locally graded groups.

Theorem E. *If G is a finitely generated locally graded group and there is a positive integer N such that there are at most N proper inclusions in the series $H = H_0 \geq H_1 \geq \cdots$ of (2), then G is polycyclic-by-finite.*

Proof. For any elements a, b in G, let $H = \langle a^{b^i}, i = 0, \dots N \rangle$. Let $H_0 = H$ and $H_i = H_{i-1} \cap H^{b^i}$. Then for some $r \leq N$, $H_{r-1} = H_r$. Thus $a^{b^{r+N}}$ lies in $\langle a^{b^i}, i = 0, \dots N + r - 1 \rangle$. Hence a^{b^N} lies in $\langle a^{b^i}, i = 0, \dots N - 1 \rangle$. Likewise a lies in $\langle a^{b^i}, i = 1, \dots N \rangle$. This shows that G is uniformly restrained and, by Theorem A, it is polycyclic-by-finite.   □

References

[1]   B. Bruno and R. E. Phillips, On Minimal Conditions Related to Miller–Moreno Type Groups, Rend. Sem. Mat. Univ. Padova 69 (1983), 153–168.

[2]   R. Botto Mura and A. H. Rhemtulla, Orderable Groups, Dekker, 1977.

[3]   J. T. Buckley, J. C. Lennox, B. H. Neumann, H. Smith and J. Wiegold, Groups with all subgroups normal-by-finite, J. Austral. Math. Soc., to appear.

[4]   S. N. Cernikov, Infinite nonabelian groups with an invariance condition for infinite nonabelian subgroups, Dokl. Akad. Nauk SSSR. 194 (1970), 1280–1283.

[5]   L. Fuchs, Partially Ordered Algebraic Systems, Pergamon Press 1963.

[6]   N. D. Gupta and S. Sidki, Some infinite p-groups, Algebra i Logika 22 (5) (1983), 584–589.

[7] R. I. Grigorchuk, On the growth degrees of p-groups and torsion-free groups, Math. Sbornik 126 (1985), 194–214; English transl. in Math. USSR Sbornik 54, (1986), 347–352.

[8] E. Hecke, Lectures on the Theory of Algebraic Numbers, translated by George U. Brauer and Jay R. Goldman, Springer-Verlag, 1981.

[9] Y. K. Kim and A. H. Rhemtulla, Weak maximality condition and polycyclic groups, Proc. Amer. Math. Soc., to appear.

[10] A. I. Kokorin and V. M. Kopytov, Fully Ordered Groups, John Wiley & Sons, 1974.

[11] J. C. Lennox, P. Longobardi, M. Maj, H. Smith and J. Wiegold, Some finiteness conditions concerning intersections of conjugates of subgroups, to appear.

[12] P. Longobardi, M. Maj and A. H. Rhemtulla, Groups with no free subsemigroups, Trans. Amer. Math. Soc., to appear.

[13] B. H. Neumann, Twisted Wreath Products of Groups, Arch. Math. 14 (1963), 1–6.

[14] A. H. Rhemtulla, A Minimality Property of Polycyclic Groups, J. London Math. Soc. 42 (1967), 456–462.

[15] J. M. Rosenblatt, Invariant measures and growth conditions, Trans. Amer. Math. Soc. 197 (1974), 33–53.

[16] J. F. Semple and A. Shalev, Combinatorial conditions in residually finite groups I, J. Algebra 157 (1993), 43–50.

[17] A. Shalev, Combinatorial conditions in residually finite groups II, J. Algebra 157 (1993), 51–62.

[18] H. Smith, On homomorphic images of locally graded groups, Rend. Sem. Mat. Univ. Padova, to appear.

[19] H. Smith and J. Wiegold, Locally Graded Groups with all Subgroups Normal-by-Finite, preprint.

[20] J. S. Wilson, Two-generator conditions for residually finite groups, Bull. London Math. Soc. 23 (1991), 239–248.

[21] E. I. Zelmanov, On some problems of group theory and lie algebras, Math. USSR Sbornik 66 (1) (1990), 159–168.

Groups with Ordered Structures

Yangkok Kim* and Akbar H. Rhemtulla

Abstract. Many well known group theoretic properties which give nice structures to soluble groups or residually finite groups also give similar nice structures to groups with full orders. Recent works on groups with ordered structures will be surveyed in this note, raising new problems.

1. Groups with Full Orders

For orderable and right orderable groups, many well known group theoretic properties pose interesting questions regarding the structure of such groups. For orderable groups, the challenge in some cases is to decide whether relatively convex subgroups are normal; in other cases normality of relatively convex subgroups is easily established reducing the problem to residually torsion-free nilpotent groups. For right orderable groups the challenge in some cases is to figure out if the group admits a right order under which the convex subgroups form a system with torsion-free abelian factors. Such a right order is called a C-order. In other cases it is easy to establish that all right orders are C-orders and the problem then becomes one in locally indicable groups. We look at several finiteness conditions in this paper. A group G is said to be **orderable** (or an O-group) if there exists a full order relation $\leq$ on the set G such that $a \leq b$ implies $xay \leq xby$ for all a, b, x, y in G. The term "ordered group" is used to denote a group endowed with a fixed order. We shall denote the identity element of G by e. A subgroup C of an ordered group G is said to be **convex** under $\leq$ if $x \in C$ whenever $e \leq x \leq c$ for some $c \in C$. An O-group G was characterized in terms of the system of its subgroups. So convex subgroups of an ordered group G play important roles in studying the whole group G. For example, if the convex subgroups of G are all normal, then G' has a central system with torsion-free factors.

A group G is called **right-orderable** (or an RO-group) if there exists a full order relation $\leq$ on the set G such that $a \leq b$ implies $ac \leq bc$ for all a, b, c in G. If $a \geq e$, a is called positive. A group G is called **locally indicable** if every finitely generated non-trivial subgroup of G has an infinite cyclic quotient. For a right orderable group

*   The first author was partly supported by KOSEF postdoctoral fellowship.

G with a fixed order, just as with orderable groups, we define a subgroup N of G to be **convex** if for every $g \in G$ and $c \in N$, $e \leq g \leq c$ implies $g \in N$. Then the set of all convex subgroups of orderable or right-orderable group forms a chain. For two convex subgroups $C \subsetneq D$ which have no proper convex subgroup between them, we call $C \to D$ a convex jump in G. For a convex jump $C \to D$, if D/C is order-isomorphic with a subgroup of additive group of real numbers, we say that this convex jump is of c-type. Clearly every element $g (\neq e)$ of G determines a jump at g, taking two convex subgroups, the largest convex subgroup containing g, the smallest convex subgroup not containing g. A group G is a **C-group** if it admits a C-order which is a right-order satisfying the following property:

If $C \to D$ is a convex jump in G, then $C \leq D$ and D/C is isomorphic to a subgroup of the additive group of the real numbers.

Let C^* denote the class of those right orderable groups in which every right-order has the above property. In [C], P. Conrad showed that the above property is equivalent to the following one:

For each pair of positive elements a, b in G there exists a positive integer n such that $a^n b > a$.

Clearly the class of orderable groups is properly contained in the class of C-groups. Every C-group is locally indicable. And locally indicable groups are RO-groups. For some time it was an open question whether the classes of right orderable groups and locally indicable groups coincide. However, in [B1], G. Bergman gave examples of RO-groups which are not locally indicable: one such group is the Seifert fibered 3-manifold group G with presentation

$$G = \langle a, b, c \, ; \, a^2 = b^3 = c^7 = abc \rangle.$$

Problem 1. Does every locally indicable group have a series from 1 to G with torsion-free abelian factors?

Now we list some sufficient conditions for RO-groups to have a C-order. The following theorem was proved in [CK]. This generalises a result of Rhemtulla [RH2] that a polycyclic group G admits a right-order if and only if it has a series with torsion-free abelian factors.

Theorem 1.1. *Let G be a soluble-by-finite group. Then G is right orderable if and only if G is locally indicable.*

Definition 1. A non-empty subset X of a group G is called **G-paradoxical** if and only if it can be expressed as a finite disjoint union of non-empty subsets, $X = A_1 \cup A_2 \cup \cdots \cup A_m \cup B_1 \cup \cdots \cup B_n (m, n \geq 1)$, in such a way that there are group elements $g_1, \ldots, g_m, h_1, \ldots, h_n$ so that $X = A_1 g_1 \cup \cdots \cup A_m g_m$ and $X = B_1 h_1 \cup \cdots \cup B_n h_n$ are disjoint unions. A group G is called **superamenable** if and only if it has no paradoxical subsets.

In [K2], Kropholler showed that if for a right-ordered group $(G, \leq)$ and $g (\neq e) \in G$, the half-open interval $[e, g)$ is not G-paradoxical, then the convex jump at g is of c-type. The following is an easy consequence of this result.

Theorem 1.2. *Every right ordered supramenable group is c-ordered and hence locally indicable.*

Note that every exponentially bounded group is superamenable. Since finitely generated nilpotent groups have polynomial growth, every locally nilpotent-by-finite group is superamenable. Thus we at once obtain that if G is a locally nilpotent-by-finite group, or more generally a exponentially bounded group then every right order on G is a c-order. The following result was shown in [LMR]. Since superamenable groups contain no free semigroups ([W1], page 189), this result extends the above result. We shall mention groups with no free subsemigroups in section 3.

Theorem 1.3. *Let $(G, \geq)$ be a right ordered group and let $P = \{g \in G; g > e\}$. If P has no free subsemigroup then G is a C-group and hence G is locally indicable.*

2. Finiteness Conditions

Engel Condition

Let n be a positive integer. A group G is called an ***n-Engel*** group if $[x, {}_n y] = 1$ for all x, y in G, where $[x, {}_n y]$ is defined inductively as follows: $[x, {}_1 y] = [x, y]$ and $[x, {}_i y] = [[x, {}_{i-1} y], y]$ for $i > 1$. G is a ***bounded Engel*** group if it is an n-Engel group for some n. For basic properties of this class of groups see [RO].

Lemma 2.1. *Let C be a convex subgroup of an ordered group $(G, \leq)$. If G is an Engel group, then C is normal.*

Proof. Suppose not. Then $C \lneq C^x$ or $C^x \lneq C$ for some x in G because the set of convex subgroups forms a chain. Let $C \lneq C^x$ and $a^x \in C^x \backslash C$ where a is in C. Since G is an Engel group, there is an integer $n > 0$ such that $[a, {}_n x] = 1$. Hence $\langle a, a^x, a^{x^2}, \ldots \rangle = \langle a, a^x, \ldots, a^{x^{n-1}} \rangle$. Thus $a^{x^n} \in C^{x^{n-1}}$, and it follows that $a^x \in C$. This is a contradiction. $\qquad\square$

Lemma 2.2. *Let $(G, \leq)$ be an ordered group in which every convex subgroup is normal. If D is a convex subgroup of G and D is finitely generated as a G-group, then there exists a convex subgroup C such that $C \to D$ is a jump.*

Proof. By hypothesis, $D = \langle X^G \rangle$ for some finite subset $X = \{x_1, \ldots, x_n\}$ of G. Suppose $e \leq x_1 \leq \cdots \leq x_n$. Let C be the union of all convex subgroups of G that do not contain x_n. Then $C \to D$ is a jump since D is the smallest convex subgroup that contains x_n. $\square$

Theorem 2.3. *A bounded Engel O-group G is nilpotent.*

Proof. Let G be finitely generated and let $G = G_0$. Then there exists a convex subgroup G_1 such that $G_1 \to G_0$ forms a jump since G_0 is finitely generated. Here G/G_1 is a finitely generated abelian group and so finitely presented. By a theorem of P. Hall G_1 is finitely generated as a G-group. Moreover all convex subgroups are normal in G by Lemma 2.1. Hence the convex subgroup G_2, where $G_2 \to G_1$ forms a jump, exists by Lemma 2.2. Now G/G_2 is nilpotent since it is a finitely generated soluble Engel group. Since finitely generated nilpotent groups are finitely presented, G_2 is finitely generated as a G-group and hence the convex subgroup G_3, where $G_3 \to G_2$ is a jump, also exists. This produces a descending central series $G = G_0 > G_1 > \cdots$, with torsion-free factors. Let N be the intersection of the G_i's so that G/N is a residually finite p-group for all primes p. Pick any odd prime p. Since G is an n-Engel group, it can not have any section isomorphic to the wreath product of a cyclic group of prime order p and a cyclic group of order greater than p^n. Thus it follows that G/N is linear. Now since nonabelian free groups are not Engel groups, it follows by Tits' Alternative [T1] that G/N is soluble-by-finite and so nilpotent. Thus $N = G_m$ for some m and hence $N = 1$ and G is nilpotent. Since G is torsion-free, the n-Engel condition implies that G is nilpotent of class depending only on n, independent of the number of generators of G, as shown by Zel'manov in [Z]. Thus an n-Engel orderable group is nilpotent of class $c = c(n)$. $\square$

Collapsing Condition

A group G is said to be ***n-collapsing*** if for any n-element subset S of G, we have $|S^n| < n^n$. We say that G is ***collapsing*** if it is n-collapsing for some n. Note that G is n-collapsing if and only if for every $x_1, \ldots, x_n$ in G, we have $x_{f(1)} \cdots x_{f(n)} = x_{g(1)} \cdots x_{g(n)}$ for some distinct functions $f, g : \{1, \ldots, n\} \to \{1, \ldots, n\}$. In [S1], J. Semple and A. Shalev introduced this condition. The following was proved in [S2].

Theorem 2.4. *Let G be a finitely generated residually finite group. Then G is collapsing if and only if G is nilpotent-by-finite.*

Let G be a collapsing O-group and C a convex subgroup of G. Suppose that $y^x \in C^x \backslash C$, for some $x, y \in G$. Then by considering a subset $\{yx^{-1}, \ldots, yx^{-n}\}$ and two distinct functions on $\{1, \ldots, n\}$, we get easy contradiction. So all convex subgroups of collapsing orderable groups are normal. With this property and Theorem 2.4, we get the following theorem. The proof is similar to the one of Theorem 2.3.

Theorem 2.5. *A group G is torsion-free nilpotent if and only if it is orderable and collapsing.*

Finite Rank and Finite Base

Recall that a group G is said to have *finite rank* if there is a positive integer d such that every finitely generated subgroup of G can be generated by d elements. The structure of soluble groups of finite rank is reasonably well known (See [RO]). More recently, A. Lubotzky and A. Mann have shown in [LM2] that *a residually finite group of finite rank has a locally soluble subgroup of finite index*. The structure of solvable O-groups of finite rank is known (see [B2]; section 3.3). The relatively convex subgroups of such a group are precisely the isolated normal subgroups. Is this true for O-groups of finite rank in general? This is equivalent to asking if the relatively convex subgroups of an O-group of finite rank are all normal.

A property closely related to that of finiteness of rank for O-groups is that of finiteness of base. If $H \leq K$ are subgroups of G, then we define the isolator $I_K(H)$ of H in K as the union of subgroups H_i, $i = 1, 2, \ldots$, where $H_0 = H$ and $H_i = \langle x \in K ; x^m \in H_{i-1}$ for some $m > 0 \rangle$. Thus $I_K(H)$ is the smallest isolated subgroup of K that contains H.

Definition 2. A group G has *finite base n* if $H \leq G$ implies $I_G(H) = I_G(X)$ for some n-generator subgroup $X \leq H$.

It is easy to show that if G is an O-group with finite base, then the relatively convex subgroups are all normal and G has finite rank. Thus the question whether O-groups of finite rank are solvable is equivalent to the question whether every relatively convex subgroup of such a group is the isolator of a finitely generated subgroup in G. The following result is one of the instances that the finite rank condition gives a nice structure to certain orderable groups.

Theorem 2.6. *Let G be an ordered group of finite rank. If the convex subgroups of G are all normal then G is nilpotent-by-abelian.*

Problem 2. Let $\leq$ be a total order on a group G of finite rank. Is it true that every convex subgroup of G is the isolator of a finitely generated subgroup?

WPSP Condition

Let G be a group. If there exists an integer $n > 1$ such that for each n-tuple $(H_1, \ldots, H_n)$ of subgroups of G, there are two distinct permutations σ and τ in Σ_n such that the two complexes $H_{\sigma(1)} H_{\sigma(2)} \cdots H_{\sigma(n)}$ and $H_{\tau(1)} H_{\tau(2)} \cdots H_{\tau(n)}$ are equal, then G is said to have the property of *weak permutable subgroup products*, or G is a *WPSP-group*. In particular, if we fix σ as an identity permutation, then we get a *PSP*-group which was discussed in [RW].

By applying similar arguments which were for *PSP*-groups in [RW], we get the following property.

Theorem 2.7. *A finitely generated soluble group G is a WPSP-group if and only if it is finite-by-abelian.*

Just like other conditions which we already mentioned, we can show the following lemma by choosing suitable subgroups of G.

Lemma 2.8. *Let C be a convex subgroup of an ordered group $(G, \leq)$. If G is a WPSP-group, then C is normal.*

Theorem 2.9. *An orderable group G is a WPSP-group if and only if it is abelian.*

Proof. Let $G = \langle x_1, \ldots, x_n \rangle$, where $e < x_1 < \cdots < x_n$. Then we have a convex jump $G \to G_1$ at x_n. Here G/G_1 is a finitely generated abelian group and so finitely presented. By a theorem of P. Hall $G_1 = \langle y_1, \ldots, y_m \rangle^G$, where $e < y_1 < \cdots < y_m$. Moreover all convex subgroups are normal in G by Lemma 2.8. y_m determines a convex jump $G_1 \to G_2$. Now G/G_2 is abelian since it is a finitely generated torsion-free soluble *WPSP*-group. Since finitely generated abelian groups are finitely presented, the convex subgroup G_3, where $G_3 \to G_2$ is a jump, also exists. This produces a descending central series $G = G_0 > G_1 > \cdots$, which should stop after finitely many steps. The converse is clear.   □

Finite Width

We say that a group G has *finite width* if for every pair (x, y) of elements in G, there is some integer $n = n(x, y)$ such that $\langle x, y \rangle = (\langle x \rangle \langle y \rangle)^n$, where $(\langle x \rangle \langle y \rangle)^n$ denotes the set $\{x^{r_1} y^{s_1} \cdots x^{r_n} y^{s_n}; r_i, s_j \in \mathbb{Z}\}$. A finitely generated soluble group of finite width is an extension of a finite group by a nilpotent group ([RW1]). The same is true for finitely generated residually finite p-groups of finite width ([RH4]).

Theorem 2.10. *Let G be an ordered group of finite width. If the convex subgroups are all normal then G is torsion-free locally nilpotent.*

Problem 3. Is every relatively convex subgroup of an orderable group G of finite width normal in G?

Weak Maximal Condition

Recall that a group G has Max or maximality condition for subgroups if every subset of the set of subgroups of G has a maximal element. This is equivalent to saying that every subgroup of G is finitely generated. One major question for this class of groups is the following: If G is an O-group with Max is it solvable? It is easy to show that every relatively convex subgroup of G is normal so that the commutator subgroup G' of G has a descending central series with torsion-free factors. One may thus assume that G is residually torsion-free nilpotent with Max. Unfortunately there are no tools so far that can deal with such groups.

Recall that G is called a **restrained group** if $\langle x^{\langle y \rangle} \rangle$ is finitely generated for all x, y in G. G is called n-restrained if $\langle x^{\langle y \rangle} \rangle$ can be generated by n elements for all x, y and G is called **uniformly restrained** if it is n-restrained for some n.

Clearly the class of restrained groups is a larger class than groups with Max. It also includes all Engel groups. Bounded Engel groups and n-collapsing groups are uniformly restrained.

It is known that for a finitely generated restrained group G if H is a normal subgroup of G such that G/H is cyclic, then H is finitely generated. By using this property, we have that for a finitely generated uniformly restrained group G, if it is a locally graded group, then G is polycyclic-by-finite.

If G is an RO-group with at least one C-order, then every finitely generated non-trivial subgroup of G has a non-trivial cyclic quotient. Thus we get the following. Let G be an RO-group with at least one C-order. If G is uniformly restrained, then it is locally polycyclic.

Note that polycyclic Engel groups are nilpotent. Hence a locally graded group G is locally nilpotent, when it is an n-Engel group for some n. Notice that if G is torsion-free, then it is nilpotent of class depending only on n, independent of the number of generators of G, as shown by Zel'manov in [Z]. In particular, an RO-group that has at least one C-order is nilpotent if it is a bounded Engel group.

The following questions remain open in this area.

Problem 4. If G is an n-Engel RO-group, does it have a C-order?

Problem 5. If G is a uniformly restrained RO-group, does it have a C-order?

Lemma 2.11. *If G is a collapsing right-orderable group then every right order on G is a C-order.*

Proof. Let P be the positive cone of a given right-order on G and a, b in P. Suppose, if possible, that $a^m b < a$ for all integer $m > 0$. Consider the set $S = \{ba, ba^2, \ldots, ba^n\}$ where n is such that $|S^n| < n^n$. Since G is n-collapsing, there exist two distinct functions f, g on the set $\{1, 2, \ldots, n\}$ such that

$$\prod_{i=1,\ldots,n} ba^{f(i)} = \prod_{i=1,\ldots,n} ba^{g(i)}.$$

Hence for some $0 < r \le n$ we have $ba^{f(1)} ba^{f(2)} \cdots ba^{f(r)} = ba^{g(1)} ba^{g(2)} \cdots ba^{g(r)}$ and $f(r) \ne g(r)$. Say $f(r) < g(r)$ and let $s = g(r) - f(r)$. Then we have $ba^{f(1)} \cdots b = ba^{g(1)} \cdots ba^s$. Now $a^m b < a$ for all $m > 0$ implies $ba^{f(1)} ba^{f(2)} \cdots b < a^{f(1)+1} ba^{f(2)} \cdots b < a^{f(2)+1} b \cdots b < \cdots < a$. On the other hand, $ba^{g(1)} \cdots b \ge e$ so that $ba^{g(1)} \cdots ba^s \ge a^s \ge a$, giving the required contradiction. $\qquad \square$

Theorem 2.12. *A finitely generated collapsing RO-group is nilpotent-by-finite.*

Proof. Let G be a finitely generated collapsing RO-group. By Lemma 2.11 it has a C-order so that it is locally indicable. Since G is uniformly restrained, G is polycyclic-by-finite. By Theorem in [S1], G is nilpotent-by-finite. $\qquad \square$

3. Groups with No Non-Abelian Free Subsemigroups

In this section we look at groups G which contain no non-abelian free subsemigroup. In other words, for every pair (a, b) of elements of G, the subsemigroup $S(a, b)$ generated by a and b has a relation of the form

$$a^{r_1} b^{s_1} \cdots a^{r_j} b^{s_j} = b^{m_1} a^{n_1} \cdots b^{m_k} a^{n_k} \tag{1}$$

where r_i, s_i, m_i, n_i are all non-negative and r_1 and m_1 are positive integers. We shall call G a group with no free subsemigroup if it has no free non-abelian subsemigroup; thus taking "free" to mean "free non-abelian." We shall call $j + k$ the width of relation (1) and the sum $r_1 + \cdots + r_j + n_1 + \cdots + n_k$ the exponent of a in the relation.

Lemma 3.1. *Let G be a group with no free subsemigroups. Then G is restrained. Moreover,*
(i)   *if G is an O-group, then relatively convex subgroups are normal;*
(ii) *if G is an RO-group, then every right order is a C-order.*

Proof. We need to show that $\langle x^{\langle y \rangle} \rangle$ is finitely generated. Substitute y for a and xyx^{-1} for b in (1) to obtain the equality

$$y^{r_1} xy^{s_1} x^{-1} \cdots y^{r_j} xy^{s_j} x^{-1} = xy^{m_1} x^{-1} y^{n_1} \cdots xy^{m_k} x^{-1} y^{n_k}.$$

If some positive power of y lies in $\langle x^{\langle y \rangle} \rangle$, then $\langle x^{\langle y \rangle} \rangle$ is of finite index in $\langle x, y \rangle$ and hence it is finitely generated. So assume otherwise. Then $r_1 + s_1 + \cdots + r_j + s_j = m_1 + n_1 + \cdots + m_k + n_k$ and x lies in the group generated by $\{x^{y^i} \; ; \; i > 0\}$. Similarly, replacing y by its inverse we get x to lie in the group generated by $\{x^{y^i} \; ; \; i < 0\}$. This then shows that $\langle x^{\langle y \rangle} \rangle$ is finitely generated. If $\langle x^{\langle y \rangle} \rangle$ is finitely generated, then x and x^y must be comparable in an O-group. This proves (i). Finally if $\leq$ is a right order on G and a, b are positive elements in G, then equation (1) shows that $a^n b \geq a$ for some n. $\qquad\qquad\qquad\square$

Theorem 3.2. *Suppose G is a group and there is a bound N such that for all pairs (a, b) of elements in G there is a relation of the form (1) whose width is at most N. Then G is nilpotent if it is residually torsion-free nilpotent.*

Proof. Since G is residually torsion-free nilpotent, there exist a descending central series $G = G_0 > G_1 > \cdots$ where $\bigcap_{i=1}^{\infty} G_i = 1$ and G/G_i is torsion-free nilpotent for all i. Now note that for a torsion-free nilpotent group H, if there is a bound M such that for all pairs (a, b) of elements in H there is a relation of the form (1) whose width is at most M, then H is nilpotent whose class is bounded by a function of M (see [LMR]). Thus there is some integer $f(N)$ such that $\gamma_{f(N)+1} \subseteq G_i$ for all i and hence $\gamma_{f(N)+1}(G) = 1$. $\qquad\qquad\qquad\square$

It is known that if G is a finitely generated soluble group with no free subsemigroup, then G is nilpotent-by-finite. Conversely, nilpotent-by-finite groups have no free subsemigroups. Thus a locally solvable O-group has no free subsemigroup if and only if it is torsion free locally nilpotent. In general, we do not have an answer for the following problem.

Problem 6. Is a finitely generated O-group with no free subsemigroups nilpotent?

However, we know that a certain restriction on the width of expression (1) gives us a positive solution as we see in the following theorem (see [LMR]).

Theorem 3.3. *Suppose G is an O-group and there is a bound N such that for all pairs (a, b) of elements in G, there is a relation of the form (1) whose width is at most N. Then G is nilpotent of class bounded by a function of N.*

Proof. Let $\geq$ be a total order on G. Let $C \to D$ be a jump in the set of convex subgroups of G under $\geq$. Then the centralizer $C_G(D/C)$ of D/C in G contains G' and is an isolated subgroup of G. Let J be the isolator of G' in G. Then $[D, J] \subseteq C$. Suppose that G is finitely generated. Then J is finitely generated. Now order J by taking the restriction of order $\geq$. Then the convex subgroups of J are $C \cap J$, where C is convex in G under $\geq$. This order on J is a G-order in that the positive cone is invariant under conjugation by elements of G. We use this order on J and extend it to an order on G by making $J \to G$ a convex jump. Since J is finitely generated, there is $J_1 \leq G$ such that $J_1 \to J$. Similarly J_1 is finitely generated and there is a jump $J_2 \to J_1$. Continue this process and let $K = \bigcap_{i=1}^{\infty} J_i$. Then J/K is residually torsion-free nilpotent-by-finite. Since a nilpotent-by-finite O-group is nilpotent. Hence G/K is nilpotent and the nilpotency class is bounded by a function N. Thus $K = J_m$ for some m and $J_m = 1$.   □

The structure of an RO-group G can be quite complicated even when the group has no free subsemigroup of bounded width. Let p be a prime, F the free group of rank two and F/R be isomorphic to the Gupta–Sidki p-group constructed in [GS]. Then F/R is infinite, residually finite p-group. Thus $G = F/R'$ is residually torsion-free solvable group (hence an RO-group). For all pairs (a, b) of elements in G, there is a relation of type (1) with $j = k = 1$. But G is not even soluble. As we however see in the following theorems, the restriction on the exponent on a in the expression (1) is on our side.

Theorem 3.4. *Suppose that G is a locally indicable group and there is a bound N such that for all ordered pairs (a, b) of elements of G there is a relation of the form (1) where the exponent of a is at most N. Then G is locally nilpotent-by-finite.*

Proof. We assume that G is finitely generated and show that it is nilpotent-by-finite. Let R be the torsion-free solvable residual of G. Then $R = \bigcap_{i=1}^{\infty} G_i$ where G/G_i is torsion-free solvable. Here the exponent of $G/\operatorname{Fitt}(G)$ and the nilpotency class of $\operatorname{Fitt}(G)$ are bounded by a function of N (for detail, see [LMR]). Thus $G_i G^m / G_i$ is nilpotent of class n with n and m depending only on N. Thus G/R is solvable and so R is finitely generated. If $R \neq 1$ then, by hypothesis, R/R' is infinite so that G/J is torsion-free solvable where J is the isolator of R' in R, and $J \neq R$ which is a contradiction.   □

Theorem 3.5. *If G is a finitely generated RO-group and there is a bound N such that for all ordered pairs (a, b) of elements in G, there is a relation of the form (1) where the exponent of a is at most N, then G is nilpotent-by-finite.*

Proof. This follows from Lemma 3.1 and Theorem 3.4.   □

Now let G be a finitely generated O-group with no free subsemigroup. Let $\leq$ be a total order on G. Then the convex subgroups are normal and hence all jumps are centralized by G' which is also finitely generated; all by Lemmas 2.2 and 2.11. Also, if $C \to D$ is a jump and D is finitely generated, then so is C so that there exists a subgroup B such that $B \to C$ is a jump. Thus if $G_0 = G$, then for all positive integers i, there exist subgroups G_i such that $G_i \to G_{i-1}$ is a jump. If G is nilpotent, then $G_n = 1$ for some n. If G is not nilpotent, then let $N = \bigcap G_i$, $i = 1, 2, \ldots$.

Then G/N is a finitely generated residually torsion-free nilpotent group with no free subsemigroup. Thus Problem 6 reduces to the following.

Problem 7. Let G be a finitely generated, residually torsion-free nilpotent group with no free subsemigroup. Is G nilpotent?

References

[B1] G. M. Bergman, Right orderable groups that are not locally indicable, Pacific J. Math. 147 (1991), 243–248.

[B2] R. Botto Mura and A. H. Rhemtulla, Orderable Groups, Marcel Dekker, 1977.

[B3] R. Botto Mura and A. H. Rhemtulla, Ordered solvable groups satisfying the maximal condition on isolated subgroups and groups with finitely many relatively convex subgroups, J. Algebra 36 (1975), 38–45.

[CK] I. M. Chiswell and P. H. Kropholler, Soluble right orderable groups are locally indicable, Canad. Math. Bull. 36 (1993), 22–29.

[C] P. F. Conrad, Right ordered groups, Michigan Math. J. 6 (1959), 267–275.

[D] M. R. Darnel, A. M. W. Glass and A. H. Rhemtulla, Groups in which every right order is two sided, Arch. Math. 53 (1989), 538–542.

[E] J.H. Evertse, On sums of S-units and linear recurrences, Compositio Math. 53 (1984), 224–244.

[F1] L. Fuchs, On orderable groups, in: Proc. Internat. Conf. Theory of Groups, Austral. Nat. Univ. Canberra, August 1965, Gordon and Breach Science Publishers Inc., 1967, 89–98.

[F2] L. Fuchs, Partially Ordered Algebraic Systems, Pergamon Press, 1963.

[G] A. M. W. Glass, Automorphisms of the ordered multiplicative group of positive rational numbers, preprint.

[GS] N. Gupta and S. Sidki, Some infinite p-groups, Algebra i Logika 22(5) (1983), 584–589.

[HR] R. J. Hursey Jr. and A. H. Rhemtulla, Ordered groups satisfying the maximal condition locally, Canad. J. Math. 22 (1970), 753–758.

[KK] A. I. Kokorin and V. M. Kopytov, Fully Ordered Groups, John Wiley & Sons, 1974.

[K1] P. H. Kropholler, On finitely generated soluble groups with no large wreath product sections, Proc. London Math. Soc. (3) 49 (1984), 155–169.

[K2] P. H. Kropholler, Amennbility and right orderable groups, Bull. London Math. Soc. 25 (1993), 347–352.

[KR] Y. K. Kim and A. H. Rhemtulla, Weak maximality condition and polycyclic groups, Proc. Amer. Math. Soc., to appear.

[LMR]   P. Longobardi, M. Maj and A. H. Rhemtulla Groups with no free subsemigroups, to appear.

[L]   A. Lubotzky, A group theoretic characterization of linear groups, J. Algebra 113 (1988), 207–214.

[LM1]   A. Lubotzky and A. Mann, Powerful p-groups I and II, J. Algebra 105 (1987), 484–515.

[LM2]   A. Lubotzky and A. Mann, Residually Finite Groups of Finite Rank, Proc. Cambridge Phil. Soc. 106 (1989), 385–388.

[RH1]   A. H. Rhemtulla, Right-ordered groups, Canad. J. Math. 24 (1972), 891–895.

[RH2]   A. H. Rhemtulla, Polycyclic right ordered groups, in: Algebra Carbondale 1980 (R. K. Amayo, ed.), Lecture Notes in Math. 848, Springer Berlin, 1981, 230–234.

[RH3]   A. H. Rhemtulla Periodic extensions of ordered groups, in: Ordered Algebraic Structures, Kluwer Academic Publishers, 1989, 65–69.

[RH4]   A. H. Rhemrulla Groups with many elliptic subgroups, in: Proceedings of Groups-Korea 1988, Lecture Notes in Math. 1398, Spinger-Verlag, 1988 156–162.

[RW]   A. H. Rhemtulla and A. R. Weiss, Groups with permutable subgroup products, in: Group Theory: Proceedings of the 1987 Singapore conference (K. N. Cheng and Y. K. Leong, eds.), Walter de Gruyter, Berlin, 1989, 485–495.

[RW1]   A. H. Rhemtulla and J. S. Wilson On elliptically embedded subgroups of soluble groups, Canad. J. Math. 39 (1987), 956–968.

[RO]   D. J. S. Robinson, Finiteness Conditions and Generalized Soluble Groups, Part I and II, Springer-Verlag, New York, 1972.

[S1]   J .F. Semple and A. Shalev, Combinatorial conditions in residually finite groups I, J. Algebra 157 (1993), 43–50.

[S2]   A. Shalev, Combinatorial conditions in residually finite groups II, J. Algebra 157 (1993), 51–62.

[T1]   J. Tits, Free subgroups in linear groups, J. Algebra 20 (1972), 250–270.

[W1]   S. Wagon, The Banach–Tarski paradox, Cambridge University Press, 1985.

[W2]   J. S. Wilson, Two-generator conditions for residually finite groups, Bull. London Math. Soc. 23 (1991), 239–248.

[Z]   E. I. Zelmanov, On some problems of group theory and Lie algebras, Math. USSR Sbornik 66(1) (1990), 159–168.

Finite Groups with Trivial
Multiplicator and Large Deficiency

L. G. Kovács

Abstract. Generalizing examples of Swan and Wiegold, this note shows how to construct more finite groups (including some perfect groups) whose Schur multiplicator is trivial but whose abelianized deficiency is arbitrarily large.

1991 Mathematics Subject Classification: 20F05

1. Introduction

Given a finite presentation for a group in terms of generators and defining relations, one obtains the deficiency of that presentation by subtracting the number of generators from the number of relations. The deficiency of a finitely presentable group G is the minimum of the deficiencies of its (finite) presentations; we write it as def G. If G is finite, then def $G \geqslant 0$.

B. H. Neumann [11] asked whether the deficiency of a finite group has to be zero whenever the Schur multiplicator of the group is trivial. The first examples to show that the answer is negative were made by Swan [12]: he gave an infinite set of finite groups whose multiplicators are trivial but whose deficiencies admit no upper bound. Later Wiegold [15] produced a different construction to the same end (and, as was reported in [15], Neumann immediately added a slight modification to reduce the number of generators).

Our aim here is to generalize both constructions. This will yield new examples which can be tailored to various purposes. In particular, it will be seen that there exist finite perfect groups with trivial multiplicator but arbitrarily large deficiency. As is the nature of generalizations, the process will direct our attention to various features of the two constructions, separating similarities from differences.

The author is indebted to the organizers of Groups—Korea 1994 for the financial support which enabled his attendance and for their warm hospitality. Work on this paper started during that conference, and its direction was influenced by continuing correspondence with C. I. Wotherspoon.

When a finite group G is written as F/R with F free of finite rank, conjugation in F yields a G-module structure on R/R'. The number obtained by subtracting the rank of F from the minimum of the cardinalities of the G-module generating sets of R/R' has been called the abelianized deficiency of G (p. 149 in Wiegold [15]). In symbols, we write this as

$$\text{abdef } G = d_G(R/R') - d(F).$$

(It was proved by Gruenberg in [7], and again as part (ii) of Corollary 7.9 in [8], that $d_G(R/R') - d(F)$ does not depend on the choices involved in writing G as F/R, so in this context we do not have to take the minimum of all possibilities.) As $\text{def } G = d_F(R) - d(F)$ for some choice of F/R and of course $d_F(R) \geqslant d_G(R/R')$, it follows that $\text{def } G \geqslant \text{abdef } G$. It is easy to adapt the arguments of Swan [12] and Wiegold [15] to show that in their examples there is no upper bound on the abelianized deficiencies either. The same will be true of the generalizations.

We shall write the Schur multiplicator of a finite group G as $\text{M}(G)$. The connection between deficiency and the Schur multiplicator may be viewed as follows. In the situation considered above, the quotient $R/[R, F]$ is the direct sum of $\text{M}(G)$ with a free abelian group of rank $d(F)$. On the other hand, $R/[R, F]$ is a G-homomorphic image of R/R', and the action of G on $R/[R, F]$ is trivial, so $d_G(R/R') \geqslant d(R/[R, F])$. Thus

$$\text{def } G \geqslant \text{abdef } G \geqslant d(\text{M}(G)).$$

(In different notation, this was (3) in Wamsley's survey [14]).

Before the fundamental paper [12] of Swan and the developments that flowed from it, it was not known that $d_G(R/R') - d(F)$ is an invariant of G, and in any case there was no method for calculating this number, so the only lower estimate of $\text{def } G$ came from the inequality $\text{def } G \geqslant d(\text{M}(G))$. A group G is called efficient if $\text{def } G = d(\text{M}(G))$ (beware: for infinite G, the definition is different). In the light of the present discussion, G cannot be efficient unless the minimum number of generators of R/R' can be read off the largest G-trivial quotient $R/[R, F]$ of this module. In general, modules with this property are fairly rare, and so one can expect that inefficient groups are far more common than efficient ones. It is not the examples of Swan and Wiegold, and the generalizations to be presented here, that are the rare exceptions, but the groups that have been found to be efficient.

As usual in such pseudo-statistical speculations, other views are also quite plausible. While modules with this property are rare in general, they may not be so rare among the modules which arise as R/R', and they may be even less rare among the R/R' that arise with $\text{M}(F/R) = 0$. Or: rarity is determined by the examples known at any one time, and by the skills available to the beholder. I, for one, would not know where to look for an efficient finite group that is not already in the literature, but would be happy to attempt to 'make to measure' any number of further examples with $\text{abdef } G > d(\text{M}(G))$.

One cannot close such a discussion without mentioning that the existence of a finite G with def G > abdef G is still an open problem and looks as intractable as ever. (This was Question 6 in Wamsley's survey [14].) Recall that Gruenberg proved that if a finite group G is written as F/R and if T is a normal subgroup of F contained in R with R/T finite or soluble, then $d_F(R/T) \leqslant d_G(R/R')$ (Propositions 2.8 and 7.10 in [8]; note that in the second line of p. 47 'finite soluable' should be replaced by 'soluble'). Thus in attempting to prove def G > abdef G for some particular G, it is a waste of time to investigate the F/T with R/T finite or soluble.

2. The Constructions

Both of the constructions that we generalize are based on a nontrivial module of order 7 for the group of order 3. All the examples are semidirect products $Q \ltimes N$ where Q is the group of order 3 (or, in the Wiegold–Neumann case, the cyclic group of order 6) and N ranges through certain groups made from copies of the module. Swan's N are simply the direct sums of copies of the module. Wiegold first takes a free product of two copies of the module, and his N are the quotients of that free product over suitable terms of the lower central series.

Perhaps the most important common property of the two constructions is that the orders of the group and module are coprime, so the orders of the semidirect factors will also be coprime. The action of Q on N (used in defining $Q \ltimes N$) yields an action of Q on M(N) (see the next section). Let M(N)^Q denote the subgroup of of M(N) consisting of the elements fixed by Q. The key to ensuring that M($Q \ltimes N$) = 0 is the following.

Lemma 1. *If the orders of Q and N are coprime, then*

$$\text{M}(Q \ltimes N) = \text{M}(Q) \oplus \text{M}(N)^Q.$$

The relevant special case of this lemma is explicitly mentioned in Swan [12] and justified by reference to the spectral sequence of the extension $N \rightarrow Q \ltimes N \rightarrow Q$. In its present generality, it can be read off as a special case from Theorem 2 of Tahara [13], who did not use spectral sequences. A proof without cohomology will be given in the next section. The reader who prefers the spectral sequence proof may still find some interest in the implicit group theoretic interpretation provided by our argument for some ingredients of the spectral sequence. (Wiegold chose to avoid this lemma in favour of a simple and direct but somewhat *ad hoc* argument.)

Both Swan and Wiegold use a simple consequence of the Reidemeister–Schreier Theorem, namely that if H is a subgroup of a group G then

$$1 + \text{def } G \geqslant (1 + \text{def } H)/|G : H|,$$

to ensure that the groups G they construct have large deficiency. The lower estimate they use for def H is $d(\mathrm{M}(H))$, so in effect they are using that

$$1 + \mathrm{def}\, G \geqslant (1 + d(\mathrm{M}(H)))/|G : H|.$$

One can equally easily deduce from the foregoing that

$$1 + \mathrm{abdef}\, G \geqslant (1 + \mathrm{abdef}\, H)/|G : H| \geqslant (1 + d(\mathrm{M}(H)))/|G : H|,$$

so in fact both constructions yield groups of arbitrarily large *abelianized* deficiency. We shall also rely on this argument.

In the generalizations we are about to describe, the task is to choose first a group Q and then infinitely many groups N on which Q acts, in such a way that $(|N|, |Q|) = 1$, $\mathrm{M}(N)^Q = 0$, and the set of the numbers $d(\mathrm{M}(N))$ is not bounded above. By the discussion so far, we know that we shall have $\mathrm{M}(Q \ltimes N) = \mathrm{M}(Q)$ with the set of the numbers $\mathrm{abdef}(Q \ltimes N)$ unbounded. In particular, if also $\mathrm{M}(Q) = 0$, then our semidirect products will have trivial multiplicator and unbounded abelianized deficiency.

For the generalization of Swan's construction, the critical property of Q is to have an element that is not conjugate to its inverse. (Thus for this purpose the group of order 2 and $SL(2, 5)$ are bad, but all larger finite cyclic groups and $SL(2, 7)$ are good.) Such a Q always has an irreducible complex character, χ say, that is different from its complex conjugate $\overline{\chi}$ (see V.13.7a in Huppert [10]). Then χ^2 does not involve the trivial character (because the scalar product of χ^2 with the trivial character is also the scalar product of χ with $\overline{\chi}$). Equivalently, for the simple module U that affords χ we have $(U \otimes U)^Q = 0$. Next, choose any prime p congruent to 1 mod $|Q|$. Over the field of p elements, which we write as $\mathbb{F}_p$, the representation theory of Q is the same as over the complex field, so there is an simple $\mathbb{F}_p Q$-module V such that $(V \otimes V)^Q = 0$. For each positive integer k, let N_k be the direct sum of k copies of V. Of course then also $(N_k \otimes N_k)^Q = 0$, so we can appeal to the completely elementary V.25.4 in [10] for the conclusion that the Sylow p-subgroup of $\mathrm{M}(Q \ltimes N_k)$ is trivial: in view of Lemma 1, this yields that $\mathrm{M}(N_k)^Q = 0$ and $\mathrm{M}(Q \ltimes N_k) = \mathrm{M}(Q)$. Alternatively, we can follow (p. 186 of) Beyl and Tappe [1] and deduce $\mathrm{M}(N_k)^Q = 0$ from $\mathrm{M}(N_k) \cong N_k \wedge N_k \leqslant N_k \otimes N_k$. Since N_k is elementary abelian and $d(N_k) = k \dim V \geqslant k$, we know that $d(\mathrm{M}(N_k)) \geqslant \frac{1}{2}k(k - 1)$, so the $d(\mathrm{M}(N_k))$ have no upper bound.

It may be worth noting here that the way Neumann made 2-generator examples from Wiegold's 3-generator groups can also be used to make 2-generator examples from Swan's many-generator groups. Choose a group Q with $\mathrm{M}(Q) = 0$ and a module V of characteristic p, as above. For each positive integer k that is prime to $p|Q|$, set $Q_k = C_k \times Q$ (where C_k is a cyclic group of order k) and let N_k be the Q_k-module induced from V (so $Q_k \ltimes N_k$ is the twisted wreath product of V by Q_k, the action of Q on V providing the twisting). Then $d(Q_k \ltimes N_k) = \min\{2, d(Q)\}$ and $\mathrm{M}(Q_k \ltimes N_k) = 0$ for all k, while the abelianized deficiencies tend to infinity.

(Note that the subgroups $Q \ltimes N_k$ form a subsequence in the corresponding sequence of generalized Swan examples.) If Q is perfect and $p > 3$, one can replace the C_k by the $SL(2, q)$ with q running through the primes larger than $|Q|$ that are congruent neither to 1 nor to -1 modulo p, and the same conclusions will still hold, while the $Q_k \ltimes N_k$ will remain perfect. (To see that $d(Q_k \ltimes N_k) \leqslant \min\{2, d(Q)\}$, argue first that Q_k can be generated by $\min\{2, d(Q)\}$ elements, then write Q_k as F/R with F a free group of this rank and appeal to a theorem of Gaschütz [6] for the fact that N_k, like every one-generator Q_k-module of characteristic p, is a quotient of $R/R'R^p$ where R^p stands for the subgroup $\langle r^p \mid r \in R \rangle$.)

For the generalization of the Wiegold–Neumann construction, the critical property of Q is to have a nontrivial central element, z say, which does not generate all of Q. (Thus here all groups of prime order and all nonabelian simple groups are bad, but cyclic groups of composite order and perfect groups like $SL(2, q)$ for odd q, $q \neq 3$, are good.) Let R be any subgroup of Q such that $R < \langle R, z \rangle < Q$, and let p be a prime which does not divide $|Q|$ and is congruent to 1 modulo $|Rz|$ (that is, modulo the order of the element Rz in the factor group $\langle R, z \rangle/R$). Choose a 1-dimensional $\mathbb{F}_p\langle R, z \rangle$-module, U say, on which R acts trivially and z acts by a scalar of multiplicative order $|Rz|$. Then z acts as that scalar also on the $\mathbb{F}_p Q$-module V induced from U, and V can be written as a direct sum of $|Q : \langle R, z \rangle|$ subspaces, each of dimension 1, that are permuted by Q. It follows that the free product of $|Q : \langle R, z \rangle|$ groups of order p, which we shall call P, admits an action by Q which is such that V is Q-isomorphic to P/P'. Write the lower central series of P as $P = P_1 > \cdots > P_i > \cdots$; for each positive integer k such that $k + 1$ is not divisible by $|Rz|$, set $N_k = P/P_{k+1}$. From Theorem 2.6 of Haebich [9] we know that $\mathrm{M}(N_k)$ is isomorphic to P_{k+1}/P_{k+2}. It follows that P/P_{k+2} is a Schur covering group of N_k. The action of Q on P yields an action on N_k, and we form the semidirect product $G = Q \ltimes N_k$ with reference to this. The action of Q on P also yields actions on the covering group P/P_{k+2} and on the copy P_{k+1}/P_{k+2} of $\mathrm{M}(N_k)$ in that covering group. These actions match in the way that is necessary to ensure that when $\mathrm{M}(N_k)$ is viewed as a Q-module in the sense of Lemma 1, it is Q-isomorphic to P_{k+1}/P_{k+2} (see the last two paragraphs of the next section). It is easy to calculate, following Wiegold [15], that P_{k+1}/P_{k+2} is an elementary abelian p-group and z has nontrivial powering action on P_{k+1}/P_{k+2}: thus $\mathrm{M}(N_k)^Q = 0$ as required. Further, Wiegold asserted in [15] that the rank of P_{k+1}/P_{k+2} tends to infinity with k, and this yields that our set of the $d(\mathrm{M}(N_k))$ has no upper bound.

However, I must admit that I have not been able to locate or devise a proof for the proposition that the rank of P_{k+1}/P_{k+2} tends to infinity. (I am grateful for a last-minute suggestion that it may be possible to deduce it from Gaglione [5].) To this extent, the justification of the construction given above is incomplete.

One way to sidestep this difficulty is by exploiting the freedom to choose p. By Dirichlet's Theorem, there are infinitely many primes satisfying the present require-ments. Write P_k and N_k as $P_{k,p}$ and $N_{k,p}$ to indicate that our groups depend on this choice as well. If $p > k+1$, then $N_{k+1,p}$ is a regular p-group (because its nilpotency class is small) and so it has exponent p (because it is generated by elements of order p). It follows that in this case $N_{k+1,p}$ is a free group of the variety $\mathfrak{B}_p \cap \mathfrak{N}_{k+1}$ of groups of exponent p and class at most $k+1$, and $P_{k+1,p}/P_{k+2,p}$ is the last term of its lower central series. Thus the rank of $P_{k+1,p}/P_{k+2,p}$ is given by Witt's Formula, from which one can see that this rank is independent of p and does tend to infinity with k. For each k such that $k+1$ is not divisible by $|Rz|$, choose N_k as one of the $N_{k,p}$ with $p > k+1$: then the sequence of groups $Q \ltimes N_k$ will have all the properties we claimed.

In the generalized Wiegold–Neumann construction, V had to be a monomial module in order that action on it should lift to action on a free product P of cyclic groups with $P/P' \cong V$. The other important point was that some element of Q act on V as a nontrivial scalar, so one can deduce that this element, and therefore also Q, acts fixed-point-free on certain lower central factors of P. If instead we choose P as a free group of $\mathfrak{B}_p \cap \mathfrak{N}_{p-1}$, then V does not have to be monomial: then any action of the p'-group Q on P/P' can be lifted to an action on P. Witt's Formula is just a special case of the character formula of Brandt [2], which enables one to compute the action on the lower central factors of these P. Using this, it may be possible to verify that the action of Q is fixed-point-free even when no individual element of Q is known to act without fixed points. Of course these P are finite and so no single one of them will yield an infinite family of examples.

3. A Proof of Lemma 1

The aim of this section is to present an elementary proof of Lemma 1.

In preparation, we establish two simple propositions. They concern the (right) action of a finite group Q on a finitely generated abelian group A. It will be convenient for the moment to write A additively. Let A^Q denote the subgroup consisting of the fixed points of Q in A, and A_0 the torsion subgroup of A. Write $[A, Q]$ for the subgroup of A generated by the elements $a(x - 1)$ with $a \in A$ and $x \in Q$.

(a) *If the index $|A : A^Q|$ is prime to $|Q|$, then $[A, Q] \cap A^Q = 0$.*

(b) *If $|A_0|$ is prime to $|Q|$, then $[A, Q] \cap A_0 = [A_0, Q]$.*

For the proof, let η denote the element $\sum_{y \in Q}(1 - y)$ of $\mathbb{Z}Q$, and note that $(x - 1)\eta = |Q|(x - 1)$ whenever $x \in Q$: thus

$$a\eta = \begin{cases} 0 & \text{if } a \in A^Q, \\ |Q|a & \text{if } a \in [A, Q]. \end{cases}$$

It follows that $[A, Q] \cap A^Q$ has exponent dividing $|Q|$. If $|A : A^Q|$ is prime to $|Q|$, then $A = A^Q + |Q|A$ and hence $[A, Q] = [|Q|A, Q] = |Q|[A, Q]$. A finitely generated abelian group that is divisible by $|Q|$ has no nonzero element of order dividing $|Q|$, so in this case $[A, Q] \cap A^Q = 0$. On the other hand, if $|A_0|$ is prime to $|Q|$ then there is a positive integer k such that $|Q|^k \equiv 1 \pmod{|A_0|}$, and then $a \in [A, Q] \cap A_0$ implies that $a = |Q|^k a = a\eta^k \in A_0\eta \leqslant [A_0, Q]$. This completes the proof of (a) and (b).

We are now ready to start on the proof of Lemma 1.

Let π denote the set of the prime divisors of $|Q|$, and π' the set of all other primes. Then $M(Q)$ is a π-group and $M(N)$ is a π'-group, so to prove Lemma 1 it will be sufficient to show that $M(Q \ltimes N)$ has a subgroup isomorphic to $M(N)^Q$ with quotient isomorphic to $M(Q)$.

The definition of multiplicator that we use is this: given a finite group G, write it as F/R with F free of finite rank, and set $M(G) = (F' \cap R)/[F, R]$. (Accordingly, in this section we use multiplicative rather than additive language even where the multiplicator is concerned.) As the discussion on pp. 29–31 of Beyl and Tappe [1] explains, there is in fact a functor M from finite groups to abelian groups; so in particular there is, for each G, a distinguished homomorphism

$$\operatorname{Aut} G \to \operatorname{Aut} M(G), \quad \alpha \mapsto M(\alpha).$$

Composition with this homomorphism converts any action on G into an action on $M(G)$. Given a semidirect product $Q \ltimes N$ formed with respect to some action of Q on N, it is in this sense that we have an action of Q on $M(N)$. In terms of normal subgroups in a free group, this comes to the following.

Write $Q \ltimes N$ as F/S with F free of finite rank, and let R/S correspond to the normal subgroup N of $Q \ltimes N$. Then $F/R = Q$ and R is also a free group of finite rank, so we have

$$M(Q \ltimes N) = (F' \cap S)/[F, S],$$
$$M(Q) = (F' \cap R)/[F, R],$$
$$M(N) = (R' \cap S)/[R, S].$$

Moreover, the relevant action of Q on $M(N)$ is that which comes from conjugation in F. To justify this last claim, recall that if $\alpha \in \operatorname{Aut} N$ then $M(\alpha)$ is defined by choosing any endomorphism, ε say, of R such that $(rS)\alpha = (r\varepsilon)S$ for all r in R, and setting $M(\alpha)\colon r'[R, S] \mapsto (r'\varepsilon)[R, S]$ for each r' in $R' \cap S$. (The last sentence of Lemma 3.1 in [1] shows that the $M(\alpha)$ so defined depends only on α and not on the choice of ε.) The α that are relevant to our claim come from conjugation in F; the same conjugation can also be used to define the corresponding ε, and then our claim follows. Note also that, by Maschke's Theorem, $M(N)^Q$ is isomorphic to the largest Q-trivial quotient of $M(N)$, so in the present terms

$$M(N)^Q \cong (R' \cap S)/[R, S][F, R' \cap S].$$

A modular lattice generated by two chains is always distributive: thus in the normal subgroup lattice of F, the sublattice generated by the two chains

$$F \geqslant R \geqslant S \geqslant [F, S] \geqslant [F, R' \cap S] \quad \text{and} \quad F' \geqslant [F, R] \geqslant R' \geqslant [R, S]$$

is distributive. The figure shows the Hasse diagram of the distributive lattice $\mathcal{L}$ defined on these nine generators by the displayed inclusions and

$$F \geqslant F', \quad R \geqslant [F, R] \geqslant [F, S], \quad R' \geqslant [F, R' \cap S]$$

as defining relations: its verification is an elementary lattice theory exercise.

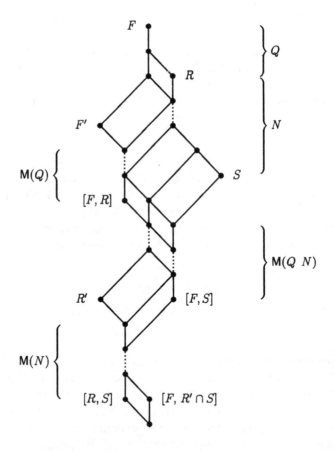

The sublattice in the normal subgroup lattice of F is a homomorphic image of this $\mathcal{L}$. Dotted lines have been used to denote edges which are present in $\mathcal{L}$ but, as we shall see, always collapse in F (in the sense that the two endpoints of the edge are equal in F). Note that once those collapses are established, the planned proof of

Lemma 1 will be complete, with $(R' \cap S)[F, S]/[F, S]$ as the required subgroup in $M(Q \ltimes N)$.

By one of the elementary isomorphism theorems,

$$(F' \cap R)/([F' \cap [F, R]S) \cong (F' \cap R)S/[F, R]S.$$

The left hand side is a quotient of the π-group $M(Q)$ and the right hand side is a section of the π'-group R/S, so both must have order 1. This proves the first pair of collapses. (It was noted in [1], at the top of p. 31, that the functorial nature of M very directly yields that $M(Q)$ is not only a quotient but even a module direct summand of $M(Q \ltimes N)$, and no coprimality was used there. However, the translation of that argument into the present setting would involve a more complicated lattice.)

The proposition (a) established at the beginning of this section may now be applied with $A = R'[F, S]$: since $R'S/R'[F, S] \leqslant A^Q$, the coprimality condition is satisfied. The conclusion gives that $[F, R] \cap R'S \leqslant R'[F, S]$, and the second pair of collapses follows. Next, apply (b) with $A = S/[R, S]$: then $A_0 = (R' \cap S)/[R, S]$, so the co-primality condition holds. The conclusion gives that $R' \cap [F, S] \leqslant [R, S][F, R' \cap S]$, the last collapse we wanted to show. This completes the proof of Lemma 1.

We conclude this section by sketching two justifications for a claim we used implicitly in the discussion of the generalized Wiegold–Neumann construction, namely that if some action on a group N lifts to an action on a covering group of N, then by restriction to the copy of $M(N)$ in that covering group we get the same action as by composition with the homomorphism Aut $N \to$ Aut $M(N)$ provided by the functor M. Given a covering group of N, this time write that as F/S (where F is free of finite rank), with R/S as the copy of $M(N)$ in F/S: then

$$F/R = N, \quad F' \cap S = [F, R], \quad (F' \cap R)S = R,$$

so that there is a natural isomorphism

$$M(N) = (F' \cap R)/[F, R] \cong R/S.$$

Let α be an automorphism of F/R that lifts to an automorphism of F/S, say, to α^*, and let us use our freedom to define $M(\alpha)$ in terms of an endomorphism ε of F that lifts α because it lifts α^*. Of course then ε and α^* agree on R/S, so the action of $M(\alpha)$ on $(F' \cap R)/[F, R]$ and the action of α^* on R/S are intertwined by the natural isomorphism $F' \cap R)/[F, R] \cong R/S$ above. This is just the rigourous form of what we had to establish.

The claim we have just proved could not serve as a general description on how to convert an action on N to an action on $M(N)$ (for example, if N is elementary abelian of order 2^3, the natural action of Aut N does not lift to any Schur covering group of N). However, it is possible to give such a description, in terms of covering groups rather than free groups, by reference to the familiar result that any two covering groups of of a finite group N are isoclinic. Strictly speaking, what one proves there is that, given any two surjective maps $\sigma_i : N_i \to N$ with ker $\sigma_i \leqslant N_i' \cap Z(N)$ and ker $\sigma_i \cong M(N)$, there is a unique isomorphism $N_1' \to N_2'$ with a certain property. If σ_1 is fixed and σ_2 ranges through the $\sigma_1\alpha$ with $\alpha \in \operatorname{Aut} N$, then to each α we get a uniquely determined automorphism, α_1 say, of N_1'. This α_1 maps $\ker \sigma_1$ onto itself, and the restriction of α_1 to that kernel 'is' $M(\alpha)$. (In fact, $\alpha \mapsto \alpha_1$ defines a homomorphism $\operatorname{Aut} N \to \operatorname{Aut} N_1'$.) The second justification promised lies in checking that if α lifts to an automorphism α^* of N_1 then the restriction of α^* to N_1' also has the property which characterizes α_1, so that restriction must be α_1.

4. Some Explicit Examples

In the perfect group $SL(2, 7)$, the elements of order 7 are not conjugate to their inverses; over any $\mathbb{F}_p$ with $p \equiv 1 \bmod 168$, this group has an simple module V of dimension 3 such that $V \otimes V$ contains no fixed point other than 0. In fact, Section 232 of Burnside [3] (see particularly its last sentence) explicitly describes such a representation over any field containing a square root of -7, so we can even choose $p = 11$. Of course the centre is nontrivial, the multiplicator is trivial, and the deficiency is zero (see for example Campbell and Robertson [4]), and therefore this group can play the role of Q in both constructions. For the generalized Wiegold–Neumann construction, $SL(2, 5)$ presents a smaller perfect alternative in the role of Q. The $Q \ltimes N$ obtained with these choices seem to be the first perfect finite groups proved to have trivial multiplicator but positive deficiency.

To provide a better understanding of the constructions and to illustrate further aspects of the available methods, we explore here some of the smallest perfect examples that we can make and determine the exact abelianized deficiency of these groups. The latter calculations will be based on the following result, whose proof is deferred to the next section and in which $\mathbb{F}_p$ stands for the trivial Q-module of p elements.

Lemma 2. *If Q is a finite p'-group acting on a finite p-group N, then*

$$\operatorname{abdef}(Q \ltimes N) = \max\{\operatorname{abdef} Q,\ d_Q(\mathbb{F}_p \oplus M(N)) - 1\}.$$

Take the generalized Swan construction first, with $G_k = Q \ltimes N_k$ where $Q = SL(2, 7)$ and N_k is the direct sum of k copies of the Q-module V of order 11^3 mentioned above, so

$$|G_k| = 336 \times 11^{3k}.$$

Though we do not need it, for orientation we note that it can be seen using the results of Gaschütz [6] that

$$d(G_k) = 1 + \lceil k/3 \rceil$$

where by $\lceil x \rceil$ we denote the unique integer such that $x \leqslant \lceil x \rceil < x + 1$. We shall prove that

$$\text{abdef } G_k = \left\lceil \tfrac{1}{6} k(k+1) \right\rceil - 1. \tag{1}$$

In particular, abdef G_1 = abdef G_2 = 0 but abdef G_3 = 1.

Towards the proof of (1), we already know that abdef $Q = 0$ (because even def $Q = 0$). By the general rules of multilinear algebra,

$$M(N_k) = N_k \wedge N_k = (V^{\oplus k}) \wedge (V^{\oplus k}) = (V \wedge V)^{\oplus k} \oplus (V \otimes V)^{\oplus \frac{1}{2} k(k-1)}.$$

As for every 3-dimensional simple module, the exterior square $V \wedge V$ of V is also 3-dimensional and simple. Any simple module of prime dimension on which the derived group acts nontrivially is in fact absolutely simple, and this applies to both V and $V \wedge V$. For this particular V, it is also easy to see that the direct complement of $V \wedge V$ in $V \otimes V$ is absolutely simple. Call that 6-dimensional module W: then

$$M(N_k) = (V \wedge V)^{\oplus \frac{1}{2} k(k+1)} \oplus W^{\oplus \frac{1}{2} k(k-1)}.$$

The way for counting the minimum number of generators of a semisimple module can be found in Lemma 7.12 of [8]; using that and the fact that the multiplicity of an absolutely simple module of dimension d in the largest semisimple quotient of the regular module is d, we obtain that

$$d_Q(\mathbb{F}_{11} \oplus M(N_k)) = \left\lceil \tfrac{1}{6} k(k+1) \right\rceil,$$

and so (1) follows from Lemma 2.

Next we turn to the generalized Wiegold–Neumann construction, this time writing $H_k = Q \ltimes N_k$, still with $Q = SL(2, 7)$. There is then only one nontrivial element in $Z(Q)$, and that must be our z. The largest subgroup R that does not contain z is of order 21, and then the index of $\langle R, z \rangle$ is 8; the smallest p we can choose is 5, and the smallest permitted value of k is 2. Now N_4 is the free group of rank 8 in the variety $\mathfrak{B}_5 \cap \mathfrak{N}_4$ which consists of all groups of exponent 5 and nilpotency class at most 4, so the ranks of the first four lower central factors of P can be calculated by Witt's Formula: they are 8, $\tfrac{1}{2}(8^2 - 8)$, $\tfrac{1}{3}(8^3 - 8)$, and $\tfrac{1}{4}(8^4 - 8^2)$. We only need the first three to deduce that

$$|H_2| = 336 \times 5^{36} \quad \text{and} \quad |M(N_2)| = 5^{168},$$

with $M(N_2)$ elementary abelian.

Notice that 168 is also the dimension of the unique largest $\mathbb{F}_5 Q$-module which can be generated by a single element and on which z acts as the scalar -1. There is no reason to expect that $M(N_2)$ is that particular module, but calculation shows that it is, and hence

$$\text{abdef } H_2 = 0$$

follows by Lemma 2. This coincidence was first established by a direct computer calculation that one could not hope to perform by hand. I am greatly indebted to Dr M. F. Newman for this, for his other contributions that will be mentioned below, and for many illuminating discussions on related matters.

There is another argument which justifies this conclusion and which does not rely on a machine; it can be sketched as follows. Groups of exponent p and nilpotency class at most $p-1$ are the same, in the strongest possible sense, as Lie algebras over $\mathbb{F}_p$ that are nilpotent of class at most $p-1$. In particular, N_3 is just the rank 8 free Lie algebra in the variety of the Lie algebras over $\mathbb{F}_5$ that are nilpotent of class at most 3. Call this Lie algebra $L(8, \mathbb{F}_5, 3)$, or briefly L. Also in this sense, $M(N_2)$ is L^3, and both are $\mathbb{F}_5 Q$-modules. The connection between the natural actions of $GL(8, 5)$ on L/L^2 and on L^3 was described by Brandt in [2]. That connection can be restricted to the action of Q, it survives the 'extension of scalars' which replaces $\mathbb{F}_5$ by its algebraic closure, and also the 'reduction of constants' which establishes the connection between the representations of Q over that algebraic closure and the representations of Q over the complex field $\mathbb{C}$. Thus the connection between the actions of Q on L/L^2 and on L^3 does not change if we replace L by $L(8, \mathbb{C}, 3)$. Let χ denote the complex character corresponding to the action of Q on P/P', that is, the character afforded by the complex L/L^2: by Brandt's Formula, the character afforded by L^3 is then given by the rule that its value at an element x of Q is $\frac{1}{3}(\chi(x) - \chi(x^3))$. Here χ is the character of Q induced from the character of $\langle R, z \rangle$ which is 1 on R and -1 on the rest of $\langle R, z \rangle$, so it is easy to calculate, and therefore so is $L_3(\chi)$. On the other hand, the character corresponding to the unique largest module which can be generated by a single element and on which z acts as the scalar -1 has value 168 at the identity element, -168 at the central involution, and 0 at every other element of Q. It remains to observe that this character and $L_3(\chi)$ coincide, and the argument is complete.

The next example in this sequence is H_4. A slight extension of the argument above yields first that

$$|H_4| = 336 \times 5^{1212} \quad \text{and} \quad |M(N_4)| = 5^{6552}.$$

Further, it justifies that $d_Q(M(N_4))$ is the minimum number of generators of the $\mathbb{C}Q$-module that affords the character $L_5(\chi)$ whose value at x is $\frac{1}{5}(\chi(x) - \chi(x^5))$. Here χ is the same as above. Using the character table of Q and the orthogonality relations, an easy hand calculation shows that $L_5(\chi)$ is the sum of

158 copies of each of two irreducible characters of degree 4,
234 copies of each of two irreducible characters of degree 6, and
310 copies of an irreducible character of degree 8.

It follows that $d_Q(M(N_4)) = 40$ and so, by Lemma 2,

$$\text{abdef } H_4 = 39.$$

To attempt to confirm this by the kind of direct machine calculations that were used above would be stretching the resources available to us at this time.

For a related example with positive deficiency which is not as far out of reach as this, we exploit the observations of the last paragraph of Section 2: one can imitate the construction of H_2 with P/P' the other 8-dimensional simple $\mathbb{F}_5 Q$-module on which the central involution acts nontrivially, even though that module is not monomial. This new version of H_2 also has order 336×5^{36}, but now the role of χ goes to the faithful simple character, ψ say, of degree 8, and $L_3(\psi)$ turns out to be the sum of

3 copies of each of two irreducible characters of degree 4,
6 copies of each of two irreducible characters of degree 6, and
9 copies of an irreducible character of degree 8.

The conclusion is that the abelianized deficiency of this version of H_2 is 1. This group is, of course, much larger than G_3, which remains the smallest perfect group we know with trivial multiplicator and positive deficiency.

The second smallest is also built on this last pattern, with $Q = SL(2, 5)$, $p = 11$, and P/P' the direct sum of two isomorphic 2-dimensional simple modules. (There are two such simple modules, but they are interchanged by an automorphism of Q, so the isomorphism type of $Q \ltimes N_2$ does not depend on which one we use). The resulting $Q \ltimes N_2$ has order 120×11^{10}, and its abelianized deficiency is 2.

The last two examples were discovered by Dr M. F. Newman and first justified by machine computations. I am grateful for his permission to include them here.

5. The Proof of Lemma 2

Set $G = Q \ltimes N$. In the notation already used in the proof of Lemma 1, abdef $G = d_G(S/S') - d(F)$, so the problem is to determine $d_G(S/S')$. By the theorem of Gruenberg [7] (see Theorem 7.8 in [8]), $d_G(S/S') = \max_q d_G(S/S'S^q)$ where q ranges through the prime divisors of $|G|$ and S^q stands for the subgroup $\langle s^q \mid s \in S \rangle$.

For simplicity, put $d = d(F)$ and assume that $d \geqslant 2$. It follows from a fundamental result of Gaschütz [6] (see Lecture 2 in [8]) that there is an exact sequence

$$0 \to S/S'S^q \to (\mathbb{F}_q G)^{\oplus d} \to \mathbb{F}_q G \to \mathbb{F}_q \to 0 \qquad (2)$$

of G-homomorphisms (where, as usual in this context, $\mathbb{F}_q$ stands both for the field of q elements and for the trivial G-module of q elements). In view of Schanuel's Lemma and the Krull-Schmidt Theorem (used as in the deduction of Corollary 2.5 in [8]), the existence of such a sequence may in fact be taken as the definition of $S/S'S^q$. Similarly, $R/R'R^q$ is characterized by the existence of an exact sequence

$$0 \to R/R'R^q \to (\mathbb{F}_q Q)^{\oplus d} \to \mathbb{F}_q Q \to \mathbb{F}_q \to 0. \qquad (3)$$

Consider first the case $q \neq p$. Given that N is a q'-group, every $\mathbb{F}_q G$-module W can be written canonically as $W^N \oplus [W, N]$, with

$$d_G(W) = \max\{d_G(W^N), d_G([W, N])\}.$$

In this manner, (2) splits into two sequences, and the first of those is

$$0 \to (S/S'S^q)^N \to (\mathbb{F}_q Q)^{\oplus d} \to \mathbb{F}_q Q \to \mathbb{F}_q \to 0$$

because, in the relevant sense, $(\mathbb{F}_q G)^N = \mathbb{F}_q Q$. Comparing this sequence with (3), we conclude that $(S/S'S^q)^N \cong R/R'R^q$. The second sequence resulting from (2) is

$$0 \to [S/S'S^q, N] \to [\mathbb{F}_q G, N]^{\oplus d} \to [\mathbb{F}_q G, N] \to 0 \to 0.$$

Since $[\mathbb{F}_q G, N]$ is a direct summand of $\mathbb{F}_q G$, it is projective, and therefore this sequence splits; hence by the Krull-Schmidt Theorem $[S/S'S^q, N] \cong [\mathbb{F}_q G, N]^{\oplus(d-1)}$. It follows that $d_G(S/S'S^q) = \max\{d_Q(R/R'R^q), d - 1\}$, whence one readily sees that, with q ranging over the prime divisors of $|Q|$,

$$\max_q (d_G(S/S'S^q) - d) = \max_q (d_Q(R/R'R^q) - d) = \text{abdef } Q.$$

It remains to consider $d_G(S/S'S^p)$. This is the same as the minimum number of generators of the largest semisimple quotient of $S/S'S^p$. On that quotient, N acts trivially (because every normal p-subgroup acts trivially on every simple module of characteristic p); conversely, the largest quotient on which N acts trivially is semisimple (by Maschke's Theorem). Thus the largest semisimple quotient of $S/S'S^p$ is $S/[R, S]S^p$, and $d_G(S/S'S^p) = d_Q(S/[R, S]S^p)$.

The torsion subgroup of the finitely generated abelian group $S/[R, S]$ is $(R' \cap S)/[R, S]$, isomorphic to the p-group $\mathrm{M}(N)$. It follows that (in additive terminology) $S/[R, S]S^p$ is a direct sum of two summands, namely of $S/(R' \cap S)S^p$ and of the Frattini factor group of $\mathrm{M}(N)$. Since $|Q|$ is prime to p, there is such a direct decomposition of $S/[R, S]S^p$ as Q-module, and so

$$d_Q(S/[R, S]S^p) = d_Q\big(S/(R' \cap S)S^p \oplus \mathrm{M}(N)\big).$$

The proof of Lemma 2 will therefore be complete if we show that

$$S/(R' \cap S)S^p \cong \mathbb{F}_p \oplus (\mathbb{F}_p Q)^{\oplus(d-1)}.$$

Here the right hand side is familiar, for by the easy (coprime) case of the Gaschütz theory quoted above we know that $R/R'R^p \cong \mathbb{F}_p \oplus (\mathbb{F}_p Q)^{\oplus(d-1)}$: thus what we need is that $S/(R' \cap S)S^p \cong R/R'R^p$. As $S/(R' \cap S)$ and $R'S/R'$ are Q-isomorphic, so are their largest exponent-p quotients: $S/(R'\cap S)S^p \cong R'S/R'R^p$. We have reduced our target to $R'S/R'R^p \cong R/R'R^p$, and this much can be seen as follows.

If A is any finitely generated $\mathbb{Z}$-free $\mathbb{Z}Q$-module and B is a maximal submodule of p-power index, then

$$A/pA \cong A/B \oplus B/pA \cong pA/pB \oplus B/pA \cong B/pB.$$

Repeating this argument shows that $A/pA \cong B/pB$ whenever B is a submodule of p-power index. This may then be applied with $A = R/R'$ and $B = R'S/R'$ because, being a quotient of N, $R/R'S$ has p-power order. The proof of Lemma 2 is now complete.

References

[1] Beyl, F. Rudolf, Tappe, Jürgen, Group extensions, representations, and the Schur multiplicator, Lecture Notes in Math. 958, Springer-Verlag, Berlin–Heidelberg–New York 1982.

[2] Brandt, Angeline J., The free Lie ring and Lie representations of the full linear group, Trans. Amer. Math. Soc. 56 (1944), 528–536.

[3] Burnside, W., Theory of groups, 2nd ed., Cambridge 1911, reprinted by Dover, New York 1955.

[4] Campbell, C. M., and Robertson, E. F., A deficiency zero presentation for $SL(2, p)$, Bull. London Math. Soc. 12 (1980), 17–20.

[5] Gaglione, Anthony M., Factor groups of the lower central series for special free products, J. Algebra 37 (1975), 172–185.

[6] Gaschütz, W., Über modulare Darstellungen endlicher Gruppen, die von freien Gruppen induziert werden, Math. Z. 60 (1954), 274–286.

[7] Gruenberg, K. W., Über die Relationenmoduln einer endlichen Gruppe, Math. Z. 118 (1970), 30–33.

[8] Gruenberg, Karl W., Relation modules of finite groups, Regional Conference Series in Mathematics No. 25, Amer. Math. Soc., Providence 1976.

[9] Haebich, William, The multiplicator of a regular product of groups, Bull. Austral. Math. Soc. 7 (1972), 279–296.

[10] Huppert, B., Endliche Gruppen I, Springer-Verlag, Berlin–Heidelberg–New York 1967.

[11] Neumann, B. H., Some groups with trivial multiplicator, Publ. Math. Debrecen 4 (1955), 190–194.

[12] Swan, R. G., Minimal resolutions for finite groups, Topology 4 (1965), 193–208.

[13] Tahara, Ken-ichi, On the second cohomology groups of semidirect products, Math. Z. 129 (1972), 365–379.

[14] Wamsley, J. W., Minimal presentations for finite groups, Bull. London Math. Soc. 5 (1973), 129–144.

[15] Wiegold, J., The Schur multiplier: an elementary approach, in: Groups—St Andrews 1981 (C. M. Campbell and E. F. Robertson, eds.), London Math. Soc. Lecture Note Series 71, Cambridge University Press, Cambridge 1982, 137–154.

TI-Subgroups of Finite Groups

*Alexandre A. Makhnev**

Abstract. This note contains a number of results on 2-groups appearing TI-subgroups of a finite group.

1991 Mathematics Subject Classification: 20D05.

1. Introduction

A subgroup H of even order of a finite group G is called tightly embedded (in G), if an intersection of H with any subgroup conjugated to H has odd order. If a 2-group A is a TI-subgroup of G ($A \cap A^g = 1$ for every $g \in G \setminus N(A)$) then A is tightly embedded.

Tightly embedded subgroups played an important role in the Program of Classification of Finite Simple Groups. This concept arose in connection with the investigation of groups of component type (M. Aschbacher [1]). It is very useful to create a common theory of tightly embedded subgroups (in particular for Revision).

Reduction Theorem (A. Makhnev [4]). *Let a 2-group A be a nonsubnormal TI-subgroup of a finite group G. Then G is a known group or A is a cyclic or elementary group.*

Thus the main interest has the case of cyclic TI-subgroups of order 4.

2. Cyclic TI-Subgroups of Order 4

Let a cyclic group $A = \langle a \rangle$ of order 4 be a TI-subgroup of a finite group G with $O(G) = 1$, $G^* = \langle A^G \rangle$. It is clear that A normalises every component of G. If G^*

---

*   This research is partially supported by Russian Fund for Fundamental Investigations (Grants 93-01-01529 and 94-01-00802a).

does not contain components then $F^*(G) = O_2(G)$ and for $\overline{G} = G/O_2(G)$ a group $\overline{G}^*$ is generated by a set $\overline{a}^G$ of $\{3, 5\}$-transpositions [2]. Thus the next situation is of interest.

(*) $G = LA$, $L = F^*(G)$ be a quasisimple group.

The first step in the investigation of (*) was the case when L is a simple group of characteristic 2-type [2]. The important intermediate step is the case when L is a known group. It is clear that for alternating and sporadic groups we have that G is a symmetric group or HiS or $O'N$.

Exceptional groups of odd characteristic were investigated by A. Makhnev and N. Zyulyarkina [6].

Theorem 1. *Let in the situation* (*) *L be a covering group for an exceptional group of Lee type of odd characteristic. Then*
(1) $G = L = E_7(q)$, $q^2 \equiv 1$ (mod 16) *or*
(2) $G = L = E_6(q)$ *or* $\mathrm{Cov}\, E_6(q)$, $q \equiv 1$ (mod 4) *or*
(3) $G = L = {}^2E_6(q)$ *or* $\mathrm{Cov}^2\, E_6(q)$, $q \equiv 3$ (mod 4)
(4) $|G : L| = 2$, $L = E_7(q)$, $q^2 \equiv 9$ (mod 16), A *an induced inner-diagonal automorphisms on L.*

The most difficult case is the case when L is a covering group for a classical group of odd characteristic. This case was investigated by N. Zyulyarkina [7]. There are 5 theorems in [7], considering special linear, symplectic, unitary, orthogonal groups and also spinor groups.

Let H be a tightly embedded subgroup of a finite group G, let a Sylow 2-subgroup A of H be a cyclic group of order 4. In [5] the classification of groups with the next property is given:

If a_0 is an involution from H, $S \in \mathrm{Syl}_2(G)$ then a weak closure of a_0 in S is abelian.

3. Elementary Abelian TI-Subgroups

Finally we consider the elementary abelian case. Let a noncyclic elementary abelian 2-group A be a TI-subgroup of a finite group G, $G^* = \langle A^G \rangle$. A very strong result was the description of groups with "large" TI-subgroup (F. Timmesfeld and others).

Theorem 2. *Let G be a finite group, M be a maximal 2-local subgroup of G with $F^*(M) = O_2(M)$. If a maximal elementary abelian normal subgroup A of M is a TI-subgroup of G then G is a known group.*

In fact the condition G is a group of characteristic 2-type was demand in the case when A is weakly closed in $O_2(M)$ and $|A| \geq 4$.

The other interesting result is the classification of groups satisfying the next condition:

(+) For any set $A_1, \cdots, A_k$ of distinct commuting subgroups of A^G the subgroup $A_2 \cdots A_k$ does not contain A_1.

Theorem 3 (A. Makhnev [3]). *Let an elementary abelian TI-subgroup A of order 2^n, $n \geq 2$, satisfy the condition (+). Then $G^* = \langle A^G \rangle$ is the central product of subgroups conjugated in G to $J = \langle A^{G^*} \rangle$ and one of the following statements holds:*

(1) $N = \langle (O_2(G) \cap A)^G \rangle$ *is a nonidentity abelian group and $\bar{J} = J/(J \cap N)$ is either an extension of an elementary 2-group of rank k by $L_k(2)$ or a covering group for $L_k(2^m)$, $Sz(2^m)$, $U_3(2^m)$, $\hat{A}_6$ ($\bar{A}$ is weakly closed in $C_{\bar{J}}(\bar{A})$), $Sp_k(2)$, $U_k(2)$, $O_k^{\pm}(2)$ or Σ_k (and for an involution a from $A \setminus N$, a^J is a class of 3-transpositions in J) or $J/(J \cap N)$ has a Sylow 2-subgroup of order 2.*

(2) *J is an extension of an elementary 2-group V of rank k by $L_k(2)$ and A is the group of all transvections centralising a hyperplane in V.*

(3) *J is a covering group of $L_k(2^m)$, $Sz(2^m)$, $U_3(2^m)$, A_6, A_7, A_9, M_{22}, M_{23}, M_{24}.*

(4) *J is an alternating group degree at least 8, and A is conjugate to the subgroup $\langle (12)(34), (13)(24) \rangle$.*

In conclusion I would like to suggest the next

Problem. Let A be an elementary TI-subgroup of order 4 of a finite group G, $G^* = \langle A^G \rangle$ and some involution a of A be weakly closed in A. Is it true that $a \in O_2(G)$ or $\langle A^{G^*} \rangle = \text{Aut } M_{12}$?

References

[1]   Ashbacher M., Tightly embedded subgroups of finite groups, J. Algebra 42 (1976), 85–101.

[2]   Makhnev A. A., TI-subgroups in groups of characteristic 2-type, Matem. Sb. 127 (1985), 239–244.

[3]   ———, On TI-subgroups of finite groups, Izv. Acad. Nauk SSSR Ser. Mat. 50 (1986), 22–36.

[4]   ———, A reduction theorem for TI-subgroups, Izv. Acad. Nauk SSSR Ser. Mat. 55 (1991), 303–317.

[5]   ———, Zyulyarkina N. D., Tightly embedded subgroups with abelian fusion, Trans. Inst. Math. Mechanics UB RAS, Ekaterinburg 2 (1993), 19–26.

[6]   ———, Cyclic TI-subgroups of order 4 in exceptional Chevalley groups, Trans. Inst. Math. Mechanics UB RAS, Ekaterinburg 3 (1994), to appear.

[7]   ———, Cyclic TI-subgroups of order 4 in classical Chevalley groups, to appear.

On the Fibonacci Groups, the Turk's Head Links and Hyperbolic 3-Manifolds

Alexander Mednykh and Andrei Vesnin

Abstract. This paper is devoted to the study of geometrical properties of the Fibonacci groups started by Helling, Kim and Mennicke [13]. We continue the investigation of compact Fibonacci manifolds uniformized by these groups. It is shown that their hyperbolic volumes correspond to limit ordinals in Thurston–Jørgensen's theorem. Moreover, these volumes are equal to the volumes of non-compact hyperbolic 3-manifolds arising as the complements of the Turk's head links. As well it is proved that Fibonacci manifolds are two-fold coverings of the three-dimensional sphere branched over the Turk's head links.

1. The Fibonacci Groups and Fibonacci Manifolds

The Fibonacci groups,

$$F(2, m) = \langle x_1, x_2, \ldots, x_m \mid x_i x_{i+1} = x_{i+2}, \ i \bmod m \rangle,$$

were introduced by J. Conway [3]. The natural algebraic question arises: whether the groups $F(2, m)$ are infinite or not. The answer is known due to [4], [6], [7] and [20]. The group $F(2, m)$ is finite if and only if $m = 1, 2, 3, 4, 5, 7$.

Algebraic generalizations of $F(2, m)$ were considered in [5].

A new stage in the investigation of the Fibonacci groups is connected with a remarkable work of H. Helling, A. C. Kim and J. Mennicke [7]. They remarked that the Fibonacci groups $F(2, 2n)$ with the even number of generators are interesting from the geometrical point of view. They showed that the group $F(2, 2n)$, $n \geq 4$, can be realized as a discrete co-compact subgroup of $\mathrm{PSL}_2(\mathbb{C})$, the full group of orientation-preserving isometries of the hyperbolic space $\mathbb{H}^3$. Therefore the group $F(2, 2n)$, $n \geq 4$, is isomorphic to the fundamental group of a closed orientable three-dimensional hyperbolic manifold. The group $F(2, 6)$ is isomorphic to the fundamental group of the compact orientable three-dimensional Euclidean Hantzsche–Wendt manifold [23]. The group $F(2, 4)$ is the cyclic group of order 5 isomorphic to the fundamental group of the lens space $L(5, 2)$ which is the compact orientable three-dimensional spherical manifold.

Summarizing we have that any group $F(2, 2n), n \geq 2$, can be realized as a discrete co-compact fixed point free subgroup of the isometries group of a simply connected 3-space X_n of a constant sectional curvature, where X_n is the 3-sphere S^3 for $n = 2$, the Euclidean 3-space $\mathbb{E}^3$ for $n = 3$ and the hyperbolic 3-space $\mathbb{H}^3$ for $n \geq 4$.

For any $n \geq 2$ we define a **Fibonacci manifold** as a corresponding manifold $M_n = X_n/F(2, 2n)$.

Various properties of the Fibonacci manifolds were studied in [8], [9], [12].

It was shown by H. Hilden, T. Lozano and J. Montesinos [9] that the Fibonacci manifold M_n can be obtained as an n-fold cyclic covering of S^3 branched over the figure-eight knot. In other words, if $\mathcal{O}(n)$ is the orbifold with the underlying space S^3 and the figure-eight knot with index n as its singular set (see Figure 1), then M_n is the n-fold covering of the orbifold $\mathcal{O}(n)$.

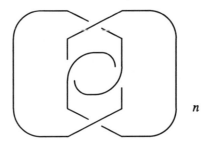

Figure 1. The singular set of the orbifold $\mathcal{O}(n)$.

The orbifold $\mathcal{O}(n)$ has a rotation symmetry of order 2 whose set of fixed points is disjoint from the singular set of $\mathcal{O}(n)$. The quotient $\mathcal{O}(n)$ by this symmetry is an orbifold $6_2^2(2, n)$ with underlying space S^3. Its singular set is a two-component link with can be described as the link 6_2^2 from Rolfsen's table [18]. Components of the singular set of $6_2^2(2, n)$ have indices 2 and n respectively (see Figure 2).

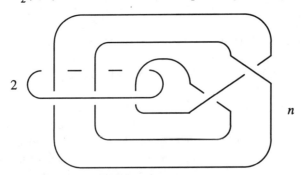

Figure 2. The singular set of the orbifold $6_2^2(2, n)$.

In the general case by an orbifold $6_2^2(m, n)$ with $m, n \in \mathbb{N}$ we mean an orbifold with the underlying space S^3 whose singular set is the two-component link 6_2^2 with indices m and n. This orbifold can be obtained by generalized Dehn surgery on two components of the link 6_2^2 with parameters $(m, 0)$ and $(n, 0)$ respectively.

For the Fibonacci manifolds M_n and the orbifolds $\mathcal{O}(n)$ and $6_2^2(2, n)$, we have the following diagram of covers:

$$M_n \xrightarrow{\ n\ } \mathcal{O}(n) \xrightarrow{\ 2\ } 6_2^2(2, n).$$

2. The Turk's Head Links

Now we consider one family of knots and links. We need this family for the description of the geometrical properties of the Fibonacci manifolds. Let us define a **Turk's head link** Th_n, $n \geq 2$, as a closed 3-strings braid $\left(\sigma_1 \sigma_2^{-1}\right)^n$. We note that Th_n is a three-component link if n is divisible by three and it is a knot otherwise. Members of the family Th_n are well-known. In particular, Th_2 is the figure-eight knot, Th_3 are the Borromean rings, Th_4 is the Turk's head knot 8_{18} (see Figure 3) and Th_5 is the knot 10_{123} from Rolfsen's table [18].

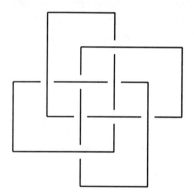

Figure 3. The Turk's head knot Th_4.

Various properties of this links were studied. Virtual Betti numbers of compact manifolds obtained by Dehn surgery on Turk's head links Th_n were studied in [10]. The symmetry groups of Th_n were calculated in [19].

We will be interesting in the consideration of a geometrical structure on knot complement $S^3 \setminus Th_n$.

We recall that three-dimensional hyperbolic manifold can be defined as a quotient space $M^3 = \mathbb{H}^3 / \Gamma$, where Γ is a discrete isometry group of the hyperbolic space $\mathbb{H}^3$ acting without fixed points. The concept of hyperbolic volume in $\mathbb{H}^3$ is carried over naturally to M^3. Henceforth we deal with hyperbolic manifolds of the finite volume.

Let $6_2^2(n, \infty)$ be an orbifold with the underlying space as a complement in S^3 to the component with the index ∞, and the singular set as the component with the index n.

It was shown by Thurston [21] that the manifold $S^3 \setminus Th_n$, $n \geq 2$ is hyperbolic and can be represented as an n-fold cyclic covering over the orbifold $6_2^2(n, \infty)$:

$$S^3 \setminus Th_n \xrightarrow{\ n\ } 6_2^2(n, \infty).$$

In particular, hyperbolic volumes of the following manifolds can be find in [1]

$$\mathrm{vol}\left(S^3 \setminus Th_2\right) = \quad 2.02988321\ldots$$
$$\mathrm{vol}\left(S^3 \setminus Th_3\right) = \quad 7.32772475\ldots$$
$$\mathrm{vol}\left(S^3 \setminus Th_4\right) = \quad 12.35090620\ldots$$
$$\mathrm{vol}\left(S^3 \setminus Th_5\right) = \quad 17.08570948\ldots$$

3. Volumes of the Hyperbolic Fibonacci Manifolds

We recall some properties of volumes of hyperbolic manifolds. The following remarkable theorem is due to Thurston and Jørgensen: *the set of volumes of three-dimensional hyperbolic manifolds form a well-ordered subset of the real line of type ω^ω*. Therefore we have the following picture for hyperbolic volumes (see Figure 4), where we show some known values:

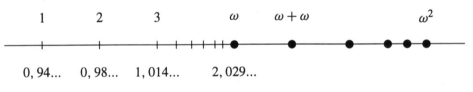

Figure 4.

In particular, it follows from the Thurston–Jørgensen theorem that there exists a three-dimensional hyperbolic manifold with the smallest volume. The minimal known value of volumes of compact manifolds is the volume of the Weeks–Matveev–Fomenko manifold [11], [22]. In [11] Matveev and Fomenko conjectured the structure of the initial part of the volume sequence of compact manifolds.

The volume of any non-compact hyperbolic manifold corresponds to the limit ordinal of the set ω^ω. The minimal known value of volumes of non-compact manifolds is equal to the double volume of the regular ideal tetrahedron and it is the volume of the complement of the figure-eight knot.

In [21] Thurston constructed an example of two non-compact manifolds with different numbers of cusps, but with the same volume. But it was unknown if there are compact manifolds which volumes correspond to limit ordinals.

From the following theorem one can see that the compact Fibonacci manifolds have this property.

Theorem 1. *For $n \geq 2$ we have*

$$\text{vol}(M_{2n}) = \text{vol}\left(S^3 \setminus Th_n\right).$$

In particular, the following assertions hold.

Corollary 1. *The volume of the manifold M_4 is equal to the volume of the complement of the figure-eight knot.*

Corollary 2. *The volume of the manifold M_6 is equal to the volume of the complement of the Borromean rings.*

Some properties of hyperbolic manifolds are defined by arithmeticity or non-arithmeticity of their fundamental group [2]. As proven in [7] and [9] the manifold M_n is arithmetic for $n = 4, 5, 6, 8, 12$ and non-arithmetic otherwise. It is well-known [17] that Th_2, the figure-eight knot, is the only arithmetic knot. It is shown in [21] that Th_3, the Borromean rings, is also arithmetic.

Corollary 3. *We have the following table of manifolds of equal volume:*

n	M_{2n}	$S^3 \setminus Th_n$
2	*arithmetic*	*arithmetic*
3	*arithmetic*	*arithmetic*
4	*arithmetic*	*non-arithmetic*
5	*non-arithmetic*	*non-arithmetic*

Therefore the Fibonacci manifold M_8 and the Turk's head knot complement $S^3 \setminus Th_4$ are an arithmetic compact manifold and a non-arithmetic non-compact manifold with the same volume. A. W. Reid kindly informed us about his unpublished results where he constructed compact arithmetic and non-compact arithmetic manifolds with the same volume by number-theoretical methods.

The proof of the Theorem 1 is based on the following ideas. Let $\Lambda(x)$ be the Lobachevsky function defined by the formula

$$\Lambda(x) = -\int_0^x \ln|2 \sin \zeta| d\zeta.$$

It is well-known [15] that the volume of the ideal tetrahedron in $\mathbb{H}^3$ with dihedral angles α, β and γ, where $\alpha + \beta + \gamma = \pi$, is equal to

$$\Lambda(\alpha) + \Lambda(\beta) + \Lambda(\gamma).$$

Theorem 1 is the consequence of the following two lemmas.

Lemma 1. *For $m \geq 4$ the orbifold $6_2^2(2, m)$ is hyperbolic and*

$$\text{vol}(6_2^2(2, m)) = \Lambda(\beta + \delta) + \Lambda(\beta - \delta),$$

where $\delta = \frac{\pi}{m}$, $\beta = \frac{1}{2} \arccos(\cos(2\delta) - \frac{1}{2})$.

Lemma 2. *For $n \geq 2$ the orbifold $6_2^2(n, \infty)$ is hyperbolic and*

$$\text{vol}(6_2^2(n, \infty)) = 4\left(\Lambda(\alpha + \gamma) + \Lambda(\alpha - \gamma)\right),$$

where $\gamma = \frac{\pi}{2n}$, $\alpha = \frac{1}{2} \arccos(\cos(2\gamma) - \frac{1}{2})$.

According to the diagram (1) the manifold M_{2n} is the $4n$-fold covering of the orbifold $6_2^2(2, 2n)$. Therefore from Lemma 1 for $m = 2n$ we obtain

$$\text{vol}(M_{2n}) = 4n \, \text{vol}(6_2^2(2, 2n)) = 4n\left(\Lambda(\alpha + \gamma) + \Lambda(\alpha - \gamma)\right),$$

where α and γ are the same as in Lemma 2.

According to the diagram (2) the manifold $S^3 \setminus Th_n$ is the n-fold covering of the orbifold $6_2^2(n, \infty)$. Therefore from Lemma 2 we have

$$\text{vol}(S^3 \setminus Th_n) = n \, \text{vol} \, 6_2^2(n, \infty) = 4n\left(\Lambda(\alpha + \gamma) + \Lambda(\alpha - \gamma)\right).$$

Comparing (5) and (6) we finished the proof of Theorem 1.

4. Fibonacci Manifolds As Two-Fold Coverings

The following theorem gives one more topological relation between Fibonacci manifolds M_n and Turk's head links Th_n.

Theorem 2. *For any $n \geq 2$ the manifold M_n can be represented as a two-fold covering of the S^3 branched over Th_n.*

Consider the orbifold $Th_n(2)$ with underlying space S^3 and singular set as the link Th_n with index 2. Then we have the following covering diagram (Figure 5).

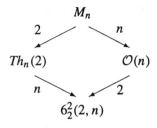

Figure 5. The diagram of coverings.

We recall that Meyerhoff and Neumann [14] have obtained a compact hyperbolic manifold $N = W(3, -2; 6, -1)$ by means of Dehn surgery on the Whitehead link whose volume approximately (up to 10^{-50}) equals to the volume of the regular ideal tetrahedron. They asked if this volumes are strictly equal and if the manifold N is arithmetic over the field $\mathbb{Q}$. As a consequence of the above results we have affirmative answers on this questions.

Theorem 3. *The Meyerhoff–Neumann manifold is arithmetic over the field* $\mathbb{Q}(\sqrt{-3})$ *and its volume equals to the volume of the regular ideal tetrahedron.*

For the demonstration of this theorem we remark that by Theorem 2 the manifold M_4 can be realized as a two-fold covering of the three-sphere branched over the knot Th_4 (see Figure 3).

The orbifold $Th_4(2)$ has an evident rotational symmetry ρ of order 4. The quotient $Th_4(2)$ by a symmetry ρ^2 is an orbifold $\Omega(2, 2)$ with the underlying space S^3. The singular set of the orbifold $\Omega(2, 2)$ consists of the two-component link Ω (see Figure 6) whose components are labelled by 2.

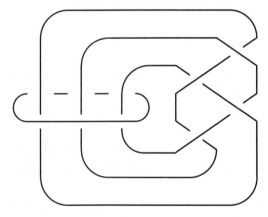

Figure 6. The link Ω.

By Montesinos theorem [16] any manifold obtained by a Dehn surgery on a strongly invertible link can be represented as a two-fold covering of S^3 branched over some link. This theorem also gives an algorithm for constructing such link. Using the Montesinos algorithm for N we get that N is the two-fold covering of the three dimensional sphere branched over the same link Ω. Therefore we have the following diagram of covers (see Figure 7):

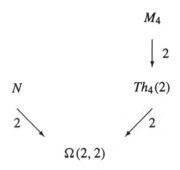

Figure 7. The diagram of coverings.

So manifolds N and M_4 are commensurable. Therefore the manifold N is arithmetic over the field $\mathbb{Q}(\sqrt{-3})$, because it is true for M_4 [7]. Moreover by Corollary 1 we have: $\mathrm{vol}(N) = \frac{1}{2}\,\mathrm{vol}(M_4) =$ the volume of the ideal regular tetrahedron $= 1.01494\ldots$.

We also remark that the manifold N does not appear in the list of Matveev and Fomenko in [11].

The second named author acknowledges financial support from Russian Foundation of Fundamental Investigations and from Korea Science and Engineering Foundation. Cordial thanks go to organizers of the Conference.

References

[1] Adams C., Hildebrand M., Weeks J., Hyperbolic invariants of knots and links, Trans. Amer. Math. Soc. 326 (1991), 1–56.

[2] Borel A., Commensurability classes and hyperbolic volumes, Annali Sci. Norm. Pisa 8 (1991), 1–33.

[3] Conway J., Advanced problem 5327, Amer. Math. Monthly 72 (1965), 915.

[4] Conway J. et al., Solution to Advanced problem 5327, Amer. Math. Monthly 74 (1967), 91–93.

[5] Johnson, D. L., Wamsley, J. W., Wright, D., The Fibonacci groups, Proc. London Math. Soc. 29 (1974), 577–592.

[6] Havas G., Computer aided determination of a Fibonacci group, Bull. Austral. Math. Soc. 15 (1976), 297–305.

[7] Helling H., Kim A. C., Mennicke J., A geometric study of Fibonacci groups, SFB 343 Bielefeld, Diskrete Strukturen in der Mathematik, preprint (1990).

[8] Hempel J., The lattice of branched covers over the figure–eight knot, Topology and its Appl. 34 (1990), 183–201.

[9] Hilden H. M., Lozano M. T., Montesinos J. M., The arithmeticy of the figure-knot orbifolds, preprint (1989).

[10] Kojima S., Long D. D., Virtual Betti numbers of some hyperbolic 3-manifolds, in: A Fete of Topology, Academic Press Inc., 1988, 417–442.

[11] Matveev V. S., Fomenko A. T., Constant energy surfaces of Hamiltonian systems, enumeration of three-dimensional manifolds in increasing order of complexity, and computation of volumes of closed hyperbolic manifolds, Russian Math. Surveys 43 (1988), 1, 3–24.

[12] Mednykh A. D., Vesnin A. Ju., On limit ordinals in the Thurston–Jorgensen theorem on the volumes of three-dimensional hyperbolic manifolds, Dokl. RAS. 336 (1994), 7–10 (in Russian).

[13] Mennicke J. L., On Fibonacci groups and some other groups, in: Proceedings of "GROUPS – KOREA 1988", Lecture Notes in Math. 1062, 117–123.

[14] Meyerhoff R., Neumann W. D., An asymptotic formula for the eta invariants of hyperbolic 3-manifolds, Comment. Math. Helv. 67 (1992), 28–46.

[15] Milnor J., Hyperbolic geometry: the first 150 years, Bull. Amer. Math. Soc. 6 (1982), 9–25.

[16] Montesinos J. M., Whitten W., Constructions of two–fold branched covering spaces, Pacific J. Math. 125 (1986), 415–446.

[17] Reid A., Arithmeticity of knot complements, J. London Math. Soc. 43 (1991), 171–184.

[18] Rolfsen D., Knots and Links, Publish of Perish Inc., Berkely Ca., 1976.

[19] Sakuma M., Weeks J., Examples of canonical decompositions of hyperbolic link complements, Preprint, 1993.

[20] Thomas R. M., The Fibonacci groups $F(2, 2m)$, Bull. London Math. Soc. 21 (1989), 463–465.

[21] Thurston W. P., The Geometry and Topology of Three-Manifolds, Lecture Notes, Princeton University 1980.

[22] Weeks J., Hyperbolic structures on 3-manifolds, Princeton Univ. Ph. D. Thesis, 1985.

[23] Zimmermann B., On the Hantzche–Wendt manifold, Monatsh. Math. 110 (1990), 321–327.

Alternating and Symmetric Groups As Quotients of $G^{5,6,36}$

Q. Mustaq and N. A. Zafar

Abstract. The study of groups $G^{5,6,m}$ is of interest for a number of reasons. For instance, they are symmetric groups of the map $\{5, 6\}_m$ which is constructed from the tessellation $\{5, 6\}$ of the hyperbolic plane by identifying two points, at a distance m apart, along a Petrie path. In this paper, by using a diagrammatic argument, we show that all but finitely many A_n and S_n are factor groups of $G^{5,6,36}$.

1. Introduction

It is well known that the modular group $\Gamma = \langle x, y : x^2 = y^3 = 1 \rangle$ has the property that every alternating and symmetric group is a homomorphic image of Γ except A_6, A_7, A_8, S_5, S_7, or S_8. G. Higman has questioned which discrete reflective hyperbolic groups also exhibit this type of behavior. Let G be an infinite, finitely presented group. We say that G has **Property H** if there is an integer $N > 1$ such that either A_n or S_n is a homomorphic image of G for all $n > N$. For $[p, q]$ we will use the presentation $[p, q] = \langle x, y, t : x^2 = y^p = (xy)^q = t^2 = (xt)^2 = (yt)^2 = 1 \rangle$. If we let $x = R_1 R_2$, $y = R_2 R_3$ and $t = R_2$ then $[p, q]$ has the presentation $\langle R_1, R_2, R_3 : R_1{}^2 = R_2{}^2 = R_3{}^2 = (R_1 R_2)^2 = (R_2 R_3)^p = (R_1 R_3)^q = 1 \rangle$. It is an infinite Coxeter group generated by the reflections in the sides of the right hyperbolic triangle forming the fundamental region of the tessellation. Let us assume without loss of generality that $p \leq q$. Geometrically, x and y are rotations at the vertices of the fundamental region having angles $\pi/2$ and π/p respectively, and t is the reflection on the side joining them. Let $[p, q]^+$ denote the index 2 subgroup of orientation preserving isometries in $[p, q]$. For more details see [2].

It has been shown in [1] that a factor group $G^{6,6,6}$ of $[3, 6]$ has Property H for all $q > 6$ and in [5] that $[p, 5p - 3]$ has Property H for all $p \geq 5$. In this paper we shall show that a factor group $G^{5,6,36}$ of $[5, 6]$ has Property H. The Property H has been proved for other useful values of p, q in [3] and [4].

2. Coset Diagrams

The method we use in proving the results in this paper is mainly of [1], [3], [4], [5] and [6].

For each integer $q > 4$ we constructed coset diagram for the groups $[5, q]^+ = \langle x, y : x^2 = y^5 = (xy)^q = 1 \rangle$. A coset diagram with n vertices depicts the transitive permutation representation of $[5, q]^+$ on the cosets of some subgroup of index n. Fixed points of y are denoted by heavy dots, and 5-cycles of y by pentagons whose vertices are permuted anti-clockwise by y. Any two points and the fixed points of x are also denoted by heavy dots. If the diagram possesses a vertical axis of symmetry, then it may be regarded as a cosets diagram for the extended group $[5, q] = \langle x, y, t : x^2 = y^5 = (xy)^q = t^2 = (xt)^2 = (yt)^2 = 1 \rangle$ where $q > 4$. In this case the action of t is given by reflection in the axis of symmetry.

For example, the following diagram depicts a (transitive) permutation representation of $[5, 7]^+$ of degree 15, in which x acts as (3 7) (4 11) (5 6) (13 14), y acts as (1 2 3 4 5) (6 7 8 9 10) (11 12 13 14 15) and t acts as (1 2) (3 5) (6 7) (8 10) (12 15) (13 14).

In an arbitrary permutation representation of the group $[5, q]$, any two points λ, μ which are fixed by x and are such that both $(xy)^j$ and t map λ to μ, will be said to form a (j)-handle. We shall write such a (j)-handle as $[\lambda, \mu]_j$. When $j = 1, 2$ or 3, the configurations appear on the central axes of the corresponding diagrams as follows:

(1)-handle $[a, b]_1$

(2)-handle $[\alpha, \beta]_2$

(3)-handle $[A, B]_3$

3. Composition of Coset Diagrams

We require a method (propounded by W. W. Stothers [6]) of stitching coset diagrams together, to give transitive permutations of the group $[p, q]$ of larger degree.

Remark 3.1. (i) Suppose $D(m)$ and $D(n)$ are coset diagrams for [5,6] both containing handles $[a, b]_1$ and $[a', b']_1$ (of the same type) respectively. Place the two diagrams on a common vertical axis of symmetry, one above the other, and add two new x-edges, the first joining a to a' and the second b to b'.

(ii) It is easy to see that the resulting diagram is also a coset diagram for [5,6]. To begin with, the relation $x^2 = y^5 = t^2 = (xt)^2 = (yt)^2 = 1$ is still satisfied. Secondly, as in [1], if $(a, b, c_1, c_2, c_3, c_4)$ and $(a', b', c'_1, c'_2, c'_3, c'_4)$ are the appropriate 6-cycles of xy in the representation of $G^{5,6,36}$ depicted by V and V' respectively, then since $(b, b') (a, a') (a, b, c_1, c_2, c_3, c_4) (a', b', c'_1, c'_2, c'_3, c'_4){=}(a, b', c_1, c_2, c_3, c_4)$ $(a', b, c'_1, c'_2, c'_3, c'_4)$, the two later 6-cycles will be cycles of xy in the new diagram. Other cycles of xy are unchanged, so xy still has order 6.

(iii) The similar thing happens to the cycles of xyt as a result of this composition.

(iv) For later use, we notice also that cycles of the permutation(s) induced by xy^2t and xy^3t are affected in the same sort of way: provided a' and b' lie in distinct cycles of xy^2t and xy^3t, the cycles ending in a and a' will be juxtaposed to form a single cycle, and then those ending in b and b' will be similarly combined.

(v) The action of xyt (or, indeed, for any element of the group $[5, q]$) can be determined easily from the inspection of the coset diagram.

4. Main Result

We now prove that for all but finitely many positive integer n, the alternating group A_n and symmetric group S_n are homomorphic images of $G^{5,6,36}$. But before we come to this we state Jordan's Theorem (13.9, [7]).

Theorem 4.1. *Let p be a prime number and G a primitive group of degree $n = p + k$ with $k \geq 3$. If G contains an element of degree and order p, then G is either alternating or symmetric.*

To prove Theorem 4.6, we use a diagrammatic argument very similar to the one adopted in [1], [3], [4] and [5]. Specifically, we use coset diagrams for the group $G^{5,6,36}$, and the method of composition [6] for joining the diagram together, in order to obtain a transitive permutation representation of $G^{5,6,36}$ of arbitrary large degree.

Proof of the theorem consists of two parts. We shall use four diagram as *basic diagrams* in one part and three diagrams in the second part of the proof. In order to facilitate composition, each of them will possess on its vertical axis at least one of the handles discussed in Section 2.

We will need four diagrams which we will connect together to form the required one. Each of these diagrams is given a specification, consisting of the degree of the corresponding permutation representation of the group $G^{5,6,36}$, the number of handles that will be used, the parity of the action t, and the cycle structure of xyt and xy^2t and xy^3t.

Remark 4.2. For instance, the specification for the diagram $D(17)$ is : 1(1)-handle, t is even, xyt: b(a7)6.2 and xy^2t: (a2)(b6)5.2, which means that on the 17 points of the diagram, t is even, and xyt has four cycles, one of length 8 containing the point a, one of the length 1 containing the point b, one of length 6, and one is of length 2, and xy^2t has also four cycles, one of length 3 containing the point a, second of length 7 containing point b, third cycle is of length 5 and the last one is of length 2.

Lemma 4.3. *The diagram $D(n) = pD(35) + qD(12) + rD(20) + D(17)$ is a connected coset diagram for $G^{5,6,36}$ acting on n points.*

Proof. Consider the following coset diagrams.

$D(17)$, 1(1)-handle, t is even
$xyt : b(a7)6.2,\ xy^2t : (a2)(b6)5.2$

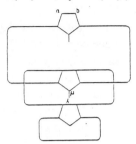

$D(20)$, 2(1)-handle, t is even
$xyt : b(a5)a'(b'5)3.3,\ xy^2t : (a5)(b2)(a'2)(b'5)1.1,\ xy^3t : (a2)(b5)(a'5)(b'2)1.1$

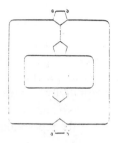

$D(35)$, 2(1)-handle, t is odd
$xyt : (a11)b(a'3)b'2.12.3,\ xy^2t : (a5)(b2)(a'5)(b'8)6.3.2,$
$xy^3t : (a2)(b5)(a'8)(b'5)6.3.2$

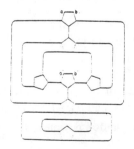

$D(12)$, 2(1)-handle, t is even
$xyt : (a4)b(b'4)a',\ xy^2t : (a2)(b2)(a'2)(b'2),\ xy^3t : (a2)(b2)(a'2)(b'2)$

Suppose the positive integer n is expressible as a linear combination of $pD(35) +$ $qD(12) + rD(20) + D(17)$, where p, q and r are positive integers. We take p number of $D(35)$ diagrams, q number of $D(12)$ diagrams, r number of $D(20)$ diagrams and one $D(17)$ diagram, and connect them together by the order as explained in Section 3. Join two copies of $D(35)$ with one copy of $D(12)$. The method of joining will be such that the handle of length 4 of $D(35)$ is joined with one handle of $D(12)$. Now with one copy of $D(35)$ join one copy of $D(20)$ and in the similar way process can be continued. Finally we connect one copy of $D(17)$ with $D(35)$. Until we get diagram with n vertices. The resulting diagram, by virtue of Remarks 3.1(i) and 3.1(ii), will be a diagram for the group $G^{5,6,36}$. The reflection t acts as an even or odd permutation depending upon the values of p, q and r.   □

Lemma 4.4. *The representation of $G^{5,6,36}$ is primitive on the n vertices of $D(n)$.*

Proof. Suppose the representation of $G^{5,6,36}$ is imprimitive on the n vertices of $D(n)$. Then, since $(xy^2t)^{144}$ fixes $n - 5$ vertices, the 5 vertices of the cycle must lie in the same block of imprimitivity. In this cycle, apart from other vertices there are the ones labelled λ and μ. Also, the fixed vertices λ and μ of y are such that $x(\lambda) = \mu$ and $x(\lambda) = \lambda$. Thus, the block of imprimitivity is preserved by all three x, y and t. By transitivity this implies that the block has n number of vertices, which is a contradiction to the fact $D(n)$ has n vertices. Hence the representation is primitive.   □

Theorem 4.5. *For all but finitely many values of n, either A_n or S_n is a homomorphic image of $G^{5,6,36}$.*

Proof. The length of every cycle of xyt will be divisor of 36 and so this diagram $D(n)$ gives a permutation representation of the group $G^{5,6,36} = \langle x, y, t : x^2 = y^5 = t^2 = (xy)^6 = (xt)^2 = (yt)^2 = (xyt)^{36} = 1 \rangle$.

Note that the cycles of xy^2t are all of length 1, 2, 3, 5, 6, 9, 12, 16. Moreover, as only one diagram $D(17)$ is used, there will be only one cycle of length 5 (Note that this cycle contains the fixed vertices λ and μ). All these numbers, expect 5, are divisors of 144. So the element $(xy^2t)^{144}$ yields a power of the cycle, fixing the remaining $n - 5$ vertices.

By the Theorem 4.1 and Lemmas 4.3 and 4.4, it now follows that the permutation representation is alternating or symmetric of degree n. Thus the result.

Since t is even or odd depending upon the values of p, q and r, if we take p to be even, and for all values of q and r, t gives odd permutation, so we get S_n.

To obtain A_n in the same way, we express n as a linear combination of $pD(35) +$ $qD(12) + rD(20)$, from the same sort of chain using p copies of $D(35)$, q copies of $D(12)$ and r copies of $D(20)$. It should be noted that we are not using any copy of $D(17)$. In this case we will take the element xy^3t, whose every cycle is of length 1, 2, 3, 6, 9, 12. Here one cycle is of prime length 3. Take $g = (xy^3t)^2$,

since $(xy^3t)^2 = xy^3txy^3t = xy^3(tx)y^3t = xy^3x^{-1}t^{-1}y^3t = xy^3x^{-1}(t^{-1}y)y^2t = xy^3x^{-1}(y^{-1}t^{-1})yt = xy^3x^{-1}y^{-1}y^{-1}t^{-1}yt = xy^3x^{-1}y^{-2}y^{-1}t^{-1}t = xy^3x^{-1}y^{-3}$,
so the resultant diagram again gives a transitive representation of $\langle x, y \rangle$ on n points, but this time x, y and t all will be even for any values of q and r, taking p even. In this case the cycles of xy^3t are all of length 1, 2, 3, 6, 9, 12, and one cycle is of length 3 containing the points λ, μ and τ. Since we know that xy^2t and xy^3t are conjugate to each other so we can do so. Now, by a similar argument as in the proof of Lemma 4.4, we note that $G^{5,6,36}$ is primitive on the n vertices of our chain. Using the previous argument we say that x, y and t generate A_n. This completes the proof of the theorem. $\quad\square$

We can rephrase the theorem in the following form.

Corollary 4.6. *For all but finitely many values of n, the alternating group A_n has the presentation $\langle x, y, t : x^2 = y^5 = (xy)^6 = (x^{-1}y^{-1}xy)^{18} = 1 \rangle$.*

Acknowledgement. The first author wishes to thank University Brunei Darussalam for providing financial assistance for participation in the conference. He wishes also to thank the organizers of the conference Groups—Korea 1994, for their hospitality.

References

[1] M. D. E. Conder, On the group $G^{6,6,6}$, Quart. J. Math. Oxford Ser. (2), 39 (1988), 175–183.

[2] H. S. M. Coxeter, The abstract group $G^{m,n,p}$, Trans. Amer. Math. Soc. 45 (1939), 73–150.

[3] Q. Mustaq, M. Ashiq and T. Maqsood, The symmetric group as a quotient of $G^{3,8,720}$, Math. Japonica 37 (1) (1992), 9–16.

[4] Q. Mustaq, M. Ashiq and T. Maqsood, Factor groups of $G^{5,5,120}$, Comm. Algebra 12 (20) (1992), 3759–3767.

[5] Q. Mustaq and G.-C. Rota, Alternating groups as quotients of two generator groups, Adv. in Math. 96 (1992), 113–121.

[6] W. W. Stothers, Subgroups of the (2,3,7) Triangle group, Manuscripta Math. 20 (1977), 323–334.

[7] H. Wielandt, Finite Permutation Groups, Academic Press, London 1964.

Covering Groups by Subgroups

B. H. Neumann

Abstract. A number of questions about covering a group by finitely many subgroups is discussed.

A colleague asked me, not long ago, what I knew about covering a group irredundantly by 5 subgroups; but when after a while I came with an answer, he no longer needed it and had, in fact, forgotten he had asked the question.

However, this had revived my interest in a subject I had first worked on 40 years ago, see [1]. In that paper I had defined Δ_n^* as the least upper bound of the index of the intersection of n subgroups in a group they cover irredundantly, and I had shown that $\Delta_3^* = 4$ and $\Delta_4^* = 9$; for greater values of n the paper contained only very large upper bounds.

In what follows I give lower bounds for Δ_n^*, which I conjecture, on experimental evidence that I do not reproduce here, to be the exact bounds.

For $n = 5$ all non-cyclic abelian groups of order 16 can be covered irredundantly by 5 subgroups with trivial intersection, and thus

$$\Delta_5^* \geq 16.$$

If this bound is, as I believe, sharp, it suggest that generally

$$\Delta_n^* = (n - 1)^2;$$

and that is what I at first conjectured on the feeble evidence of $n = 3, 4, 5$. This is, however, false. In fact

$$\Delta_n^* \geq 2^{n-1},$$

as is seen by considering the elementary abelian group of order 2^{n-1}, written additively, which is covered irredundantly and with trivial intersection by the $n - 1$ subgroups obtained from some fixed basis by putting the coefficient of each basis element equal to zero in turn, with all the other coefficients ranging independently over $\{0, 1\}$, and additionally the "diagonal" subgroup which has equal coefficients 0 or 1 for all basis elements.

However, for $n \geq 6$ one can do even better, namely

$$\Delta_n^* \geq 9 \times 2^{n-4}.$$

To see this, we consider the vectors $(x, y, z_1, z_2, \ldots, z_{n-4})$ under addition, where x and y are counted *modulo* 3, and the z_i *modulo* 2. They form an abelian group of order $9 \times 2^{n-4}$. In this group one takes the $n - 2$ subgroups consisting of the vectors that have a 0 in one place and everywhere else all possible values [*modulo* 3 or 2, as the case may be], and in addition the two "diagonal" subgroups consisting of the vectors $(x, x, z, z, \ldots, z)$ and $(x, 2x, z, z, \ldots, z)$, respectively, [again *modulo* 3 or 2, as the case may be]. It is easy to verify that these subgroups cover the whole group, that they do it irredundantly, and that their intersection is trivial [in fact the intersection of the first $n - 2$ subgroups is clearly trivial].

In the lecture I raised the question of finding an irredundant covering of a group by subgroups whose intersection is not normal in the whole group. Professor Noboru Ito quickly produced an example, namely the symmetric group of degree 4 with a cover of 5 subgroups that intersect in a non-normal subgroup of order 2. A slightly smaller example is the elementary abelian group of order 9 extended by the automorphism of order 2 that inverts all elements. The 4 subgroups of index 3 in the abelian subgroup of order 9, each with the involutory automorphism adjoined, cover the whole group irredundantly, and their intersection is the subgroup of order 2 generated by that automorphism, which is clearly not normal in the whole group.

An open problem, not mentioned in my talk, is the following: is it true that among the groups of some fixed finite order that can be covered irredundantly with proper subgroups with trivial intersection, the abelian groups need no more such subgroups than any of the non-abelian groups? I guess that this is not the case.

References

[1] B. H. Neumann, Groups covered by finitely many cosets, Publicationes Mathematicae 3 (1954), 227–242 [=Selected Works of B. H. Neumann and Hanna Neumann, Winnipeg, Canada, 1988, Volume V, 954–969].

Kleinian Groups Generated by Rotations

Walter D. Neumann

Abstract. We discuss which Kleinian groups are commensurable with Kleinian groups generated by rotations, with particular emphasis on Kleinian groups that arise from Dehn surgery on a knot.

Introduction

In a problem session, organized by A. Kim, of the 1991 German–Korean–SEAMS Conference on Geometry, E. Vinberg asked for a cocompact Kleinian group which is not commensurable with a group generated by rotations (elements of finite order). Examples are, in fact, not hard to find. In this note we describe in some detail which Kleinian groups have this property among the Kleinian groups that occur as fundamental groups of Dehn surgeries on knots. We also briefly discuss some related questions.

In the following Γ and Λ will always denote Kleinian groups of finite covolume, that is, discrete subgroups of $PSL(2, \mathbb{C}) = Isom^+(\mathbb{H}^3)$ such that the orbifold $\mathbb{H}^3 / \Gamma$ or $\mathbb{H}^3 / \Lambda$ has finite volume. They are *commensurable* if they have isomorphic subgroups of finite index. By Mostow–Prasad rigidity this is equivalent to the condition that they can be conjugated within $PSL(2, \mathbb{C})$ so their intersection has finite index in each.

Kleinian groups commensurable with groups generated by reflections (rather than rotations) have been studied by E. Vinberg [V] and E. M. Andreev [An]. They are a very restricted class of Kleinian groups. For example, as is pointed out in [NR], the invariant trace field of such a Kleinian group — indeed, of any Kleinian group commensurable with a non-orientation-preserving subgroup of $Isom(\mathbb{H}^3)$ — has to be preserved by complex conjugation (and the same for the invariant quaternion algebra), which is a rare occurrence. For a Kleinian group generated by rotations I know of no similar restriction on the invariant trace field, although the invariant trace field severely restricts the possible orders of the rotations involved (cf. Theorem 3 below).

In the following section we discuss Dehn surgery on knots and state and prove our main theorem (Theorem 1). In Section 2 we discuss the case of knot complements themselves: a hyperbolic knot complement is commensurable with a $\mathbb{H}^3 / \Lambda$ with Λ generated by rotations if the knot is invertible and conjecturally in just one other case (Theorem 2). In Section 3 we make some additional comments about what rotations can be contained in Kleinian groups commensurable with a given Kleinian group Γ. If such rotations do not occur in Γ itself, it is reasonable to call them *hidden rotations* of Γ. We show that "most" Kleinian groups have no hidden rotations of orders other than 2, 3, 4, or 6.

1. Dehn Surgery on Knots

Let (S^3, K) be a hyperbolic knot (i.e., $M = S^3 - K$ admits a complete finite-volume hyperbolic structure). Thurston's hyperbolic Dehn surgery theorem ([T], [N-Z]) implies that by excluding finitely many integer pairs (p, q) we may assume the result $M(p, q)$ of (p, q)-Dehn-Surgery on (S^3, K) is hyperbolic. Say $M(p, q) = \mathbb{H}^3/\Gamma(p, q)$. Group-theoretically, $\Gamma(p, q)$ is the result of factoring $\pi_1(S^3 - K)$ by the normal closure of the element $m^p l^q$, where $m, l \in \pi_1(S^3 - K)$ are represented by a meridian and a longitude in a torus neighbourhood boundary of K in S^3.

Theorem 1.
(i)   If (S^3, K) is non-invertible and does not branched cyclic cover a torus knot, then for all but finitely many values of p/q the above group $\Gamma(p, q)$ is not commensurable with a group generated by rotations.
(ii) If (S^3, K) is an invertible knot, then each $\Gamma(p, q)$ has index 2 in a group generated by rotations.
(iii) If (S^3, K) is a d-fold cyclic cover of the unknot, then each $\Gamma(p, q)$ is commensurable with a group generated by rotations.
(iv) If (S^3, K) is not in one of the above cases, then infinitely many $\Gamma(p, q)$ are commensurable with groups generated by rotations and infinitely many are not.

Example. Figure 1 shows a knot 3-fold cyclic covering an unknot. This knot (and its obvious generalizations) provides an example for case (iii). However, in some sense "most" knots have no symmetries at all, so they satisfy case (i).

Proof. It is well known (cf. [Ar]) that $\Lambda \subset Isom^+(\mathbb{H}^3)$ is generated by rotations if and only if the space $\mathbb{H}^3/\Lambda$ is simply connected.

Let $M(p, q) = \mathbb{H}^3/\Gamma(p, q)$ be as in the theorem. We shall abbreviate $\Gamma = \Gamma(p, q)$. We first assume (S^3, K) is non-invertible.

Thurston's Dehn surgery theorem (loc. cit.) also says $\text{vol}(M(p, q)) < \text{vol}(M)$, and since, by Borel [B], only finitely many hyperbolic orbifolds with volume below a given bound are arithmetic, we may, by excluding finitely many (p, q), assume $M(p, q)$ is non-arithmetic. Then by Margulis (cf. [Z], ch. 6) there exists a maximal element Γ_0 in the commensurability class of Γ. Denote $M_0(p, q) = \mathbb{H}^3/\Gamma_0$. The map

$M(p, q) \rightarrow M_0(p, q)$ is a covering map of orbifolds. We shall show first that for (p, q) sufficiently large it is a cyclic covering.

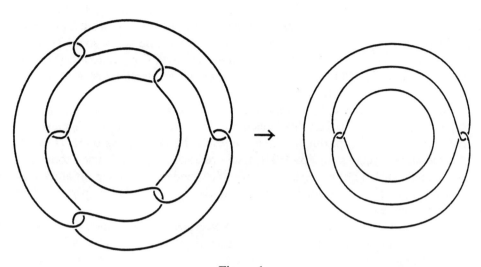

Figure 1

There is a bound on the degree of $M(p, q) \rightarrow M_0(p, q)$, independent of (p, q), since a complete hyperbolic orbifold has volume above some fixed positive bound (Margulis). If (p, q) is sufficiently large, then the geodesic γ added to M by Dehn surgery is much shorter than the other closed geodesics of $M(p, q)$. It follows that, for (p, q) sufficiently large, the image γ_0 of γ in $M_0(p, q)$ is the shortest closed geodesic of $M_0(p, q)$, and that no other geodesic of $M(p, q)$ could cover γ_0. Then $M = M(p, q) - \gamma$ covers $M_0(p, q) - \gamma_0$. We exclude the finitely many (p, q) for which this is not the case.

Now, consider this covering restricted to the boundary T of a solid torus neighborhood of γ. The image T_0 in $M_0(p, q) - \gamma_0$ is an orbifold covered by the torus. Let A and A_0 be the fundamental groups of T and T_0. If A is not normal in A_0, that is the covering $T \rightarrow T_0$ is not regular, then T_0 must be a triangle orbifold of type $(2, 4, 4)$, $(2, 3, 6)$, or $(3, 3, 3)$, since these are the only orbifolds covered by a torus for which the orbifold fundamental group A_0 contains non-normal torsion-free subgroups. However, such a covering cannot extend to a solid torus and the covering $T \rightarrow T_0$ extends to a tubular neighborhood of γ. Hence the covering $T \rightarrow T_0$ is regular. As described for instance in [R], it follows that the covering $M = M(p, q) - \gamma \rightarrow M_0(p, q) - \gamma_0$ is regular, since $\pi_1(T)$ normally generates $\pi_1(M)$ (M is a knot complement); we repeat the argument for completeness.

We show Γ is normal in Γ_0 by showing that it is the normal closure of a suitable subgroup. If we identify A_0 with its image in Γ_0 then $A = A_0 \cap \Gamma$. The degree of our covering is on the one hand equal to $|A_0/A| = |A_0/(A_0 \cap \Gamma)| = |A_0 \Gamma / \Gamma|$ and on the other hand equal to $|\Gamma_0/\Gamma|$, so $A_0\Gamma = \Gamma_0$. The normal closure of A in $\Gamma_0 = A_0\Gamma$ therefore equals the normal closure of A in Γ, which is Γ itself, as pointed out above.

The covering transformation group G for $M \to M_0(p, q) - \gamma_0$ is $G = \Gamma_0/\Gamma = A_0/A$, so it is the same as for $T \to T_0$. Since the longitude of the knot complement $M(p, q) - \gamma = S^3 - K = M$ generates the kernel of $H_1(T) \to H_1(M)$, it is preserved by the G-action up to sign. Hence the same is true for the meridian, so the G-action extends to an action on S^3. By the solution to the Smith Conjecture (cf. [MB]) G is cyclic or dihedral. But dihedral is excluded by our assumption that $K \subset S^3$ is a non-invertible knot, so G is cyclic.

Now suppose that $\Lambda \subset \Gamma_0$ exists so that Λ is generated by rotations, i.e., the space $\mathbb{H}^3/\Lambda$ is simply connected (cf. first sentence of this section). Then $\mathbb{H}^3/\Gamma_0$ must have finite fundamental group. But $\mathbb{H}^3/\Gamma_0$ is the result of a Dehn filling of M/G, where G is the above cyclic group. By [BH], since we assumed M/G is not a torus knot complement, such a Dehn surgery can give finite fundamental group for at most 24 values of p/q. Thus by excluding these (p, q) we can avoid this, so part (i) of the theorem is proved.

(We note that this final exclusion may be necessary. For example, if (S^3, K) is the $(-2, 3, 7)$-pretzel knot then $(r, 1)$-Dehn surgery gives a manifold with finite fundamental group for $r = 17, 18, 19$ — see [BH] — so in this case if $\Gamma(p, q)$ is a Kleinian group with $p/q = 17, 18, 19$ then it has a subgroup of finite index generated by rotations.)

For part (iii) of the theorem suppose that $(S^3, K)/G$ is the unkot. Then the underlying space of $M_0(p, q) = M(p, q)/G$ is the result of (p, dq)-Dehn-surgery on this unknot with $d = |G|$. That is, $M_0(p, q)$ has underlying space the lens space $L(p, dq)$, which has a simply-connected d-fold covering, proving part (iii). The proof of (iv) is similar: in this case $M_0(p, q)$ has underlying space equal to the result of (p, dq) Dehn surgery on a torus knot, which is a Seifert fibered manifold with infinite fundamental group for infinitely many (p, q) and with finite fundamental group for infinitely many (p, q).

Finally, for part (ii), suppose (S^3, K) is invertible. That is, there is an involution of S^3 which reverses K. The quotient S^3/C_2 by this involution, as a space, is S^3, while the quotient of a tubular neighborhood of K is a ball. Dehn surgery just replaces this ball by another ball with different orbifold structure, so $M(p, q)/C_2$, as a space, is still S^3. The corresponding C_2-extension of Γ is thus generated by rotations. □

2. Hyperbolic Knot Complements and Groups Generated by Rotations

One can also ask when a hyperbolic knot complement $M = S^3 - K$ is itself commensurable with a $\mathbb{H}^3/\Lambda$ with Λ generated by rotations. Again, this means $\mathbb{H}^3/\Lambda$ has simply connected underlying space. If the knot is invertible, then the quotient of

$M = S^3 - K$ by the inversion has underlying space an open disk, so the answer is "yes." Otherwise, M is non-arithmetic by Reid [R] (who shows that the figure-eight knot is the only knot with arithmetic hyperbolic complement; the figure-eight knot is invertible) and we can again argue that the ***orientable commensurator quotient*** M_0 of M (i.e., the quotient of $\mathbb{H}^3$ by the largest Kleinian group Γ_0 containing $\Gamma = \pi_1(M)$) would have underlying space with finite fundamental group. This could only happen if Γ_0 is larger than the normalizer $N(\Gamma)$ of Γ in $PSL(2, \mathbb{C})$ (since $N(\Gamma)/\Gamma$ is cyclic, by the same argument as in Sect. 1, so $\mathbb{H}^3/N(\Gamma)$ still has infinite homology). As described in [NR, Sect. 9], this is an exceedingly rare phenomenon which quite possibly only happens for the figure-eight knot and the two ***dodecahedral knots*** of Aitchison and Rubinstein [AR]. One of the dodecahedral knots is invertible (this knot is number 5 in the series of knots of which number 3 is shown in Fig. 1). The other dodecahedral knot is non-invertible and its $\mathbb{H}^3/\Gamma_0$ is contractible. Thus summarizing:

Theorem 2. *Let* (S^3, K) *be a hyperbolic knot. Then* $\Gamma = \pi_1(S^3 - K)$ *is commensurable with a Kleinian group generated by rotations if* (S^3, K) *is invertible or is the non-invertible dodecahedral knot. Any other example would have to have the normalizer* $N(\Gamma)$ *of* Γ *not equal to the commensurator* Γ_0 *and conjecturally there are no further examples of this.*   □

3. Hidden Rotations in Kleinian Groups

One can ask whether a group Λ commensurable with a given Kleinian group $\Gamma \subset PSL(2, \mathbb{C})$ can contain rotations than Γ does not contain — we call these ***hidden rotations*** for Γ.

Of course, Kleinian groups can have hidden rotations — any torsion-free subgroup Γ of a group Λ with torsion does, for example. But we can exclude a lot of possible hidden rotations too.

Let $k(\Gamma)$ be the invariant trace field of Γ. That is, it is the field generated by the traces of squares of elements of Γ, cf. [NR].

Theorem 3. *If a group commensurable with* Γ *contains a* $(2\pi/p)$-*rotation, then* $\cos(2\pi/p) \in k(\Gamma)$.

Proof. Suppose g is a $(2\pi/p)$-rotation in a group Λ commensurable with Γ. Then the trace of g^2 is $2\cos(2\pi/p)$, so $2\cos(2\pi/p) \in k(\Lambda)$. Since $k(\Lambda) = k(\Gamma)$ (cf. [NR]), the Theorem follows.   □

Note that $\cos(2\pi/p)$ is rational for $p \le 4$ and $p = 6$, so Theorem 2 never excludes elements of order ≤ 4 or of order 6. However, "most" fields will not contain $\cos(2\pi/p)$

for $p = 5$ or $p > 6$, so "most" Kleinian groups Γ will admit no hidden rotations of these orders.

References

[AR]   I. R. Aitchison, and J. H. Rubinstein, Combinatorial cubings, cusps and the dodec-ahedral knots, in: Topology 90, Proceedings of the Research Semester in Low Dimensional Topology at Ohio State, Walter de Gruyter Verlag, Berlin–New York 1992, 17–26.

[An]   E. M. Andreev, On convex polyhedra in Lobacevskii space, Math. USSR Sbornik 10 (1970), 413–440.

[Ar]   M. A. Armstrong, The fundamental groups of the orbit space of a discontinuous group, Proc. Cambridge Philos. Soc. 64 (1968), 299–301.

[BH]   S. A. Bleiler and C. D. Hodgson, Spherical space forms and Dehn fillings, Topology, to appear.

[B]   A. Borel, Commensurability classes and volumes of hyperbolic 3-manifolds, Ann. Scuola Norm. Sup. Pisa 8 (1981), 1–33.

[MB]   Morgan, J. W. and Bass, H, eds., The Smith Conjecture, Academic Press 1984.

[NR]   W. D. Neumann and A. Reid, Arithmetic of hyperbolic 3-manifolds, in: Topology 90, Proceedings of the Research Semester in Low Dimensional Topology at Ohio State, Walter de Gruyter Verlag, Berlin–New York 1992, 273–310.

[NZ]   W. D. Neumann and D. Zagier, Volumes of hyperbolic 3-manifolds, Topology 24 (1985), 307–332.

[R]   A. W. Reid, Arithmeticity of knot complements, J. London Math. Soc. (2) 43 (1991), 171–184.

[T]   W. P. Thurston, The geometry and topology of 3-manifolds, Mimeographed Notes, Princeton Univ. (1977).

[V]   E. Vinberg, Discrete groups generated by reflections in Lobachevskii space, Math. Sb. 114 (1967), 471–488.

[Z]   R. Zimmer, Ergodic Theory and Semi-Simple Lie Groups, Birkhäuser, Boston, 1984.

From Geometries to Loops and Groups

Markku Niemenmaa and Ari Vesanen

1. Introduction

In his famous work "Grundlagen der Geometrie", Hilbert considered a projective plane as a system of axioms for the incidence relations between points and lines. One of the beautiful features of this approach is the way it connects certain concepts of algebra with those of geometry. This delicate interplay of geometry and algebra leads to the study of algebraic structures like alternative division rings and loops. The first two major works [14, 15] were introduced by R. Moufang in 1933 and 1934. Later on (1939–1944) A. Albert [1, 2] and R. Baer [3] considered algebraic structures called quasigroups and finally in 1946 R. Bruck [5] laid the foundation of loop theory. In this article Bruck defined the concepts of the multiplication group and the inner mapping group of a loop thus creating a link between loop theory and group theory. In the following sections we give a short description of this connection and introduce some basic results on loops and their multiplication groups.

2. Quasigroups, Loops and Groups

Let Q be a quasigroup (i.e. a groupoid with unique division). For each $a \in Q$ we have two permutations L_a (left translation) and R_a (right translation) on Q define by $L_a(x) = ax$ and $R_a(x) = xa$ for every $x \in Q$. The subgroup $M(Q)$ of S_Q generated by the set of all left and right translations is called the ***multiplication group*** of Q. Clearly, $M(Q)$ is transitive on Q and the stabilizers of elements of Q are conjugated in $M(Q)$. If Q is a loop (i.e. a quasigroup with a neutral element e), then the stabilizer of e is denoted by $I(Q)$ and is called the ***inner mapping group*** of Q (note that it is the analogue for loops of the inner automorphism group of a group). In general, the inner mapping group is not a group of automorphisms; however, there are loops whose inner mappings are automorphisms. We say that a loop is a ***Moufang loop*** if it satisfies the law

$$x(y.zy) = (xy.z)y.$$

If Q is a commutative Moufang loop, then $I(Q)$ is a group of automorphisms. Several other interesting results about Moufang loops are known and we list some of these before starting to work with the general case:

1) If Q is a commutative Moufang loop, then $I(Q)$, $M(Q)/Z(M(Q))$ and $M(Q)'$ are locally finite 3-groups and are finite if Q is finitely generated (Bruck [6], Theorem 11.4 in Chapter VIII).

2) Every finite Moufang loop of odd order is soluble (Glauberman [9, 10]; this clearly is a result which is an analogue of the odd order theorem in group theory).

3) If Q is a finite simple non-associative Moufang loop, then $M(Q)$ is a non-abelian finite simple group with triality (Doro [7]).

Our analysis of the general case starts with the following observation: If Q is a loop and $A = \{L_a \mid a \in Q\}$ and $B = \{R_a \mid a \in Q\}$, then A and B are transversals to $I(Q)$ in $M(Q)$ and $[L_a, R_b] \in I(Q)$ for every $a \in Q$ and $b \in Q$. We say that A and B are $I(Q)$-connected transversals. Now the relation between multiplication groups of loops and connected transversals is given by

Theorem 2.1 ([16], Theorem 4.1). *A group G is isomorphic to the multiplication group of a loop if and only if there exist a subgroup H satisfying $H_G = 1$ (the core of H in G is trivial) and H-connected transversals A and B such that $G = \langle A, B \rangle$.*

Now one of the main problems in loop theory is to decide which groups are isomorphic to multiplication groups of loops. Another interesting problem is to study how the structure of $I(Q)$ influences on the structure of $M(Q)$ and Q. By using Theorem 2.1 we have been able to prove:

1) If $I(Q)$ is cyclic, then Q is an abelian group (and then, of course $I(Q) = 1$).

2) $I(Q)$ cannot be a Prüfer-group C_{p^∞} for a prime p.

3) If $I(Q)$ is a finite abelian group, then $M(Q)$ is soluble, $I(Q)$ is subnormal in $M(Q)$ and Q is a centrally nilpotent loop.

4) If $I(Q)$ is a dihedral 2-group, then $M(Q)$ is soluble, $I(Q)$ is subnormal in $M(Q)$ and Q is a centrally nilpotent loop.

5) If $|I(Q)| = 6$, then $M(Q)$ is soluble. However, if $I(Q)$ is non-abelian, then Q is not necessarily centrally nilpotent.

In 1946, Bruck showed that the multiplication group of a centrally nilpotent loop is soluble and if the multiplication group of a loop is nilpotent, then the loop is centrally nilpotent. A closer investigation revealed that

6) If $M(Q)$ is a finite nilpotent group, then Q is a direct product of centrally nilpotent p-loops.

For the proofs of these results, see [12, 16, 17, 18, 19, 20, 24].

Before we put an end to this section, we state two problems.

Problem 1. If $I(Q)$ is a finite nilpotent group, does it then follow that $M(Q)$ is soluble and Q is a centrally nilpotent loop?

Problem 2. If $|I(Q)| = pq$ (where p and q are different prime numbers), does it then follow that $M(Q)$ is soluble?

3. Finite Simple Groups and Multiplication Groups of Loops

In this section we sum up what is currently known about the relation between simple groups and multiplication groups of loops; here 'simple group' means a non-commutative finite simple group. We first note that if G is a simple group, then $I(G) \cong G$; thus any simple group can occur as the inner mapping group of a loop. However, relatively few simple groups are known to be multiplication groups of loops and in general it seems easier to prove non-existence results than to actually construct loops with given multiplication group. We call groups isomorphic to multiplication groups of loops *loop groups*.

Let Q be a loop such that $M(Q)$ is simple; then $I(Q)$ is maximal in $M(Q)$ (see [16]). Hence, to decide whether a given simple group is a loop group or not, we have to study transversals of its maximal subgroups. This is, however, often impossible in practice. We can put some restrictions on $M(Q)$ using representation theory as follows. Let $R_0, R_1, \ldots, R_d$ be the orbits of $M(Q)$ acting on $Q \times Q$, where R_0 is the diagonal $\{(x, x) \mid x \in Q\}$. Then each R_i is a graph and $M(Q)$ is a group of automorphisms of R_i. Let $(x, y) \in R_k$ and let p_{ij}^k denote the number of elements $z \in Q$ satisfying $(x, z) \in R_i$ and $(z, y) \in R_j$; this number is independent of choice of x and y. From the fact that $L_x(y) = R_y(x)$ and that the translations are automorphisms of graphs, it is easy to deduce that $p_{ij}^k = p_{ji}^k$ for all i, j, k. This means ([4], p. 53) that the *centralizer ring* of $M(Q)$ is commutative and we get a bound

$$|M(Q)| \leq |I(Q)|\{(C - 1)|I(Q)| + 1\}$$

that can be used to rule out maximal subgroups of small order. The fact that the centralizer ring is commutative can be proved using other techniques, too (see [11]).

We first list some non-existence results. It was shown in [22, 23] that the group $L_2(q)$ is not a loop group if $q = 2^k$ or if $q > 59$ is odd; with a little effort it is possible to prove that $L_2(q)$ is a loop group if and only if $q = 9$. We have also studied the mentioned groups as multiplication groups of quasigroups and have obtained the following result: The group $L_2(q)$ is isomorphic to the multiplication group of quasigroup if and only if $q = 2^k$ or $q \in \{3, 5, 7, 9, 11\}$. Finally we have proved that (see [25]) if G is one of the groups

1) $S_n(q)$,

2) $U_n(q)$, $n \geq 6$,

3) $O_n(q)$, where n is odd and $n \geq 7$, or

4) $O_n^\epsilon(q)$, where n is even and $n \geq 7 - \epsilon$,

acting on the set Ω of the isotropic points of the corresponding projective space, and if there exists a loop Q defined on Ω such that $M(Q) \leq G$, then $G = S_2(q)$ and we have that q is even and Q is cyclic group (of order $q + 1$) or $q = 3$ and Q is the four-group. These are also the only cases that G can be isomorphic to the multiplication group of quasigroup.

We finally list those simple groups that are known to be loop groups. It was first shown by Drapal and Kepka [8] that A_5 is not a loop group but that there exists a loop Q of order n such that $M(Q) \cong A_n$, whenever $n \geq 6$. On the other hand, there is a loop of order 15, whose multiplication group is isomorphic to A_8 (see [25]); this is the only known example of a simple group being the multiplication group of two loops of different orders. Finally, a loop Q can be defined on the points of norm 1 of the orthogonal projective space such that $M(Q) \cong O_8^+(q)$ ([13], pp. 34–35).

As mentioned in the previous section, the multiplication group of a nonassociative simple Moufang loop is simple group with triality. Liebeck proved in [13] that the only simple groups with triality are the groups $O_8^+(q)$ and with the help of this result he was able to classify the simple finite Moufang loops: they are exactly the finite simple groups and the loops defined by Paige in [21]. We end the paper by stating

Problem 3. Is it possible to classify the finite simple (left) Bol loops using methods similar to Liebeck? (Left Bol loops satisfy the law

$$x(y.xz) = (x.yx)z$$

so that Moufang loops are left Bol loops)

References

[1]   Albert, A. A., Quasigroups I, Trans. Amer. Math. Soc. 54 (1943), 507–519.

[2]   Albert, A. A., Quasigroups II, Trans. Amer. Math. Soc. 55 (1944), 401–419.

[3]   Baer, R., Nets and groups, Trans. Amer. Math. Soc. 46 (1939), 110–141.

[4]   Bannai, E., and T. Ito, Algebraic combinatorics I, Benjamin-Cummings Publishing Company Inc., California, 1984.

[5]   Bruck, R. H., Contributions to the theory of loops, Trans. Amer. Math. Soc. 60 (1946), 245–354.

[6]   Bruck, R. H., A survey of binary systems, Springer-Verlag, Berlin–Heidelberg–New York 1971.

[7]   Doro, S., Simple Moufang loops, Math. Proc. Camb. Phil. Soc. 83 (1978), 377–392.

[8] Drapal, A., and T. Kepka, Alternating groups and latin squares, Europ. J. Combinatorics 10(2) (1989), 175–180.

[9] Glauberman, G., On loops of odd order, J. Algebra 1 (1964), 374–396.

[10] Glauberman, G., On loops of odd order II, J. Algebra 8 (1968), 393–414.

[11] Ihringer, T., Quasigroups, loops and centralizer rings, Contributions to general algebra, 3, in: Proceedings of the Vienna Conference, 1984.

[12] T. Kepka and M. Niemenmaa, On loops with cyclic inner mapping groups, Arch. Math. 60 (1993), 233–236.

[13] Liebeck, M. W., The classification of finite simple Moufang loops, Math. Proc. Camb. Phil. Soc. 102 (1987), 33–47.

[14] Moufang R., Alternativkörper und der Satz vom vollständigen Vierseit, Abh. Math. Sem. Hamburg Univ. 9 (1933), 207–222.

[15] Moufang R., Zur Struktur von Alternativkörpern, Math. Annalen 110 (1934), 416–430.

[16] Niemenmaa, M., and T. Kepka, On multiplication groups of loops, J. Algebra 135 (1990), 112–122.

[17] M. Niemenmaa and T. Kepka, On connected transversals to abelian subgroups in finite groups, Bull. London Math. Soc. 24 (1992), 343–346.

[18] M. Niemenmaa and T. Kepka, On connected transversals to abelian subgroups, Bull. Australian Math. Soc. 49 (1994), 121–128.

[19] M. Niemenmaa, Transversals, commutators and solvability in finite groups, Boll. Un. Math. Ital., to appear.

[20] M. Niemenmaa, On loops which have dihedral 2-groups as inner mapping groups, unpublished manuscript.

[21] Paige, J. P., A class of simple Moufang loops, Proc. Amer. Math. Soc. 7 (1956), 471–482.

[22] Vesanen, A., On connected transversals in PSL(2,q), Ann. Acad. Sci. Fenn. Ser. A I Math. Dissertationes 84, 1992.

[23] Vesanen, A., The Group PSL(2,q) is not the Multiplication Group of a Loop, Comm. Algebra 22 (4) (1994), 1177–1195.

[24] Vesanen, A., On p-groups as loop groups, Arch. Math 61 (1993), 1–6.

[25] Vesanen, A., Finite classical groups and multiplication groups of loops, Math. Proc. Camb. Phil. Soc, to appear.

A Simplification of Golod's Example

A. Yu. Ol'shanskii*

Abstract. Our talk at the Conference concerned with periodic groups consisted of 3 parts. The third one presented a simplified proof of Golod's theorem on existence of periodic finitely generated infinite groups. Below we produce this proof that could be useful for students who are interested in algebra.

1991 Mathematics Subject Classification: 20F50, 16N40, 16W50

Let $F_p = \mathbb{Z}/p\mathbb{Z}$ be a field of a prime order p. Denote $A = F_p[x, y]$ the algebra of non-commutative polynomials in x, y over F_p without constant terms. It is a direct sum $A = A_1 \oplus A_2 \oplus \cdots$ of homogeneous components A_i of degrees $i = 1, 2, \ldots$, where $A_1 = \langle x, y \rangle$, $A_2 = \langle x^2, xy, yx, y^2 \rangle$, $\ldots$, and obviously $\dim A_i = 2^i$. Recall that I is a homogeneous ideal of A if $I = I_1 \oplus I_2 \oplus \cdots \oplus I_k \oplus \cdots$, where $I_k \leq A_k$.

Lemma. *Let I be a homogeneous ideal of A generated by a set $\{r_6, r_7, \ldots\}$ where $r_i \in A_i$ for $i = 6, 7, \ldots$. Then $I_k \neq A_k$ for every $k = 1, 2, \ldots$.*

Proof. Every component I_k is a linear envelope of all products $u r_j v$ where $6 \leq j \leq k$ and u, v are monomials. There are two cases: (a) If $\deg v > 0$, i.e., $v \in A_1^s$ for some positive s and $j \leq k$, then evidently $u r_j v \in I_{k-1} A_1$. (b) If v is an empty word then $u r_j v \equiv u r_j \in A_{k-j} r_j$.

Taking both cases into account one obtains

$$I_k \leq I_{k-1} A_1 + \sum_{j=6}^{k} A_{k-j} r_j. \tag{1}$$

Choose any subspace C_{k-j} in A_{k-j} such that $A_{k-j} = I_{k-j} \oplus C_{k-j}$, i.e.

$$A_{k-j} r_j = I_{k-j} r_j + C_{k-j} r_j . \tag{2}$$

Inequality (2) and the obvious inclusion $I_{k-j} r_j \subseteq I_{k-1} A_1$ allow us to rewrite (1) as

---
*Partially supported by RFFI grant # 011 1541 and by ISF grant # MID000

follows:

$$I_k \le I_{k-1} A_1 + \sum_{j=6}^{k} C_{k-j} r_j . \tag{3}$$

All above is just Golod's argument [1]. To avoid power series and to simplify the proof of the inequality $c_k = \dim A_k - \dim I_k = \dim C_k > 0$, we observe that $c_k = \dim A_k = 2^k$ for $k = 1, \ldots, 5$ and prove below by induction on k that $c_k \ge \frac{3}{2} c_{k-1}$ for $k = 6, 7, \ldots$ Let $n_k = \dim I_k$. Then (3) gives us: $n_k \le 2 n_{k-1} + \sum_{j=6}^{k} c_{k-j}$. Here $n_k = \dim A_k - c_k = 2^k - c_k$. Therefore

$$(2^k - c_k) \le 2(2^{k-1} - c_{k-1}) + \sum_{j=6}^{k} c_{k-j}. \tag{4}$$

By induction hypothesis, $c_{k-6} \le (2/3)^5 c_{k-1}, c_{k-7} \le (2/3)^6 c_{k-1}, \ldots$. Hence inequality (4) implies:

$$c_k \ge 2 c_{k-1} - \sum_{j=6}^{k} c_{k-j} \ge 2 c_{k-1} - c_{k-1} \left((2/3)^5 + (2/3)^6 + \cdots \right)$$

$$\ge c_{k-1} \left(2 - \frac{(2/3)^5}{1 - 2/3} \right) = c_{k-1} (2 - \frac{32}{81}) \ge \frac{3}{2} c_{k-1}. \qquad \square$$

Theorem (Golod). *There exists a 2-generated infinite p-group.*

Proof. One can enumerate all elements of A, i.e. $A = \{a_1, a_2, \ldots\}$. Let $a_1^{n_1} = r_{11} + \cdots + r_{1k_1}$ where $n_1 \ge 6$ and r_{1i} is a homogeneous element of degree i. Choose $n_2 > k_1$ and denote $a_2^{n_2} = r_{21} + \cdots + r_{2k_2}, \ldots$ By the choice, each A_i contains at most one element of the set $\{r_{11}, \ldots, r_{1k_1}, r_{21}, \ldots\} = R$. So the homogeneous ideal I generated by R satisfies the condition of the lemma, and $I_k < A_k$ for every $k \ge 1$. Every element of the quotient $B = A/I$ is nilpotent by the choice of R. Thus the group $G = 1 + B$ is a p-group because $(1 + a)^p = 1 + a^p$ for $a \in B$. For any $k \ge 1$ there is a monomial $z_{1k} \cdots z_{kk} \notin I_k$ (where $z_{11}, \ldots, z_{kk} \in \{x, y\}$). Therefore $(1 + z_{1k}) \cdots (1 + z_{kk}) \ne (1 + z_{1j}) \cdots (1 + z_{jj}) \mod I$ if $k \ne j$ because the ideal I is homogeneous. Consequently, the subgroup of G, generated by $1 + x$ and $1 + y$ is infinite. $\qquad \square$

Remark. A quite similar approach can be applied for the more general theorem of Golod on the existence of an infinite d-generated p-group G for given $d \ge 2$ such that every $(d - 1)$-generated subgroup of G is finite.

References

[1] Golod, E. S., On nil-algebras and residually finite groups, Izv. Akad. Nauk SSSR Ser. Mat. 28 (1964), 273–276.

Restricted Movement for Intransitive Group Actions

Cheryl E. Praeger*

Abstract. Let G be a permutation group on Ω and let Γ be a finite subset of Ω. It was shown in 1976 by B. J. Birch, R. G. Burns, S. Oates Macdonald and P. M. Neumann that, if $\Gamma^g \cap \Gamma \neq \emptyset$ for all $g \in G$, then G has an orbit in Ω of length at most $|\Gamma|^2$. The bound $|\Gamma|^2$ is not sharp but it is very close to the best possible bound, there being examples in which the smallest orbit length is $|\Gamma|^2 - |\Gamma| + 1$. The main purpose of this paper is to provide improvements of the bound $|\Gamma|^2$ in the case where Γ intersects nontrivially more than one of the G-orbits; in this case we show that there is a G-orbit of length at most $(|\Gamma| - 1)^2$. As a corollary we deduce that in general there is a G-orbit of length at most $|\Gamma|^2 - 1$.

1991 Mathematics Subject Classification: 20B05, 20B07.

1. Introduction

In 1954, B. H. Neumann [BHN, Lemma 4.1] proved a fundamental result about the covering of a group G with a finite number of cosets of proper subgroups; he showed that at least one of the subgroups involved in the coset covering must have finite index in G. This result was shown by P. M. Neumann [PMN, Lemma 2.3] to be equivalent to a result about permutation groups, namely, for a permutation group G all of whose orbits are infinite, every finite subset of points has an image under some element of G which is disjoint from it. Alternatively this result may be viewed as asserting that the failure of a permutation group G to separate a finite subset of points completely from itself implies the existence of a finite G-orbit. It is the latter "failure of separation" form of the result on which we shall focus here. For a permutation group G on a set Ω and a finite subset Γ of Ω, we define the ***movement*** of Γ as

$$\text{mov}(\Gamma) := \max_{g \in G} |\Gamma^g \setminus \Gamma|.$$

Using this notation the Separation Theorem has the following form.

---

*   The research for this paper forms part of an ARC small grant project 3-ARCSG94-2

Theorem 1.1. *If G is a group of permutations of a set Ω such that* $\mathrm{mov}(\Gamma) < |\Gamma|$ *for some finite subset Γ of Ω, then G has a finite orbit in Ω.*

A quantitative version of the Separation Theorem was proved in 1976 by B. J. Birch, R. G. Burns, S. Oates Macdonald and P. M. Neumann [BBMN, Theorem 2]. This gave an upper bound, in terms of positive integers m and n, on the minimal orbit length of a group which failed to separate subsets of points of sizes m and n.

Theorem 1.2 [BBMN, Theorem 2]. *If G is a group of permutations of a set Ω and Γ, Δ are finite subsets of Ω, and if $\Gamma \cap \Delta^g \neq \emptyset$ for all $g \in G$, then some G-orbit in Ω has length at most $|\Gamma| \cdot |\Delta|$. In particular, if* $\mathrm{mov}(\Gamma) < |\Gamma|$, *then some G-orbit in Ω has length at most $|\Gamma|^2$.*

The upper bound given in [BBMN, Theorem 2] for the length of the smallest G-orbit is sharp in the case of separating *different* subsets Γ, Δ. For example if $|\Omega| = mn$ and $G = S_m$ wr S_n is the subgroup of the symmetric group Sym (Ω) preserving some partition $\mathcal{P}$ of Ω into n subsets of size m, and if Γ is one of the subsets of $\mathcal{P}$ while Δ consists of exactly one point from each subset of $\mathcal{P}$, then $\Gamma \cap \Delta^g \neq \emptyset$ for all $g \in G$. However, if $\Gamma = \Delta$, and if G is transitive, then it has been shown in [BPP] that the bound $|\Gamma|^2$ is not sharp.

Theorem 1.3 [BPP, Proposition D]. *If G is a transitive group of permutations of a set Ω and for some m-element subset Γ,* $\mathrm{mov}(\Gamma) < m$, *then $|\Omega| \leq m^2 - 1$.*

Although the bound $|\Gamma|^2$ in Theorem 1.2 is not the best possible for transitive groups G, it is not far from optimal, as the following family of examples shows.

Example 1. Let $m - 1$ be a prime power, let Ω be the set of points of the Desarguesian projective plane PG $_2(m - 1)$, and let Γ be the set of points on a line of PG $_2(m - 1)$. Then $G := \mathrm{PGL}(3, m-1)$ is transitive on Ω, $|\Omega| = m^2 - m + 1$, and Γ is an m-element subset with $\mathrm{mov}(\Gamma) < m$.

Thus the best possible upper bound for the minimal G-orbit length in Theorem 1.2, for a *transitive* group G and a set Γ with $\mathrm{mov}(\Gamma) < |\Gamma|$, is $|\Gamma|^2 - \delta$ for some $1 \leq \delta \leq |\Gamma| - 1$. In this paper we shall show that the transitivity assumption in Theorem 1.3 can be omitted. We obtain improvements to the bound $|\Gamma|^2$ in the case where Γ intersects nontrivially at least two G-orbits. We show in Theorem 1.4 that, if Γ meets nontrivially at least two G-orbits and $\mathrm{mov}(\Gamma) < |\Gamma|$, then the minimal G-orbit length is at most $(|\Gamma| - 1)^2$. This reduces the problem of determining the best possible upper bound in Theorem 1.2 (when $\Gamma = \Delta$) to the case where the group is transitive.

Problem 1. Determine the value of δ $(1 \leq \delta \leq m - 1)$ such that
(a)   if G is transitive on a set Ω of size greater than $m^2 - \delta$ then for every m-element subset Γ of Ω, $\mathrm{mov}(\Gamma) = m$, and

(b)   there exists a transitive permutation group G on a set Ω of size $m^2 - \delta$ and an m-element subset Γ of Ω such that mov$(\Gamma) < m$.

Several people believe that the correct value of δ is $m - 1$, and hence that Example 1 gives a class of extreme examples. For this problem, $m \geq 2$. If $m = 2$ then of course $\delta = m - 1 = 1$, while if $m = 3$ then it follows from the classification in [BPP2] that $\delta = m - 1 = 2$. The problem is open for $m \geq 4$. The papers [BPP, BPP2] contain relevant information relating $|\Omega|$ and mov(Γ). In particular, in the situation of Problem 1(b), it follows from [BPP, Theorem 2.1] that mov$(\Gamma) = m - 1$.

We now turn to the consideration of intransitive permutation groups and their separation properties. First we give an example, similar to Example 1, which shows that we may have a group with two orbits, one of length as large as $(m - 1)^2$, and an m-element subset Γ meeting each orbit nontrivially such that mov$(\Gamma) < m$.

Example 2. Let $m - 1$ be a prime power, let Ω be the set of points of the Desarguesian projective plane PG$_2(m - 1)$, let Ω_1 and Γ be sets of points on distinct lines of the plane, and let G be the setwise stabiliser of Ω_1 in PGL$(3, m - 1)$. Then the G-orbits in Ω are Ω_1 and $\Omega_2 := \Omega \setminus \Omega_1$, and we have $|\Omega_1| = m$ and $|\Omega_2| = m^2 - 2m + 1$. Moreover Γ is an m-element subset with mov$(\Gamma) = m - 1$, and $|\Gamma \cap \Omega_1| = 1$, $|\Gamma \cap \Omega_2| = m - 1$.

Note that in this example, setting $\Gamma_i := \Gamma \cap \Omega_i$, we have mov$(\Gamma_i) = |\Gamma_i|$ for $i = 1, 2$ even though mov$(\Gamma) < |\Gamma|$. This illustrates the fact that results about separation of subsets by intransitive groups may not be deduced immediately from similar results for transitive groups. The intransitive case must be considered separately. In Section 2 we prove the following result.

Theorem 1.4. *Let G be an intransitive permutation group on a set Ω, and let Γ be an m-element subset of Ω which intersects nontrivially $r \geq 2$ of the G-orbits in Ω, say $\Omega_1, \ldots, \Omega_r$. If $|\Omega_i| \geq (m - 1)^2$ for $i = 1, \ldots, r$, then either*
(a)   mov$(\Gamma) = m$; *or*
(b) $r = 2$ *and, say,* $|\Gamma \cap \Omega_1| = 1$, $|\Gamma \cap \Omega_2| = m - 1$, *and either*
      (b1) $|\Omega_2| = (m - 1)^2$ *and* $|\Omega_1| \leq |G|/(m - 1)$, *or*
      (b2) $|\Omega_2| = (m - 1)^2 + 1$, $|\Omega_1| = (m - 1)^2$ *and* $|\Omega_1| \cdot |\Omega_2| \leq |G|/(m - 1)$.

I am grateful to Dima Pasechnik for bringing to my attention techniques from [BPP] which lead to an improvement of the original version of part (b) above. The fact that the groups in Example 2 satisfy the conditions of Theorem 1.4(b1) suggests that it may be impossible to prove a better result than this. However I know of no permutation groups G satisfying the conditions of Theorem 1.4(b2). In fact I know of no examples of permutation groups G with two orbits, each of size close to m^2, say each of length greater than $m^2/2$, which fail to separate from themselves m-element subsets of points. (Arguing as in the proof of Theorem 1.4 it can be shown that such a group must separate from itself any m-element subset which intersects each of the two orbits in at least $m/4$ points.) As an immediate corollary of Theorems 1.4 and 1.3 we obtain the following improvement to the Separation Theorem.

Corollary 1.5. *Let G be a permutation group on a set Ω, and let Γ be an m-element subset of Ω such that $\mathrm{mov}(\Gamma) < m$. Suppose that Γ meets nontrivially exactly r of the G-orbits in Ω, say $\Omega_1, \ldots \Omega_r$. Then*
(a)   *at least one of the Ω_i has length at most $m^2 - 1$, and moreover*
(b)   *if $r \geq 2$, then at least one of the Ω_i has length at most $(m - 1)^2$.*

These results suggest that the number of nontrivial orbit intersections of Γ may affect significantly the bound which can be obtained on the minimal orbit length. A closer examination of the situation in Section 3 shows that, for reasonably small values (relative to $m := |\Gamma|$) of the number r of nontrivial orbit intersections of Γ with finite G-orbits, an orbit bound close to $(m - r + 1)^2$ can be proved. Our investigation shows that the sizes, as well as the number of nontrivial orbit intersections for Γ influence the bound. A technical result of this nature, Proposition 3.1, will be stated and proved in Section 3; it implies the following result giving bounds in terms of m and r only. (When $r = 2$, Theorem 1.4 gives a better result.)

Theorem 1.6. *Let G be an intransitive permutation group on a set Ω with at least two finite orbits, and let m be a positive integer. Suppose that Γ is an m-element subset of Ω which intersects nontrivially $r \geq 2$ of the finite G-orbits in Ω, and suppose that $\mathrm{mov}(\Gamma) < m$.*
(a)   *If $r < \sqrt{m} + 1/2$, then Γ intersects nontrivially a G-orbit of length less than*

$$m^2 - 2m(r - 2) - (r - 1)(r - 2).$$

(b)   *If $r > \sqrt{m}$, then Γ intersects nontrivially a G-orbit of length less than*

$$m^2 \left(1 - \frac{2(r - 1)}{r^2}\right) - (r - 1)(r - 2).$$

For r approximately $\sqrt{m}$, the upper bound given by Theorem 1.6 is approximately $m^2 - 2m^{3/2}$.

2. Subsets with More Than One Nontrivial Orbit Intersection

Suppose that G is a permutation group on Ω and that Γ is an m-element subset of Ω with $\mathrm{mov}(\Gamma) < m$. First we show that to prove Theorem 1.4 it is sufficient to deal with the case where Ω is finite. We introduce some notation which allows us to focus on the finite G-orbits and the infinite G-orbits separately. Let Ω_{fin} and Ω_{inf} denote the unions of the finite G-orbits and the infinite G-orbits respectively, and let $\Gamma_{\text{fin}} := \Gamma \cap \Omega_{\text{fin}}$ and $\Gamma_{\text{inf}} := \Gamma \cap \Omega_{\text{inf}}$.

Proposition 2.1. *If* $\text{mov}(\Gamma) < m$ *then* $m' := |\Gamma_{\text{fin}}| > 0$ *and*

$$\text{mov}(\Gamma_{\text{fin}}) = \text{mov}(\Gamma) - |\Gamma_{\text{inf}}| < m - |\Gamma_{\text{inf}}| = m'.$$

Proof. If $m' = 0$ then $\Gamma \subseteq \Omega_{\text{inf}}$, and as $\text{mov}(\Gamma) < m$, we have a contradiction by Theorem 1.1 on considering the action of G on Ω_{inf}. Thus $m' > 0$. The second assertion follows from results in [BPP]: by [BPP, Lemma 3.6], $\text{mov}(\Gamma_{\text{fin}}) = \text{mov}(\Gamma) - d(\Gamma_{\text{inf}})$ where by [BPP, Lemma 3.2] it follows that $d(\Gamma_{\text{inf}}) = |\Gamma_{\text{inf}}| = m - m'$. Thus $\text{mov}(\Gamma_{\text{fin}}) = \text{mov}(\Gamma) - m + m' < m'$. □

Corollary 2.2. *If Theorem 1.4 holds in the case where Ω is a finite set and each G-orbit intersects Γ nontrivially, then it holds in general.*

Proof. Suppose that Theorem 1.4 holds when Ω is finite and each G-orbit intersects Γ nontrivially, and suppose that G, Ω, Γ are as in the statement of Theorem 1.4. Let Ω_{fin} and Ω_{inf} be defined as above, and let Ω_{fin}' be the union of G-orbits in Ω_{fin} which intersect Γ nontrivially. Note that Ω_{fin}' is a finite set and $\Gamma \cap \Omega_{\text{fin}} = \Gamma \cap \Omega_{\text{fin}}'$. If $\Gamma \subseteq \Omega_{\text{fin}}$ then Theorem 1.4 follows on considering the action of G on Ω_{fin}'. Thus we may assume that $\Gamma_{\text{inf}} := \Gamma \cap \Omega_{\text{inf}} \neq \emptyset$. By Proposition 2.1, $\Gamma_{\text{fin}} := \Gamma \cap \Omega_{\text{fin}} \neq \emptyset$ and $\text{mov}(\Gamma_{\text{fin}}) < |\Gamma_{\text{fin}}|$. Now $m' := |\Gamma_{\text{fin}}| = m - |\Gamma_{\text{inf}}| \leq m - 1$ and again Theorem 1.4 follows on considering the action of G on Ω_{fin}'. □

We now complete the proof of Theorem 1.4. The argument in the last paragraph of the proof was suggested by Dima Pasechnik.

Proof of Theorem 1.4. By Corollary 2.2 we may assume that Ω is finite and that each of the G-orbits $\Omega_1, \ldots, \Omega_r$ intersects Γ nontrivially. Set

$$n_i := |\Omega_i|, \quad \text{and} \quad v_i := |\Gamma \cap \Omega_i|$$

for $i = 1, \ldots, r$, and relabel the Ω_i if necessary so that

$$0 < v_1 \leq \cdots \leq v_r.$$

We are assuming that each $n_i \geq (m-1)^2$, and since $\sum_{1 \leq i \leq r} v_i = m$, we have $v_r \geq m/r$. As Ω is finite there are only a finite number of distinct images of Γ under elements of G, say $\Gamma^{(1)} := \Gamma, \ldots, \Gamma^{(s)}$.

Since G is transitive on Ω_i, the number x_i of the sets $\Gamma^{(j)}$ containing a point $\alpha \in \Omega_i$ is independent of the point α. The number of pairs (α, j) such that $\alpha \in \Omega_i \cap \Gamma^{(j)}$ is equal to $n_i x_i$ and also to $s v_i$, whence $x_i = s v_i / n_i$.

Now we estimate the number N of pairs (α, j) such that $\alpha \in \Gamma \cap \Gamma^{(j)}$. Note that there are m such pairs with second entry equal to 1, and that $m \geq 2$ (since Γ meets at least two orbits nontrivially). Thus if, at some stage of the proof, we find that $N \leq s$ then we may conclude that, for some j, $\Gamma \cap \Gamma^{(j)} = \emptyset$, whence $\mathrm{mov}(\Gamma) = m$ and part (a) of Theorem 1.4 holds. To determine N we note that there are v_i points $\alpha \in \Gamma \cap \Omega_i$ and that each such point lies in x_i of the $\Gamma^{(j)}$. Thus

$$N = \sum_{1 \leq i \leq r} v_i \cdot x_i = s \cdot \sum_{1 \leq i \leq r} \frac{v_i^2}{n_i} \leq \frac{s}{(m-1)^2} \cdot \sum_{1 \leq i \leq r} v_i^2.$$

Now $\sum_{1 \leq i \leq r} v_i^2 = (\sum v_i)^2 - 2\sum_{i<j} v_i v_j = m^2 - 2\sum_{i<j} v_i v_j$, and so if $\sum_{i<j} v_i v_j \geq m - 1/2$ then we could deduce that $N \leq s$ and hence that Theorem 1.4(a) holds. Note that $\sum_{i<j} v_i v_j = v_r(m - v_r) + \sum_{i<j<r} v_i v_j$.

Suppose first that $v_r < m - m/r$. Then we must have $r \geq 3$. Since the function $f(x) = x(m - x)$ has minimum value on the interval $[m/r, m - m/r]$ equal to $f(m/r) = m^2(r - 1)/r^2$, it follows that $v_r(m - v_r) \geq m^2(r - 1)/r^2$. Then since the function $g(x) = (x - 1)/x^2$ is decreasing on the interval $[3, m]$, $(r - 1)/r^2 \geq (m - 1)/m^2$ and hence $v_r(m - v_r) \geq m - 1$. Thus $\sum_{i<j} v_i v_j \geq m - 1 + \sum_{i<j<r} v_i v_j \geq m$ (since $r \geq 3$) and again Theorem 1.4(a) holds.

Thus we may assume that $v_r \geq m - m/r = m(r - 1)/r$. In this case

$$v_r(m - v_r) = v_r(v_1 + \cdots + v_{r-1}) \geq \frac{m(r-1)^2}{r} > m(r - 2).$$

If $r \geq 3$ then we deduce that $\sum_{i<j} v_i v_j > m$ and as before Theorem 1.4(a) holds. So we may assume that $r = 2$. Then $\sum_{i<j} v_i v_j = v_1 v_2 \geq v_1 m/2$ which again is at least m if $v_1 \geq 2$. So we may assume that $v_1 = 1$ and $v_2 = m - 1$. Returning to the original expression for N we have

$$N = s \cdot \left(\frac{1}{n_1} + \frac{(m-1)^2}{n_2}\right)$$

which, if both n_1 and n_2 are at least $(m - 1)^2 + 1$, implies that $N \leq s$ and hence that Theorem 1.4(a) holds. Thus we are left with the case where at least one of the n_i is equal to $(m - 1)^2$.

To complete the proof of Theorem 1.4 we assume that $\mathrm{mov}(\Gamma) < m$; to obtain part (b) we use the techniques of the proof of [BPP, Theorem 2.1] which are a refinement of the methods used above and in [BBMN]. Let $\Gamma \cap \Omega_1 := \{\alpha\}$ and $\Gamma \cap \Omega_2 = \{\beta^{y_1}, \ldots, \beta^{y_{m-1}}\}$ where $\alpha \in \Omega_1$, $\beta \in \Omega_2$ and $y_1, \ldots, y_{m-1} \in G$. If $g \in G \setminus G_\alpha$ then $\alpha^g \in \Gamma^g \setminus \Gamma$ and, since $|\Gamma^g \setminus \Gamma| \leq \mathrm{mov}(\Gamma) < m$, there exist i, j such that $\beta^{y_i g} = \beta^{y_j}$, that is $g \in y_i^{-1} G_\beta y_j$. It follows that $G = G_\alpha \cup \left(\cup_{i,j} y_i^{-1} G_\beta y_j\right)$. Noting that G_α and $y_i^{-1} G_\beta y_i$, for each $i = 1, \ldots, m - 1$, contain 1_G, it follows that $|G| \leq$

$|G_\alpha| + (m-1)^2|G_\beta| - (m-1)$. Dividing by $|G_\alpha|$ we have

$$n_1 \leq 1 + \frac{(m-1)^2 n_1}{n_2} - \frac{m-1}{|G_\alpha|}.$$

If $n_2 = (m-1)^2$ this implies that $|G_\alpha| \geq m-1$, so $n_1 \leq |G|/(m-1)$ and (b1) holds. So assume that $n_2 > (m-1)^2$ whence $n_1 = (m-1)^2$. Then the displayed inequality above implies that

$$\frac{(m-1)^4}{n_2} \geq (m-1)^2 - 1 + \frac{m-1}{|G_\alpha|} > (m-1)^2 - 1$$

whence

$$n_2 < \frac{(m-1)^4}{(m-1)^2 - 1} = (m-1)^2 + 1 + \frac{1}{(m-1)^2 - 1} < (m-1)^2 + 2.$$

It follows that $n_2 = (m-1)^2 + 1$, and then our inequality above yields

$$\frac{m-1}{|G_\alpha|} \leq \frac{(m-1)^4}{n_2} - (m-1)^2 + 1 = \frac{1}{n_2}$$

and (b2) holds. $\qquad\square$

3. The Effect of the Sizes of Orbit Intersections

In this section we extend the ideas used in the previous section to investigate the influence of the relative sizes of the orbit intersections of the subset Γ on the orbit length bound. In the previous section the arithmetic was quite straightforward and the bound obtained was the best we could hope for in the case of two nontrivial orbit intersections of Γ. Here the arithmetic needs more care, and extra information is obtained in the case in which Γ has at least three nontrivial orbit intersections.

Suppose that G is an intransitive permutation group on Ω and that Γ is an m-element subset of Ω with $\text{mov}(\Gamma) < m$. From the proof of Proposition 2.1 and its corollary it is clear that most of the analysis needed to obtain orbit bounds may be done in the case where Ω is a finite set. Accordingly in the first part of this section we shall assume that Ω is finite and that each of the G-orbits intersects Γ nontrivially. We prove a technical proposition which we shall use in the proof of Theorem 1.6.

Proposition 3.1. *Suppose that G is an intransitive permutation group on a finite set Ω with orbits $\Omega_1, \ldots, \Omega_r$. Suppose further that Γ is an m-element subset of Ω which meets Ω_i in v_i points for $i = 1, \ldots, r$, where $0 < v_1 \leq \cdots \leq v_r$, and that $\text{mov}(\Gamma) < m$.*

(a) If $v_r \leq m - m/r$ then at least one of the Ω_i has length less than

$$m^2(1 - \frac{2(r-1)}{r^2}) - (r-1)(r-2).$$

(b) If $v_r \geq m - m/r$ then at least one of the Ω_i has length less than

$$m^2 - 2\frac{m}{r}(r-1)^2 - (r-1)(r-2).$$

Proof. Set $n_i := |\Omega_i|$, for $i = 1, \ldots, r$, and let $\Gamma^{(1)} := \Gamma, \ldots, \Gamma^{(s)}$ denote the distinct images of Γ under elements of G. As in Section 2 the number N of pairs (α, j) such that $\alpha \in \Gamma \cap \Gamma^{(j)}$ is equal to

$$N = s \cdot \sum_{1 \leq i \leq r} \frac{v_i^2}{n_i}.$$

Since there are m such pairs with second entry equal to 1, and since $m \geq 2$ (since Γ meets at least two orbits nontrivially), $N \leq s$ would imply that, for some j, $\Gamma \cap \Gamma^{(j)} = \emptyset$, contradicting the fact that $\mathrm{mov}(\Gamma) < m$. Thus $N > s$. Let M denote the minimum of $n_1, \ldots, n_r$. Then we have $s < N \leq (s/M) \sum_{1 \leq i \leq r} v_i^2$, that is, $M < \sum_{1 \leq i \leq r} v_i^2$. Now

$$\sum_{1 \leq i \leq r} v_i^2 = (\sum_{1 \leq i \leq r} v_i)^2 - 2 \sum_{i<j} v_i v_j = m^2 - 2v_r(m - v_r) - 2\sum_{i<j<r} v_i v_j.$$

Since $0 < v_1 \leq \cdots \leq v_r$ we have $v_r \geq m/r$, and $2\sum_{i<j<r} v_i v_j \geq (r-1)(r-2)$. If $v_r \leq m - m/r$ then, as $f(x) = x(m-x)$ has minimum value on the interval $[m/r, m-m/r]$ equal to $m(m-m/r)/r = m^2(r-1)/r^2$, part (a) follows. Similarly, if $v_r \geq m - m/r$ then

$$v_r(m - v_r) = v_r(v_1 + \cdots + v_{r-1}) \geq (m - \frac{m}{r})(r-1) = \frac{m}{r}(r-1)^2,$$

and part (b) follows.   □

Now we use Proposition 3.1 to prove Theorem 1.6.

Proof of Theorem 1.6. By Proposition 2.1 (and arguing as in the proof of Corollary 2.2) we may assume that Ω is finite, and that Γ intersects nontrivially each of the G-orbits $\Omega_1, \ldots, \Omega_r$. Set $v_i := |\Gamma \cap \Omega_i|$, for $i = 1, \ldots, r$. Further we may assume that $0 < v_1 \leq \cdots \leq v_r$ so that $m/r \leq v_r \leq m-r+1$. We consider the two cases of Proposition 3.1 separately. Suppose first that $v_r > m - m/r$ then $m - m/r < m - r + 1$ whence $m > r(r-1)$ and hence $r < \sqrt{m} + 1/2$. By Proposition 3.1(b), one of the $|\Omega_i|$ is less than $m^2 - 2m(r-1)^2/r - (r-1)(r-2)$ which is less than $m^2 - 2m(r-2) - (r-1)(r-2)$, as in Theorem 1.6(a). Now we assume that $v_r \leq m - m/r$. Then by Proposition 3.1(a), one of the $|\Omega_i|$ is less than $M := m^2(1 - 2(r-1)/r^2) - (r-1)(r-2)$. If $r > \sqrt{m}$, then M is the bound required for Theorem 1.6(b), so we may assume that $r \leq \sqrt{m}$.

Thus $m/r^2 \geq 1$, and so M is at most $m^2 - 2m(r-1) - (r-1)(r-2)$, which is better than is needed for Theorem 1.6(a). $\square$

Addendum. It was recently proved by P. M. Neumann and the author that, if G is a transitive permutation group on Ω, and if $\Gamma \subseteq \Omega$ is such that $\text{mov}(\Gamma) < |\Gamma| = m$, then $|\Omega| \leq m^2 - m + 1$, with equality if and only if $\{\Gamma^g \mid g \in G\}$ is the set of lines of a projective plane with point set Ω. Thus, in Problem 1, $\delta = m - 1$ whenever $m = q + 1$ for some prime power q, and in Corollary 1.5(a) we may replace $m^2 - 1$ by $m^2 - m + 1$.

References

[BBMN] Birch, B. J., Burns, R. G., Macdonald, S. O. and Neumann, P. M., On the orbit sizes of permutation groups containing elements separating finite subsets, Bull. Austral. Math. Soc. 14 (1976), 7–10.

[BPP]   Brailovsky, L., Pasechnik, D. V. and Praeger, C. E., Subsets close to invariant subsets under group actions, preprint, University of Western Austalia, 1993.

[BBP2] Brailovsky, L., Pasechnik, D. V. and Praeger, C. E., Classification of 2-quasi-invariant sets, Ars Combin., to appear.

[BHN]   Neumann, B. H., Groups covered by permutable subsets, J. London Math. Soc. 29 (1954), 236–248.

[PMN]   Neumann, P. M., The structure of finitary permutation groups, Arch. Math. (Basel) 29 (1976), 3–17.

A Class of Inefficient Groups with Symmetric Presentation

E. F. Robertson, R. M. Thomas and C. I. Wotherspoon

Abstract. We examine a class of groups, originally introduced by Coxeter, which are described by a symmetric presentation and show that they are inefficient. We also prove that every finite simple group can be embedded in a finite inefficient group.

1991 Mathematics Subject Classification: 20F05

A finite group G is said to have **deficiency n**, if n is the smallest integer such that G can be presented with d generators and $d + n$ relations. A finite group G is said to be **efficient** if G has a presentation with d generators and $d + r$ relations, where r is the rank of the Schur multiplier [7] $M(G)$ of G. If no such presentation exists, then $n > r$ and the group is said to be **inefficient**.

There are many examples of efficient groups. For example all abelian [4], metacyclic [9] and many simple groups [2] are known to be efficient. However, there seems to be essentially only one class of inefficient groups known, which was provided by Swan [8] in 1965. For a survey of the efficiency problem and a good description of the Swan groups see Wiegold [10].

In this paper we examine a class of groups, originally introduced by Coxeter [3], which are described by a symmetric presentation [5] and show that they are inefficient. We also prove that every finite simple group can be embedded in a finite inefficient group.

Definition. Let G be the n-generator group defined by the presentation

$$P = \langle a_1, a_2, \ldots, a_n \mid w_i(a_1, a_2, \ldots, a_n) = 1, \ 1 \le i \le m \rangle.$$

Then P is a **symmetric presentation** if $G = S_n G$, where $S_n G$ is the group defined by the presentation

$$\langle a_1, a_2, \ldots, a_n \mid w_i(a_{1\theta}, a_{2\theta}, \ldots, a_{n\theta}) = 1, \ 1 \le i \le m, \ \theta \in S_n \rangle.$$

We now state our main theorem.

Theorem 1. *The group presentation*

$$P_n = \langle a, b, c \mid a^2 = b^2 = c^2 = (ab)^3 = (bc)^3 = (ca)^3 = (abac)^n = 1 \rangle$$

is the symmetric presentation of a finite group G_n which is inefficient when $(n, 6) = 3$.

These groups are finite three-dimensional irreducible groups generated by three reflections; a full description of their geometrical significance is given in Coxeter [3]. In particular, the group denoted here by G_3 is in the list of transitive groups produced by Butler and MacKay [1], where it is named $t9n12$.

We prove Theorem 1 via a series of lemmas.

Lemma 1. *P_n is a symmetric presentation.*

Proof. The relations $a^2 = b^2 = c^2 = 1$ and $(ab)^3 = (bc)^3 = (ca)^3 = 1$ are clearly symmetric. The relation $(abac)^n = 1$ is symmetric because $abac = babc$ (since $ababab = 1$) and so $(abac)^n = 1 \Leftrightarrow (babc)^n = 1$. The other relations $(cacb)^n = 1$ etc. follow in the same way.

Lemma 2. *$\mid G_n \mid = 2 \cdot 3 \cdot n^2$.*

Proof. Introduce $x = ab$, $y = ac$ and eliminate b and c from the presentation P_n. G_n is now presented by

$$\langle a, x, y \mid a^2 = 1, x^a = x^{-1}, y^a = y^{-1}, x^3 = y^3 = (xy^{-1})^3 = (xy)^n = 1 \rangle.$$

Clearly G_n is the semidirect product of $H_n = \langle x, y \rangle$ by $\langle a \rangle$ and H_n has presentation

$$\langle x, y \mid x^3 = y^3 = (xy^{-1})^3 = (xy)^n = 1 \rangle.$$

It remains to show that $\mid H_n \mid = 3 \cdot n^2$. Introduce $u = xy$, $v = yx$, so that H_n is defined by

$$\langle x, y, u, v \mid x^3 = y^3 = (xy^{-1})^3 = (xy)^n = 1, u = xy, v = x^{-1}ux \rangle,$$

and then delete $y = x^{-1}u$ to obtain

$$\langle x, u, v \mid x^3 = 1, (x^{-1}u)^3 = 1, (xu^{-1}x)^3 = 1, u^n = 1, v = x^{-1}ux \rangle.$$

We see that

$$
\begin{aligned}
(x^{-1}u)^3 = 1 \quad &\Leftrightarrow \quad x^{-1}uxxux^{-1}u = 1 \quad (x^3 = 1) \\
&\Leftrightarrow \quad vxux^{-1}u = 1 \quad\quad\quad (x^{-1}ux = v) \\
&\Leftrightarrow \quad xux^{-1} = v^{-1}u^{-1} \\
&\Leftrightarrow \quad x^{-2}ux^2 = v^{-1}u^{-1} \quad (x^3 = 1) \\
&\Leftrightarrow \quad x^{-1}vx = v^{-1}u^{-1} \quad (x^{-1}ux = v),
\end{aligned}
$$

$$(xu^{-1}x)^3 = 1 \quad \Leftrightarrow \quad (xu)^3 = 1 \qquad (x^3 = 1)$$
$$\Leftrightarrow \quad xux^{-1}x^{-1}uxu = 1 \quad (x^3 = 1)$$
$$\Leftrightarrow \quad v^{-1}u^{-1}vu = 1 \qquad (xux^{-1} = v^{-1}u^{-1}, \ x^{-1}ux = v)$$
$$\Leftrightarrow \quad [v, u] = 1.$$

Also, since $x^{-1}ux = v$ and $u^n = 1$, v has order n and the presentation for H_n may now be written as

$$\langle x, u, v \mid x^3 = 1, \ u^x = v, \ v^x = v^{-1}u^{-1}, \ u^n = v^n = [u, v] = 1 \rangle,$$

so that H_n is a semidirect product of $\langle u, v \rangle = C_n \times C_n$ by $\langle x \rangle = C_3$, with the homomorphism $u \longmapsto v \longmapsto v^{-1}u^{-1}$. The order of H_n is $3 \cdot n^2$, $[G_n : H_n] = 2$, and so G_n has order $2 \cdot 3 \cdot n^2$.

Lemma 3. *The Schur multiplier of G_n is trivial when n is odd.*

Proof. The order of G_n is $2 \cdot 3 \cdot n^2$ and n is odd. If e is the exponent of the multiplier, then e^2 is a divisor of the order of G_n [6] and so the exponent of the multiplier of G_n must have odd order. If the multiplier is non-trivial, then G_n will have a stem (and therefore a central) extension by a cyclic group of odd order. We would therefore have a group $\tilde{G}_n$ with presentation

$$\langle a, x, y, z \mid a^2 = z^d, \ x^a = x^{-1}z^i, \ y^a = y^{-1}z^j,$$
$$x^3 = z^p, \ y^3 = z^q, \ (xy^{-1})^3 = z^r, \ (xy)^n = z^s, \ z^t = 1, \ z \text{ central} \rangle,$$

where t is odd.

Consider the relation $a^2 = z^d$ and choose $\alpha = az^{-u}$ where $2u \equiv d \pmod{t}$. Now

$$a^2 = z^d \quad \Leftrightarrow \quad \alpha^2 z^{2u} = z^d \quad \Leftrightarrow \quad \alpha^2 = 1.$$

Without loss of generality we can therefore assume that $a^2 = 1$ and so we have

$$\langle a, x, y \mid a^2 = 1, \ x^a = x^{-1}z^i, \ y^a = y^{-1}z^j, \ x^3 = z^p,$$
$$y^3 = z^q, \ (xy^{-1})^3 = z^r, \ (xy)^n = z^s, \ z^t = 1, \ z \text{ central} \rangle.$$

Consider the relation $x^a = x^{-1}z^i$ and choose $\chi = xz^{-v}$ where $2v \equiv i \pmod{t}$. Now

$$x^a = x^{-1}z^i \quad \Leftrightarrow \quad a^{-1}\chi az^v = (\chi z^v)^{-1}z^i \quad \Leftrightarrow \quad \chi^a = \chi^{-1}.$$

Without loss of generality we can therefore assume that $x^a = x^{-1}$ and similarly $y^a = y^{-1}$. The presentation for $\tilde{G}_n$ may now be written as

$$\langle a, x, y \mid a^2 = 1, \ x^a = x^{-1}, \ y^a = y^{-1}, \ x^3 = z^p,$$
$$y^3 = z^q, \ (xy^{-1})^3 = z^r, \ (xy)^n = z^s, \ z^t = 1, \ z \text{ central} \rangle.$$

Since $x^3 = z^p$, we have that

$$z^{-p} = x^{-3} = a^{-1}x^3a = a^{-1}z^pa = z^p,$$

so that $z^{2p} = 1$, and hence $z^p = 1$ (z has odd order). So $x^3 = 1$. Also, since $(xy^{-1})^3 = z^r$, we have that

$$z^{-r} = a^{-1}z^{-r}a = a^{-1}(xy^{-1})^{-3}a = (x^{-1}y)^{-3} = z^r,$$

so that $z^{2r} = 1$, and hence $z^r = 1$. So $(xy^{-1})^3 = 1$, and similarly, $y^3 = (xy)^n = 1$. So we have the presentation

$$\langle a, x, y \mid a^2 = 1, \ x^a = x^{-1}, \ y^a = y^{-1},$$
$$x^3 = y^3 = (xy^{-1})^3 = (xy)^n = 1, \ z^t = 1, \ z \text{ central}\rangle.$$

Hence $\tilde{G}_n = G_n \times \langle z \rangle$. We see that there is no central extension $\tilde{G}_n$ with $z \in Z(\tilde{G}_n) \cap \tilde{G}_n'$, and so G_n has trivial multiplier. We now turn to the subgroup H_n of index 2.

Lemma 4. *If H_n is defined by*

$$\langle x, y \mid x^3 = y^3 = (xy^{-1})^3 = (xy)^n = 1\rangle,$$

then it has Schur multiplier of rank 2 when $(n, 6) = 3$.

Proof. Consider the group $\overline{H}_n$ defined by the presentation

$$\langle x, y, z, t \mid x^3 = 1, \ y^3 = z, \ (xy^{-1})^3 = 1, \ (xy)^n = t, \ z^3 = t^3 = 1, \ z, t \text{ central}\rangle.$$

Clearly, $\overline{H}_n$ is a central extension of H_n by $\langle z, t \rangle$. It is also a stem extension. To see this, abelianise to obtain the presentation for $\overline{H}_n/\overline{H}_n'$ and find that now $z = t = 1$ and so $\langle z, t \rangle \leq \overline{H}_n'$

We will show that when $(n, 6) = 3$, the rank of $\langle z, t \rangle$ is 2, which shows that the multiplier of H_n has rank at least 2. H_n is expressed with a deficiency 2 presentation and so the multiplier will have rank exactly 2.

We first introduce the generators $u = xy^{-1}$, $v = y^{-1}x$, and then eliminate $y = u^{-1}x$ to get

$$\langle x, z, t, u, v \mid x^3 = 1, \ (u^{-1}x)^3 = z, \ u^3 = 1, \ (xu^{-1}x)^n = t,$$
$$z^3 = t^3 = 1, \ x^{-1}ux = v, \ z, t \text{ central}\rangle.$$

We see that

$$(u^{-1}x)^3 = z \quad \Leftrightarrow \quad u^{-1}x^{-2}u^{-1}x^2x^{-1}u^{-1}x = z \qquad (x^3 = 1)$$
$$\Leftrightarrow \quad u^{-1}x^{-1}v^{-1}xv^{-1} = z \qquad (x^{-1}ux = v)$$
$$\Leftrightarrow \quad x^{-1}vx = v^{-1}u^{-1}z^{-1} \qquad (z \text{ central}).$$

$$(xu^{-1}x)^n = t \quad \Leftrightarrow \quad (u^{-1}x^{-1})^n = t \qquad (t \text{ central}, \ x^3 = 1)$$
$$\Leftrightarrow \quad (u^{-1}x^{-1}u^{-1}xx^{-2}u^{-1}x^2)^{n/3} = t \qquad (x^3 = 1, \ (n, 6) = 3)$$
$$\Leftrightarrow \quad (u^{-1}v^{-1}x^{-1}v^{-1}x)^{n/3} = t \qquad (x^{-1}ux = v)$$
$$\Leftrightarrow \quad (u^{-1}v^{-1}uvz)^{n/3} = t \qquad (x^{-1}vx = v^{-1}u^{-1}z^{-1})$$
$$\Leftrightarrow \quad [u, v]^{n/3} = tz^{-n/3}.$$

The presentation for $\overline{H}_n$ now becomes

$$\langle x, z, t, u, v \mid x^3 = 1, \ u^x = v, \ v^x = v^{-1}u^{-1}z^{-1}, \ u^3 = 1,$$
$$[u, v]^{n/3} = tz^{-n/3}, \ z^3 = t^3 = 1, \ z, t \ \text{central}\rangle.$$

Since $u^x = v$, $v^x = v^{-1}u^{-1}z^{-1}$, $u^3 = 1$ and $z^3 = 1$, we have that $v^3 = (uv)^3 = 1$. So $\overline{H}_n$ is a semidirect product of $\langle u, v, z, t \rangle$ by $\langle x \rangle$, where $\langle u, v, z, t \rangle$ has presentation

$$\langle u, v, z, t \mid u^3 = v^3 = (uv)^3 = 1, \ [u, v]^{n/3} = tz^{-n/3}, \ z^3 = t^3 = 1, \ z, t \ \text{central}\rangle.$$

Introduce $w = tz^{-n/3}$ and eliminate t to obtain

$$\langle u, v, z, w \mid u^3 = v^3 = (uv)^3 = 1, \ [u, v]^{n/3} = w, \ z^3 = w^3 = 1, \ z, w \ \text{central}\rangle.$$

This is a presentation for the direct product of $\langle u, v, w \rangle$ and $\langle z \rangle$, which shows that z is non-trivial. We now consider the factor group $\overline{H}_n/\langle z \rangle$, and show that (the image of) t is non-trivial in this group.

$\overline{H}_n/\langle z \rangle$ has presentation

$$\langle x, y, t \mid x^3 = 1, \ y^3 = 1, \ (xy^{-1})^3 = 1, \ (xy)^n = t, \ t^3 = 1, \ t \ \text{central}\rangle.$$

Introduce $u = xy$, $v = yx$, and eliminate $y = x^{-1}u$, to obtain

$$\langle x, t, u, v \mid x^3 = 1, \ (x^{-1}u)^3 = 1, \ (xu^{-1}x)^3 = 1,$$
$$u^n = t, \ v = x^{-1}ux, \ t^3 = 1, \ t \ \text{central}\rangle.$$

We see that

$$(x^{-1}u)^3 = 1 \quad \Leftrightarrow \quad x^{-1}uxx^{-2}ux^2u = 1 \quad (x^3 = 1)$$
$$\Leftrightarrow \quad x^{-1}vx = v^{-1}u^{-1} \qquad (v = x^{-1}ux).$$

$$(xu^{-1}x)^3 = 1 \quad \Leftrightarrow \quad (xu)^3 = 1 \qquad\qquad (x^3 = 1)$$
$$\Leftrightarrow \quad xux^{-1}x^{-1}uxu = 1 \qquad (x^3 = 1)$$
$$\Leftrightarrow \quad [v, u] = 1 \qquad\qquad (xux^{-1} = x^{-1}vx = v^{-1}u^{-1})$$

$$u^n = t \quad \Leftrightarrow \quad v^n = t \qquad\qquad\qquad (x^{-1}ux = v).$$

The presentation may now be written as

$$\langle x, t, u, v \mid x^3 = 1, \ u^x = v, \ v^x = u^{-1}v^{-1}, \ [u, v] = 1,$$
$$u^n = t, \ v^n = t, \ t^3 = 1, \ t \ \text{central}\rangle,$$

which is a presentation of a semi-direct product of $\langle t, v, u \rangle$ by $\langle x \rangle = C_3$ with the homomorphism, $u \longmapsto v \longmapsto v^{-1}u^{-1}$. The group $\langle t, v, u \rangle$ is clearly abelian of order $3n^2$, and so t is non-trivial as required.

We have now shown that $\langle z, t \rangle$ has rank 2, and so the multiplier of H_n has rank 2 when $(n, 6) = 3$.

Lemma 5. *If G is a finite group with a subgroup H of index k and the Schur multiplier of H has rank r, then the deficiency of G satisfies*

$$\text{def}(G) \geq \frac{r+1}{k} - 1.$$

Proof. This follows immediately from the Reidemeister–Schreier theorem.

We are now in a position to prove that G_n is inefficient when $(n, 6) = 3$. In Lemma 4 it was shown that the subgroup H_n of index 2, has Schur multiplier of rank 2. By Lemma 5 the deficiency of G_n satisfies

$$\text{def}(G_n) \geq \frac{2+1}{2} - 1 = \frac{1}{2}.$$

The deficiency of G_n is greater than equal to one, but by Lemma 2, G_n has trivial Schur multiplier and so G_n is inefficient when $(n, 6) = 3$.

In fact we can calculate the exact deficiency of this class of inefficient groups. A proof to the next theorem will appear in [11].

Theorem 2. *When n is odd, a deficiency 1 presentation for G_n is*

$$\langle a, b, c \mid a^{-1}babab = b^{-1}cbcbc = c^{-1}acaca = a^{-1}bac(abac^{-1})^{n-1} = 1\rangle.$$

Since S_3 acts naturally on G_n, we can extend G_n by any subgroup of S_3. Some of these extensions give further classes of inefficient groups, while some give classes of efficient groups; see [11] for details.

No example is known of an inefficient simple group. However, given any non-abelian finite simple group S, it is possible to find an inefficient group G involving S, as the next part shows.

Theorem 3. *Every non-abelian finite simple group occurs as a composition factor of an inefficient group.*

Proof. If S is any finite simple group, then the covering group $\overline{S}$ of S has trivial Schur multiplier [10] and the direct product $\overline{S} \times G_n$ has S as a composition factor. We now show that when $(n, 6) = 3$, $\overline{S} \times G_n$ is an inefficient group. By the Schur–Künneth formula [7], the Schur multiplier of $\overline{S} \times G_n$ is

$$M(\overline{S} \times G_n) = M(\overline{S}) \times M(G_n) \times (\overline{S} \otimes G_n),$$

where $\otimes$ is the tensor product. $\overline{S}$ is perfect and so $M(\overline{S} \times G_n) = 1$. Consider the subgroup $\overline{S} \times H_n$ of index 2, which has multiplier

$$\begin{aligned}
M(\overline{S} \times H_n) &= M(\overline{S}) \times M(H_n) \times (\overline{S} \otimes H_n) \\
&= M(H_n).
\end{aligned}$$

The multiplier of H_n has rank 2 and so, by Lemma 5, $\overline{S} \times G_n$ is an inefficient group with the simple group S as a composition factor.

László Kovács pointed out to us that one can use this argument to embed any group in an inefficient group providing we use the Swan groups instead of the groups G_n. Each Swan group T_n is a semi-direct product of P_n by Q_n, where P_n is elementary abelian of order 7^n and Q_n cyclic of order 3. If we again invoke the Schur–Künneth formula, we have

$$M(T_n \times G) = M(T_n) \times M(G) \times (T_n \otimes G)$$

for any finite group G. Since $M(T_n) = 1$ and T_n/T_n' is cyclic of order 3 for all n, this is independent of n, and has rank q (say). Now $T_n \times G$ has $P_n \times G$ as a subgroup of index 3, and we choose n such that $M(P_n \times G)$ has rank r, where

$$\frac{r+1}{3} - 1 > q.$$

It follows immediately from Lemma 5, that $T_n \times G$ is inefficient.

Acknowledgement. The second author would like to thank Hilary Craig for all her help and encouragement.

References

[1] G. Butler and J. MacKay, The transitive groups of degree up to eleven, Comm. Algebra 11 (8) (1983), 205–221.

[2] C. M. Campbell, E. F. Robertson and P. D. Williams, Efficient presentations for finite simple groups and related groups, in: Groups—Korea 1988 (A. C. Kim and B. H. Neumann, eds.), Lecture Notes in Math. 1398, Springer-Verlag, Berlin–Heidelberg–New York 1989, 65–72.

[3] H. S. M. Coxeter, Groups generated by unitary reflections of period two, Canadian J. Math. 9 (1957), 243–272.

[4] D. Epstein, Finite presentations of groups and 3-manifolds, Quart. J. Math. Oxford Ser (2) 12 (1961), 205–221.

[5] E. F. Robertson and C. M. Campbell, Symmetric presentations, in: Group Theory (K. N. Cheng and Y. K. Leong, eds.), Walter de Gruyter, Berlin–New York 1989, 497–506.

[6] I. Schur, Über die Darstellung der endlichen Gruppen durch gebrochene lineare Substitutionem, J. Reine Angew. Math. 127 (1904), 20–50.

[7] I. Schur, Untersuchungen über die Darstellung der endlichen Gruppen durch gebrochene lineare Sustitutionen, J. Reine Angew. Math. 132 (1907), 85–137.

[8] R. G. Swan, Minimal resolutions for finite groups, Topology 4 (1965), 193–208.

[9] J. W. Wamsley, The Deficiency Of Finite Groups, Ph.D. thesis, University of Queensland, 1968.

[10] J. Wiegold, The Schur Multiplier, in: Groups—St Andrews 1981 (C. M. Campbell and E. F. Robertson, eds.), London Math. Soc. Lecture Note Ser. 71, Cambridge University Press, Cambridge 1982, 193–208.

[11] C. I. Wotherspoon, The Deficiency of Particular Finite Groups, Ph.D. thesis, University of St.Andrews, to appear.

On Quasi Left Groups

K. P. Shum, X. M. Ren and Y. Q. Guo

Abstract. Let $\overline{G}$ be a quasi group and I a left zero band. Form the direct product $\overline{G} \times I$. In this paper, a necessary and sufficient condition which leads to the quasi left group S be isomorphic to $\overline{G} \times I$ is found. Semigroups which are retract nil extensions of a left group are also characterized.

1. Introduction

A semigroup S is called quasi regular if for any $a \in S$ there exists $x \in S$ and a natural number n such that $a^n = a^n x a^n$. Clearly, the class of regular semigroups is a particular subclass of the class of quasi regular semigroups. A semigroup S is called a left group if for any $a, b \in S$ there exists $x \in S$ such that $xa = b$. According to M. Petrich ([3], p. 118), a semigroup S is a left group if and only if S is regular and E_S, the set of all idempotents of S, is a left zero band. Because of this result, we call a semigroup S a quasi left group if and only if S is quasi regular and E_S is a left zero band. Naturally, we call a quasi regular semigroup S a quasi group if $|E_S| = 1$.

Many interesting properties of left groups can be found in the text of M. Petrich ([3], p. 117–120). Among all these properties, the following property is associated with bands : S is a left group if and only if S is isomorphic to the direct product of a group and a left zero band. It is therefore natural to ask whether the corresponding result holds for quasi left groups? In fact, if $\overline{G}$ is a quasi group then, for any $a \in \overline{G}$ there exists $n \in N$ such that a^n is a completely regular element in $\overline{G}$. Hence, we deduce that $x^n = (a^n, i) \in \overline{G} \times I = S$ is a completely regular element in S as well, where I is a left zero band. In other words, the semigroup S is quasi regular. Moreover, if we let e be the identity element of $\overline{G}$, then the set $E_S = \{(e, i) \mid i \in I\}$ is trivially the left zero band of S. Consequently, $S = \overline{G} \times I$ is a quasi left group. However, the converse is not obvious. If S is a quasi left group, then S need not be expressed as the direct product of a quasi group and a left zero band. For instance, consider $S = \{a, e, f\}$ with the following Cayley table

·	a	e	f
a	f	f	f
e	e	e	e
f	f	f	f

Then S is clearly a quasi left group, where $I = \{e, f\}$ is its left zero band.

In this paper, we shall explore some conditions which lead to $S \cong \overline{G} \times I$. Retract nil extensions of left groups will also be studied. For definitions and terminology not given in this paper, the reader is referred to [2] or [3].

2. Quasi Left Groups

Throughout this section, S is assumed to be a quasi left group. The set of all regular elements of S is denoted by Reg S. Let $\mathcal{H}$ be the usual Green's relations defined on S. On the quasi regular semigroups S, define $x\mathcal{H}^*y$ if and only if there exists $n, m \in N$ such that $x^n, y^m \in$ Reg S and $x^n\mathcal{H}y^m$. It can be easily seen that $\mathcal{H}^*$ is also a equivalent relation on S and we call it the star-generalized Green's relation. Star-generalized relations, namely, the Green's relations, on quasi regular semigroups have already been studied in [4]. As usual, we denote the $\mathcal{H}^*$-class containing the element $a \in S$ by H_a^*.

The following theorem is a characterization of quasi left groups.

Theorem 2.1. *Let S be a quasi left group; $\overline{G}$ a quasi group and I a left zero band. Then $S \cong \overline{G} \times I$ if and only if the following conditions are satisfied:*
(i)   *$\mathcal{H}^*$ is a left zero band congruence on S;*
(ii) *For any $e \in E_S$, there exists an isomorphism φ_e from H_e^* onto the quasi group $\overline{G}$ such that $x\varphi_e y\varphi_f = (xy)\varphi_e$ whenever $x \in H_e^*$ and $y \in H_f^*$.*

Proof. $\Rightarrow$:   Without loss of generality, assume $S = \overline{G} \times I$. We first claim that

$$\mathcal{H}^* = \{(x, y) \mid (\exists i \in I, \ a, b \in \overline{G}), \ x = (a, i), \ y = (b, i)\}.$$

For this purpose, we let $x\mathcal{H}^*y$ for some $x = (a, k), y = (b, j) \in \overline{G} \times I$. Then, by the definition of $\mathcal{H}^*$, there exists $n, m \in N$ such that $x^n, y^m \in$ Reg S and $x^n\mathcal{H}y^m$, where $\mathcal{H}$ is the Green's relation on S. Because $\overline{G}$ is a quasi group, $\overline{G}$ is regular and $|E_{\overline{G}}| = 1$. By $x^n Hy^m$ and $S = \overline{G} \times I$, x^n is contained in a subgroup of S. This implies that x^n is a completely regular element of S. Similarly, y^m is also a completely regular element of S. Thus, there exists an idempotent $u = (\overline{e}, i) \in \overline{G} \times I$ such that $u\mathcal{H}x^n\mathcal{H}y^m$, where $\overline{e}$ is the unique idempotent of the quasi group $\overline{G}$. From $u\mathcal{H}x^n$ we deduce immediately that $x^n = x^n u = ux^n$, thereby $i = k$. Similarly, we can prove that $j = i$. In other words, we have demonstrated that $x = (a, i)$ and $y = (b, i)$. On the other hand, for each $i \in I$, let $x = (a, i)$ and $y = (b, i)$ for some $a, b \in \overline{G}$. Then, by the quasi completely regularity of the quasi group $\overline{G}$, we have $x\mathcal{H}^*y$. Our claim is thus established.

Now, if $x = (a, i) \in H_e^*$, $y = (b, j) \in H_f^*$ for some $e, f \in E_S$, then $xy = (a, i)(b, j) = (ab, ij) = (ab, i) \in H_e^*$. This means that $H_e^* H_f^* \subseteq H_e^*$ since $ab \in \overline{G}$. In other words, the star-Green's relation $\mathcal{H}^*$ is a left zero band congruence on S. Condition (i) of the theorem is satisfied.

Now for any $e \in E_S$, let $x = (a, i) \in H_e^*$. Clearly, the mapping $\varphi_e : (a, i) \longmapsto a$ is an isomorphism from H_e^* onto the group $\overline{G}$. In addition, if $x = (a, i) \in H_e^*$ and $y = (b, j) \in H_f^*$ for any idempotents $e, f \in E_S$, then $xy = (ab, i)$ since I is a left zero band. This implies that $x\varphi_e y\varphi_f = ab = (xy)\varphi_e$. Hence, condition (ii) is verified.

$\Leftarrow$: Let $\mathcal{H}^*$ be the left zero band congruence on S. Then, by the idempotent lifting theorem of quasi regular semigroups [4] and the fact that each $\mathcal{H}^*$-class contains at most one idempotent element, we deduce that each $\mathcal{H}^*$-class of S is a quasi group. By condition (ii), it is obvious that every $\mathcal{H}^*$-class is isomorphic to each other. Denote one of these $\mathcal{H}^*$-classes by $\overline{G}$ and the set of all idempotents of S by I. Then, according to our hypotheses, I is a left zero band. For any $x \in S$, define a mapping $\varphi : S \longrightarrow \overline{G} \times I$ by $x \longmapsto (x\varphi_e, e)$ where e is the idempotent of the $\mathcal{H}^*$-class which contains x. As φ_e is the isomorphism of the group H_e^* onto the group $\overline{G}$, it is easy to verify that φ is actually an isomorphism from S onto $\overline{G} \times I$. The proof is completed.

3. Retract Nil Extensions of Left Groups

Let T be an ideal of the semigroup S. Then S is called a nil extension of T if the Rees quotient S/T is a nil semigroup. Also, S is called the retract extension of T if there exists a homomorphism $\varphi : S \longrightarrow T$ such that $\varphi(t) = t$ for all $t \in T$. It was shown by S. Bogdanovic that a semigroup S is a quasi left group if and only if S is a nil extension of a left group [1]. Thus the retract nil extension of a left group is just a special class of quasi left groups. We shall call this special class of quasi left groups be retractable quasi left groups.

The following theorem is to give a description for retractable quasi left groups.

Theorem 3.1. *A semigroup S is a retractable quasi left group if and only if S is a subdirect product of a left group and a nil semigroup.*

Proof. $\Rightarrow$: Suppose S is a retractable quasi left group. Then, by definition, there exists an ideal H of S which is a left group such that the Rees quotient S/H is nil.

Consider the Rees congruence on S determined by H, namely

$$\rho_H = \{(a, b) \mid a, b \in H \text{ or } a = b \text{ for all } a, b \in S \setminus H\}.$$

Because S is retractable on H, there exists a homomorphism φ such that $\varphi(h) = h$ for all $h \in H$. Let ρ the congruence on S induced by φ. Obviously, $\rho|_H = 1_H$. Hence $\rho_H \cap \rho = 1_S$. Invoking theorem 5.9 in Howie ([2], IV.5), we know immediately that S is isomorphic to the subdirect product of the left group $H = S/\rho$ and the nil semigroup $S/_H$.

$\Leftarrow$: Suppose that $S \subseteq H \times Q$, where H is a left group and Q is a nil semigroup. Consider the product $H \times \{o\}$. By the definition of subdirect product, $H \times \{o\} \subseteq S$. Clearly, $H \times \{o\}$, is an ideal of S. Also, since every element of Q is nilpotent, for any $a = (x, q) \in S \subseteq H \times Q$ we have $a^n = (x, q)^n = (x^n, q^n) = (x^n, o) \in H \times \{o\}$ for some positive integer n. This implies that the Rees quotient $S/_{H \times \{o\}}$ is a nil semigroup. As $H \cong H \times \{o\}$, S is clearly a nil extension of the left group H. We now prove the retractness of S. By the basic property of left group ([3] p. 117, IV 3.8), we know $H = G \times I$ for some group G and some left zero band I. Hence, for any $a \in S$ $a = (g, i, q)$ for some $g \in G$, $i \in I$ and $q \in Q$. Because $H \cong H \times \{o\} = G \times I \times \{o\}$, we can define a mapping $\Phi : S \longmapsto H \times \{o\}$ by

$$\Phi : (g, i, q) \longmapsto \Phi((g, i, q)) = (g, i, q)(1_G, i, o)$$

where 1_G is the identity element of the group G.

It is now trivial to observe that $\Phi|_{H \times \{o\}}$ is an identity mapping. Moreover, for any $a = (g_1, i, q_1)$, $b = (g_2, j, q_2) \in S$, we have

$$\begin{aligned}
\Phi(a)\Phi(b) &= (g_1, i, q_1)(1_G, i, o)(g_2, j, q_2)(1_G, j, o) \\
&= (g_1 g_2, i, o)
\end{aligned}$$

and

$$\begin{aligned}
\Phi(ab) &= (g_1 g_2, i, q_1 q_2)(1_G, i, o) \\
&= (g_1 g_2, i, o)
\end{aligned}$$

Thus, Φ is verified to be a homomorphism from S onto $H \times \{o\} \cong H$. This means that, S is a retractable nil extension of a left group. The proof is completed.

References

[1] Bogdanović, S., Semigroups with a system of subsemigroups, Lecture Notes, Novi Sad University, Yugosalvia, 1986.

[2] Howie, J. M., An introduction to semigroup theory, Academic Press, London 1976.

[3] Petrich, M., Introduction to semigroups, Charles E. Merill Publishing Company, 1973.

[4] Ren X. M., Guo Yuqi and Shum K. P., On quasi Clifford regular semigroups, Chinese Ann. Math. Ser. A 15 (3) (1994), 319–325.

Generic Elements in Certain Groups

John R. Stallings

Abstract. Dold [4] found several examples of elements in free groups, which we call "generic with respect to a verbal subgroup $\mathcal{V}$". An instance which had been noted earlier by Zieschang and others is the commutator $[a, b] = aba^{-1}b^{-1}$ in the free group F on $\{a, b\}$. The "generic" property of this element is that whenever there is any homomorphism of groups $\alpha\colon G \to F$, taking an element of the commutator subgroup of G to $[a, b]$, then it follows that α is surjective. We describe some related results, how to compute whether an element is generic with respect to certain verbal subgroups, and some interesting questions.

1991 Mathematics Subject Classification: 20E10, 20F32

1. The Categorical Picture

Definition 1.1. By a *categorical subgroup $\mathcal{S}$*, is meant a functor from the category of groups and homomorphisms to itself, together with a natural transformation E from $\mathcal{S}$ to the identity functor, such that for every group G, the homomorphism $E_G\colon\mathcal{S}(G) \to G$ is injective. We generally simply consider E_G to be an inclusion, and thus consider $\mathcal{S}(G)$ to be a subgroup of G.

One example is the commutator subgroup of a group; this subgroup $\mathcal{C}(G)$ is generated by all commutators $[a, b]$, for $a, b \in G$, homomorphisms of G restrict to homomorphisms of $\mathcal{C}(G)$, and so on.

A related example it the "perfection" $\mathcal{P}(G)$; this is the largest perfect subgroup of G; where a "perfect" group is a group H such that $H = \mathcal{C}(H)$; it is an elementary exercise to show that the subgroup of G generated by any union of perfect subgroups is perfect, so that there is a maximum one $\mathcal{P}(G)$.

Another example, unrelated to these two, is the residually finite kernel $\mathcal{F}(G)$, the intersection of all subgroups of finite index in G

This work is based on research partly supported by the National Science Foundation under Grant No. DMS-9203941.

There is a difference among these categorical subgroups, in that C preserves epimorphism, while $\mathcal{P}$ and $\mathcal{F}$ do not. In other words, if $\alpha: G \to H$ is surjective (= epimorphism), then the restriction, the map $C(\alpha): C(G) \to C(H)$ must also be surjective. But, for instance, you can map a free group F onto a non-trivial perfect group G, and $\mathcal{P}(F) = \{1\}$, whereas $\mathcal{P}(G) = G$. This difference leads to a definition:

Definition 1.2. A **verbal subgroup** $\mathcal{V}$ is a categorical subgroup functor which takes epimorphisms to epimorphisms.

This definition is somewhat more general than usual [6]; but if we restrict ourselves to countable groups, we can determine $\mathcal{V}$ by what it does to a free group F of countable rank. If G is any countable group, then there is some epimorphism $\alpha: F \to G$, and then, $\mathcal{V}$ taking epimorphisms to epimorphisms, we simply have $\mathcal{V}(G) = \alpha(\mathcal{V}(F))$.

In the opposite direction, if X is any subset of F, and we define $\mathcal{V}_X(G)$ to be the subgroup of G generated by the union of all $\beta(X)$, for all homomorphisms $\beta: F \to G$, this determines such a categorical subgroup $\mathcal{V}_X$ which takes epimorphisms to epimorphisms; the main point being that if $\beta: F \to H$ is any homomorphism and $\alpha: G \to H$ any epimorphism, then β factors through a homomorphism $\beta': F \to G$. The crucial object $\mathcal{V}_X(F)$ is then the minimal "fully invariant" subgroup of F containing X; "fully invariant" meaning that it is mapped into itself by all endomorphisms.

The commutator subgroup is then the verbal subgroup determined by the singleton set $X = \{xyx^{-1}y^{-1}\}$ of the free group F having $\{x, y\}$ as a subset of a free basis.

Definition 1.3. The verbal subgroup functors $\mathcal{V}_n$, for $n = 0, 1, 2, \ldots$, are defined to be those defined by the singleton sets $\{z^n xyx^{-1}y^{-1}\}$, where $\{x, y, z\}$ is part of a free basis of F. In other words, $\mathcal{V}_n(G)$ is the subgroup of G generated by all commutators and all n-th powers. In particular, $\mathcal{V}_0(G)$ is the commutator subgroup of G; $\mathcal{V}_1(G)$ is the entire group G. The quotient group $G/\mathcal{V}_n(G)$ is the largest quotient group of G which is abelian and has exponent n.

Definition 1.4. For any categorical subgroup $\mathcal{V}$, we can define $\mathcal{V}^*(G)$ to be the subgroup of G generated by all subgroups $S \leq G$, such that $S = \mathcal{V}(S)$.

In fact, this subgroup itself has this property, so that $\mathcal{V}^*(G)$ is the maximum subgroup $T = \mathcal{V}^*(G)$ of G, such that $T = \mathcal{V}(T)$. It is easily seen that $(\mathcal{V}^*)^* = \mathcal{V}^*$. The relation between commutator subgroup C and perfection $\mathcal{P}$ is that $\mathcal{P} = C^*$.

Definition 1.5. An element h of a group H is said to be **generic in H with respect to the verbal subgroup $\mathcal{V}$**, when: First, $h \in \mathcal{V}(H)$. And, second, for every group G and for every homomorphism $\phi: G \to H$, if there exists $g \in \mathcal{V}(G)$ such that $\phi(g) = h$, then ϕ is surjective.

The word "generic" was suggested by Jon Sjogren. It has a tenuous connection with algebraic geometry, perhaps. The fact that in the free group $F(a, b)$, with basis

$\{a, b\}$, the element $[a, b] = aba^{-1}b^{-1}$ is generic goes far back in history, perhaps to Nielsen; this was generalized by Zieschang [11], and more recently by Dold [4]. There is some connection also, with the work of Brunner, Burns, and Oates-Williams [1], and with work of Shpilrain [8].

2. The Generic Elements Described by Dold

Definition 2.1. Suppose that F is free with a finite basis $A = \{a_1, \ldots, a_n\}$, and that $w \in F$. Suppose that w is cyclically reduced; that is, that the length of w as a reduced word in the basis elements and their inverses cannot be shortened by conjugation. Then the **Whitehead graph of w with respect to the basis** A is the graph $\mathrm{Wh}(w, A)$ or, abbreviated, $\mathrm{Wh}(w)$, defined thus:

There are $2n$ vertices to $\mathrm{Wh}(w)$, denoted by $A \cup \overline{A} = a_1^{+1}, \ldots, a_n^{+1}, a_1^{-1}, \ldots, a_n^{-1}$; we also denote a_i^{+1} by a_i and a_i^{-1} by $\overline{a_i}$. If w has length k in terms of the basis A, then there are k edges in $\mathrm{Wh}(w)$; for each occurrence of $a_i^\epsilon a_j^\eta$ as a subword of w in reduced form read cyclically, there is an edge in $\mathrm{Wh}(w)$ from a_i^ϵ to $a_j^{-\eta}$. — Think of a graph with one vertex and edges corresponding to the elements of A; in this graph the edge a has a direction on it, so that it has terminal and initial end bits. The terminal end is thought of as a^{+1}, and the initial end is thought of as a^{-1}. As we trace the word w as a loop in this graph, a two-letter segment of w of the form ab goes from the terminal bit of a to the initial bit of b, and thus connects a^{+1} to b^{-1}. A two-letter segment of the form $a^{-1}b$ will go from the initial bit of a to the initial bit of b, and so on. This is one way to intuit this graph. Another way, equivalently, is to attach a 2-cell to the graph along the closed path w; the boundary of a small neighborhood of the vertex (what one could call the "link of v") then has the structure of $\mathrm{Wh}(w)$.

For instance, in $F(a, b, c)$, if $w = a^2bcb^{-1}c^{-1}$, then $\mathrm{Wh}(w)$ is a hexagon with six vertices, which read around circularly: $a, \overline{a}, \overline{b}, c, b, \overline{c}$. If $u = a^3bc^3b^{-1}$, then $\mathrm{Wh}(u)$ has two connected components; one triangle with vertices $a, \overline{a}, \overline{b}$ having two edges between a and $\overline{a}$; and another triangle with vertices $b, c, \overline{c}$, with two edges between c and $\overline{c}$. As an exercise, the reader should try other examples, such as $v = abca^{-1}b^{-1}c^{-1}$.

In Whitehead's original work [10] using this graph, he found a method to determine whether an element of F is a member of some basis, by looking for cut vertices in this graph. In the subject of this paper, we are only interested in whether or not the graph is connected.

Definition 2.2. Consider, as before, the free group F with finite basis A. Let $k \geq 0$ be an integer. For $w \in F$, we say that w is an **k-word**, when: First, in the reduced word for w in terms of this basis, each element of the basis A occurs. And second, for each

$a \in A$, the occurrences of a in the reduced expression for w are either (a) two times, once as a and once as a^{-1}; or (b) exactly k times, always as a, never as a^{-1}.

For instance, in $F(a, b, c)$, the element $w = a^2bcb^{-1}c^{-1}$ is a 2-word, the element $u = a^3bc^3b^{-1}$ is a 3-word, and the element $v = abca^{-1}b^{-1}c^{-1}$ is a 0-word. A 0-word is an n-word for all n.

Dold's theorem [4] is as follows:

Theorem 2.3. *Let F be the free group with finite basis A, and let $k \geq 0$ be a chosen integer. Suppose that $w \in F$ is a k-word with respect to the basis A, and that w has a connected Whitehead graph $\mathrm{Wh}(w, A)$. Then w is generic in F with respect to the verbal subgroup $\mathcal{V}_k$.*

Proof. Let $\phi: G \to F$ be a homomorphism, and $g \in \mathcal{V}_k(G)$ such that $\phi(g) = w$. We can look at the image of ϕ as a subgroup of F and transfer the problem of surjectivity of ϕ to this image. In other words, without loss of generality we can suppose that ϕ is the inclusion of G as a subgroup of F. The basis A of F yields a description of F as the fundamental group of a one-vertex graph Γ; a subgroup G is then described via a graph Δ and a locally injective map of graphs $\Delta \to \Gamma$; see Stallings [9] for details. The element g, as an element of $\pi_1(\Delta)$, in reduced form covers a finite subgraph $\Delta_1 \subseteq \Delta$, whose fundamental group is a free factor of G, and thus $g \in \mathcal{V}_n(\pi_1(\Delta_1))$. This, plus the fact that g, considered as a loop, immerses into Γ as the k-word w, implies that the map $\Delta_1 \to \Gamma$ is a bijection on edges. A little thought now shows that the connected components of $\mathrm{Wh}(w)$ relate to the vertices of Δ_1, so that the assumption that $\mathrm{Wh}(w)$ is connected implies that Δ_1 has only one vertex, so that the map $\Delta_1 \to \Gamma$ is an isomorphism of graphs. This, plus local injectivity of $\Delta \to \Gamma$, implies that $\Delta_1 = \Delta$. And, so, on the fundamental group level, we get that $G = F$.   □

The question of whether or not it can be determined in general if some element of a free group is generic with respect to various verbal subgroups has been studied by Comerford [2]. In the case of $\mathcal{V}_k$, we can show that this question is fairly easily solvable:

Theorem 2.4. *Let F be a free group with finite basis A, and let $k \geq 0$ be a chosen integer. Then there is an algorithm to determine, given the input $w \in F$, whether or not w is generic with respect to $\mathcal{V}_k$.*

Proof. Let q be the length of w in terms of the basis A. Let Γ be the one-vertex graph with fundamental group F determined by A. The set of all connected graphs with at most q edges is finite and can be enumerated; the set of all graph-maps of these to Γ is calculable too, and thus the subset of all locally injective graph-maps $\Delta \to \Gamma$, where Δ is connected and has at most q edges, is finite and can be explicitly described. In each such Δ we can consider closed paths of length q, and determine whether such a path u maps to w as a loop in Γ; furthermore, we can, by counting how many times u traverses each edge of Δ, determine whether or not u is "null-homologous mod k", which is the exact condition needed to determine whether u belongs to $\mathcal{V}_k$ of $\pi_1(\Delta)$. If this investigation produces a graph Δ with more than one vertex, and the loop u in Δ that maps to w goes through more than one vertex of Δ, and if u is null-homologous mod k in Δ, then we have found a map $G \to F$, where $G = \pi_1(\Delta)$, with an element u in the source group's $\mathcal{V}_k$ mapping to w, such that G does not map onto (i.e., is a proper subgroup of) F; in which case, w is not generic with respect to $\mathcal{V}_k$. On the other hand, if there is no such example, then w is indeed generic.   □

This can be carried out to some extent in easy cases. For instance, in $F(a, b)$, an odd power $w = (aba^{-1}b^{-1})^n$, for n odd ≥ 3, is not generic with respect to $\mathcal{V}_0$. In fact, w belongs to the commutator subgroup of a subgroup G of F of index n. Explicitly, if $n = 2k + 1$, there is a subgroup of $F(a, b)$ determined by an action on the set $\{1, 2, \ldots, n\}$ by letting a act by the permutation $(1, 2)(3, 4) \cdots (2k - 1, 2k)$, and b by $(2, 3)(4, 5) \cdots (2k, 2k + 1)$. The subgroup S of F stabilizing 1 is of index n, and the n-th power of the commutator of a and b is in its commutator subgroup; explicitly, in fact, there is a basis of S consisting of $2k + 2$ elements $\{s_1, t_1, \ldots, s_{k+1}, t_{k+1}\}$, such that $[a, b]^n$ is the product of the $k+1$ commutators of the form $[s_i, t_i]$. — Another way to look at this is to consider $F(a, b)$ to be the fundamental group of a punctured torus T with boundary curve C; there exist irregular n-fold covering spaces $\widehat{T}$ of T, such that over C there is a single component; the subgroup of index n is the fundamental group of $\widehat{T}$, which contains the curve covering C, which is the n-th power of the commutator, as an element of the commutator subgroup of $\pi_1(\widehat{T})$. Some of these details are in Culler [3].

But, if $n = 2^k$ is a power of 2, then results by Rosenberger [7, Corollary 1] imply that $[a, b]^n$ is indeed generic with respect to $\mathcal{V}_0$.

3. Question

Generic elements in free groups are an interesting phenomenon. However, there are other cases which might be of more geometric interest. In an arbitrary group G, and with respect to an arbitrary verbal subgroup $\mathcal{V}$, one could try to search for generic elements, to determine if a given element is generic, and so on. The "hyperbolic groups" defined by Gromov [5] have some things in common with free groups, and are of considerable geometric interest in the theory of surfaces and 3-manifolds. Here is a very explicit project:

Let T be the closed orientable surface of genus 2, and let $H = \pi_1(T)$. Find elements of H which are generic with respect to $\mathcal{V}_0$. Observe geometric properties of the non-simple closed curves on T which determine these generic elements. Give an algorithm to determine if a curve on T is generic or not.

References

[1]   A. M. Brunner, R. G. Burns, S. Oates-Williams, On almost primitive elements of free groups with an application to Fuchsian groups, Canad. J. Math. 45 (1993), 225–254.

[2]   L. P. Comerford, Generic elements of free groups, preprint 1994.

[3]   M. Culler, Using surfaces to solve equations in free groups, Topology 20 (1981), 133–145.

[4]   A. Dold, Nullhomologous words in free groups which are not nullhomologous in any proper subgroup, Arch. Math. 50 (1988), 564–569.

[5]   M. Gromov, Hyperbolic groups, in: Essays in group theory, Math. Sci. Res. Inst. Publ. 8, Springer-Verlag 1987.

[6]   H. Neumann, Varieties of Groups, Ergeb. Math. Grenzgeb. 37, Springer-Verlag 1967.

[7]   G. Rosenberger, A property of subgroups of free groups, Bull. Austral. Math. Soc. 43 (1991), 269–272.

[8]   V. Shpilrain, Recognizing automorphisms of the free groups, Arch. Math. 62 (1994), 385–392.

[9]   J. R. Stallings, Topology of finite graphs, Invent. Math. 71 (1983), 551–565.

[10] J. H. C. Whitehead, On certain sets of elements in a free group, Proc. London Math. Soc. 41 (1936), 48–56.

[11] H. Zieschang, Alternierende Produkte in freien Gruppen, Abh. math. Sem. Univ. Hamburg 27 (1964), 13–31.

Geometric Understanding of
the Angle between Subgroups

John R. Stallings

Abstract. There is an interesting group-theoretical concept of "triangle of groups" [9], which has been extended by Haefliger [2] to more complex situations. In this subject, the concept of angle between subgroups of a group is defined, so that non-positive curvature can be discussed. The computation of the angle in the simplest case of a free group is not entirely solved, but results of Gilman [1] give a good idea of what is happening. In this paper, however, we apply some ideas from the theory of 3-manifolds, especially the Loop Theorem, to discuss this "angle", and give some examples.

1991 Mathematics Subject Classification: 20F32, 57M35

1. Angles and Triangles

Definition 1.1. Let G be a group with subgroups A, B, X, such that $X \leq A \cap B$. Then the amalgamated free product $A *_X B$ is defined, and there is a homomorphism $\iota : A *_X B \to G$. The elements of $A *_X B$ can be written as "reduced words", products of elements alternately from $A \setminus X$ and $B \setminus X$ (and there are also words of length one, the elements of $A \cup B$); the length of such reduced word is well-defined. The homomorphism ι might have non-trivial kernel, in which case the minimum length of elements in the kernel is even, say $2n$; in which case, we say that the *angle* $(G; A, B; X)$ is equal to π/n. If ι has trivial kernel, we make the convention that the angle is zero. — An equivalent way to say the same thing is in terms of graphs; consider the sets of right cosets G/A, etc.; there are functions as follows:

$$G/A \leftarrow G/X \to G/B$$

and we can regard this as a description of a bipartite graph, with vertex set divided into two parts G/A and G/B, and edges G/X. In the case that ι is injective, this graph is a

---

This work is based on research partly supported by the National Science Foundation under Grant No. DMS-9203941.

forest; otherwise, there exist non-trivial closed curves of even length; the shortest such closed curve has length $2n$, and the angle between these subgroups is, by definition, π/n.

Definition 1.2. Suppose that there are groups P, Q, R, A, B, C, X, and injective homomorphisms making a commutative diagram as follows:

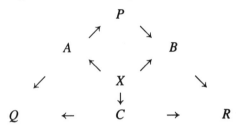

We call this a *triangle of groups*. The *group of the triangle*, G, is the "pushout" of this diagram, or its "colimit"; this is the group generated by the vertex groups with relations saying that the images in the two vertex groups, of elements of each edge group, are equal; the central group X does not come into the definition of the group of the triangle, but it is there implicitly, and is explicitly used in the notion of curvature. A *bounded* subgroup $S \le G$ is such that there exists a positive integer N, such that every element of S can be written as the product of the images of at most N elements of the vertex groups. We consider the homomorphisms in the triangle to be inclusions, and thus can form three angles at the three corners of the triangle, the angles $(P; A, B; X)$, $(Q; A, C; X)$, and $(R; B, C; X)$. If the sum of these three angles is $\le \pi$, we say that this triangle of groups is *non-positively curved*.

From [9], there are two major results;

Theorem 1.3. *In a non-positively curved triangle of groups as above, the homomorphism from each vertex group into G, the group of the triangle, is injective.*

Thus, in the non-positively curved case, we can consider the various groups of the triangle to be subgroups of the group of the triangle.

Theorem 1.4. *In a non-positively curved triangle of groups, if S is a bounded subgroup of the group G of the triangle, then S is conjugate in G to a subgroup of some vertex group.*

2. Computation of Angles in Certain Cases

Clearly, if one is given two subgroups A, B, of a finite group P, with $X = A \cap B$, then, by exhausting all possibilities of words of $A *_X B$ of lengths 2, 4, etc., one eventually gets all elements of P, and thus finds the minimal length of an element in the kernel of $A *_X B \to P$. Therefore, the angle between these subgroups is computable. Of course, in the case of moderately large-sized finite groups, it can be quite tedious. As an example, (cf. [3]) an exercise for the reader:

Exercise 2.1. Consider the non-abelian group P of order 21; in this group there are various subgroups of order 3. Then the angle between any two distinct subgroups of order 3 in P (intersecting in $X = \{1\}$), is $\pi/3$.

In the case of finitely generated subgroups A, B, of a finitely generated free group F, one can determine exactly what $X = A \cap B$ consists of, as a rational subset of F given by a finite state automaton which is basically an immersion of graphs [8].

Question 2.2. Is there an algorithm to determine, given two finitely generated subgroups A, B, of a finitely generated free group F, with $X = A \cap B$, whether or not the homomorphism $A *_X B \to F$ is injective?

If the answer to this question is "yes", then we can determine whether an angle of these free groups is zero or not. The question can be answered if we compute that X is trivial. However, when X is non-trivial, there are interesting examples where the homomorphism is injective, such as the case, when F is free on $\{x, y\}$, and A is generated by $\{x^2, y^2\}$ and B is generated by $\{x, y^2x^2yx^{-2}y^{-2}\}$.
There is an approximate answer to the question, as follows:

Theorem 2.3. *Suppose F is a finitely generated free group, with finitely generated subgroups A and B, with $X = A \cap B$. Let $\epsilon > 0$. Then there is an algorithm to determine whether or not the angle $(F; A, B; X)$ is less than ϵ.*

$$\left(\begin{array}{l} C = A \cap B \\ I \text{ talked to} \\ \text{Stallings} \end{array} \right)$$

Proof. Let n be an integer such that $\pi/n \leq \epsilon$. The sets $A \setminus \{1\}$ and $B \setminus \{1\}$ of F are rational subsets (accepted by certain finite state automata). The product of two rational subsets is again rational, by Gilman [1]. The question of whether 1 belongs to a given rational subset of F is decidable; if one goes through Gilman's method in detail, in fact, this is quite clear. Thus, we compute the rational sets $((A \setminus \{1\}) \cdot (B \setminus \{1\}))^k$, for $k = 1, \ldots, n$; if at any stage we find that 1 belongs to this set, the first such occurrence will determine exactly the angle $(F; A, B; X)$. If we do not find such an occurrence of 1 in our computed rational subsets of F, we can conclude that the angle is $< \pi/n \leq \epsilon$.

$\square$

3. The Use of the Loop Theorem to Compute Angles

The Loop Theorem of Papakyriakopoulos [5] says, roughly, that if a 3-manifold $\mathcal{M}$ has a boundary component $\mathcal{B}$ with $\pi_1(\mathcal{B}) \to \pi_1(\mathcal{M})$ not injective, then there is an embedded circle in B representing a non-trivial element of the kernel of $\pi_1(B) \to \pi_1(M)$. A proof of this, modeled on the Shapiro–Whitehead proof of Dehn's Lemma [6], occurs in [7]. For the application to the angle between subgroups, however, we need to state the theorem with more exactness and delicacy than heretofore. We shall then indicate how the [7] proof in fact proves this more delicate statement.

Theorem 3.1 (Loop Theorem). *Let M be a 3-manifold (not necessarily compact). Let B be a connected component of the boundary of M. Let N be a normal subgroup of $\pi_1(B)$ (since we are assuming N is normal, we do not have to worry about the basepoint). Let C be a closed curve in B having only double points, and only a finite number of these; let L be a curve in B which misses all the double points of C, and which intersects C transversely in exactly k points (L does not have to be connected). Suppose that the element $[C] \in \pi_1(B)$ represented by C does not belong to N, and suppose that $[C] \in \ker(\pi_1(B) \to \pi_1(M))$. — Then there is a simple closed curve (i.e., an embedded circle) C' in B such that $[C'] \in \ker(\pi_1(B) \to \pi_1(M)) \setminus N$, such that C' is transverse to L, and such that $C' \cap L$ is contained in $C \cap L$; in particular, the number of points of intersection of C' with L is at most k.*

Proof. Go through the proof in [7] carefully. The tower procedure (section 4 of [7]) needs to be modified only by pulling back the curve L at each stage, restricting L to the parts of the boundary of the 3-manifolds being kept alive each time. Thus, the manifold V_n, which is a regular neighborhood of a map of a disk into M_n, has its boundary cut back to $B'_n = B_n \cap V_n$; at the same time there is a curve L_n on the boundary of M_n, which becomes in B'_n the restriction of this curve to a curve $L'_n = L_n \cap B'_n$. We have a double covering $p_{n+1} : M_{n+1} \to V_n$, and, among the other things, we define L_{n+1} to be $p_{n+1}^{-1}(L'_n)$.

Eventually, as in [7], M_n will have no connected 2-sheeted covering space, so that all the boundary components of M_n will be spheres. What we then do is to find a simple closed curve near the lifted closed curve (denoted $|L_n|$ in [7]) above C, on one of the boundary spheres of M_n, which does not belong to the normal subgroup N_n. This closed curve bounds a disk. The extra comment to make is that this closed curve does not go across any of the arcs making up $|L_n|$ more than once; thus the curve does not hit L_n in extra points. We then push the disk downward through the tower, cutting it along double curves of singularities, until we get the final disk in M itself; in this construction, the curve maintains its good intersection qualities with L_n at every point. □

Now we give an example of this, and its application to angles and triangles of groups. Consider the closed upper half space in R^3; its boundary plane contains four points in a line; from the leftmost point draw a half-line perpendicular to the plane, and from the rightmost point do the same; between the two middle points, draw a semicircle. The 3-manifold M under consideration is the upper half space with these two lines and semicircle removed; its boundary B consists of the horizontal plane with four points removed. The Loop Theorem gives a simple closed curve C' on B which is not contractible on B but which is contractible in $\mathcal{M}$; in this case the curve C' is unique up to isotopy. This is Figure 1.

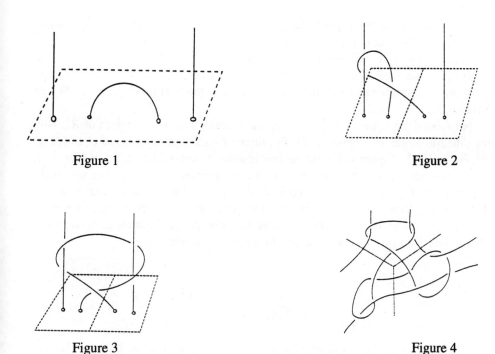

Figure 1   Figure 2

Figure 3   Figure 4

Now, imagine in the boundary B a line $\mathcal{L}$ embedded so that its endpoints are at infinity, and so that any imaginable distortion of C' will still intersect $\mathcal{L}$ in at least 6 points. By an isotopy of the picture, we can straighten out $\mathcal{L}$ and consider thus pictures such as Figure 2 or Figure 3. The situation with the fundamental groups is this: P is the fundamental group of $\mathcal{M}$, and thus is free of rank 3; the groups A and B are the fundamental groups of the closed halves of B on either side of $\mathcal{L}$; they are, in each example, free of rank 2. The group X is the fundamental group of the line $\mathcal{L}$, and hence is trivial; $X = \{1\}$. The angle $(P; A, B; X)$ is determined by the minimum length of an element in the free product $A * B = A *_X B$ which maps to 1 in P. Such an element determines a closed (non-simple) curve C on B, non-trivial in $\pi_1(B)$, but trivial in $\pi_1(\mathcal{M})$; taking the normal subgroup N to be $\{1\}$, we apply the Loop Theorem, and obtain a simple closed curve C', which has to be the curve C' already discussed above.

In fact, there is an interesting result, which can be described as an exercise; this is related to Magnus's corollary to the Freiheitssatz ([4], §6.2).

Exercise 3.2. Suppose that $\mathcal{M}$ is a connected (non-compact) 3-manifold with connected boundary B, such that both $\pi_1(B)$ and $\pi_2(\mathcal{M})$ are free groups of ranks $n + 1$

and n respectively, and such that the inclusion induces a surjection $\pi_1(\mathcal{B}) \to \pi_1(\mathcal{M})$. Then there is in $\mathcal{B}$ a simple closed curve C' which is not contractible in $\mathcal{B}$ and which is contractible in $\mathcal{M}$. The curve C' is unique up to isotopy on $\mathcal{B}$, and it represents a primitive element of the free group $\pi_1(\mathcal{B})$.

Now, examine the two figures 2 and 3. The group $\pi_1(\mathcal{B})$ is free of rank 4 represented as the free product of two free groups A, B each of rank 2, and the curve C' can be explicitly drawn and its length in terms of this free product can be computed. In the case of Figure 2, the length is 6; in Figure 3, the length is 10. This means that in Figure 2, the angle determined by the fundamental groups in the picture is $\pi/3$, and in Figure 3 it is $\pi/5$.

Now, bend the upper half space along $\mathcal{L}$, so that it occupies one third of R^3, with the dihedral angle at $\mathcal{L}$ being $2\pi/3$. Put three of these examples together to fill up R^3. For example, Figure 4. Taking the fundamental groups of the various pieces, with the figures made out of the straight lines and semicircles removed, what we obtain is a triangle of groups, whose group is the fundamental group of R^3 minus this 1-dimensional picture. By the above argument, each angle of this triangle is $\pi/3$ or $\pi/5$, and so this triangle of groups is non-positively curved. Thus the conclusions in Theorems 1.2 and 1.3 can be made for this kind of example.

4. Questions

One problem would be to extend this geometric reasoning further. In particular, the case discussed here is extremely simple, and inquiry can be made into more complex matters. The method of Gilman to determine the angle should have a simpler and more transparent variant.

Question 4.1 (Related to the Freiheitssatz). Suppose that A, B, P are finitely generated free groups, and that $\alpha: A * B \to P$ is a surjective homomorphism which is injective on A and on B; and furthermore suppose that the rank of $A * B$ exceeds that of P by exactly 1. Then the kernel of α is the normal closure of a single primitive element of $A * B$, which is uniquely determined up to conjugacy and exponent ± 1 (cf. [4], §6.2). Choose such an element which is cyclically reduced in the free product structure $A * B$. It then has a certain length $2n$. Is this the smallest length of a non-trivial element of the kernel of α? In other words, the obvious conjecture, which is true in cases such as we considered in section 3 because of geometric things and the Loop Theorem, is that this is a way to compute the angle $(P; A, B; \{1\})$ as π/n.

Question 4.2. Can a similar construction be made in case the excess of the rank of $A * B$ over that of P is greater than 1? So that one can, by a geometric construction, compute the angle.

Question 4.3. Can something be done for the case that the intersection $A \cap B$ is non-trivial?

Question 4.4. Imagine a picture similar to Figure 4, except that the sum of the angles is $> \pi$; for instance, with all angles $\pi/2$. There may be an example where the bounded subgroup theorem (Theorem 1.3) fails. Does this have some interesting geometric significance?

References

[1] R. Gilman, Computations with rational subsets of confluent groups, EUROSAM 84, Lecture Notes in Comput. Sci. 174, Springer-Verlag 1984, 207–212.

[2] A. Haefliger, Complexes of groups and orbihedra, in: Group Theory from a Geometrical Viewpoint, World Scientific, Singapore 1991, 504–540.

[3] P. Köhler, T. Meixner, M. Wester, Triangle groups, Comm. Algebra 12 (13) (1984), 1596–1626.

[4] W. Magnus, Über diskontinuierliche Gruppen mit einer definierenden Relation (Der Freiheitssatz), J. Reine Angew. Math. 163 (1930), 141–165.

[5] C. D. Papakyriakopoulos, On solid tori, Proc. London Math. Soc. (3) 7 (1957), 281–299.

[6] A. Shapiro and J. H. C. Whitehead, A proof and extension of Dehn's Lemma, Bull. Amer. Math. Soc. 64 (1958), 174–178.

[7] J. Stallings, On the loop theorem, Ann. of Math. 72 (1960), 12–19.

[8] J. Stallings, Topology of finite graphs, Invent. Math. 71 (1983), 551–565.

[9] J. Stallings, Non-positively curved triangles of groups, in: Group Theory from a Geometrical Viewpoint, World Scientific, Singapore 1991, 491–503.

Augmentation Quotients of Integral Group Rings

L. R. Vermani

Abstract. Let G be a group, ZG the integral group ring of G and $I(G)$ the augmentation ideal of ZG. Following Passi [20] we define polynomial groups $Q_n(G) = I^n(G)/I^{n+1}(G)$, $n \geq 0$, which are Abelian groups. Although the structure of the Abelian groups $Q_n(G)$ has been extensively investigated by several authors during the last twenty-five years, the structure of these groups is not yet completely determined in the general case. An excellent survey of the work done on the problem up to 1977 has been given by Passi [23] and, as such, we shall limit ourselves mainly to the later developments. However, I have tried to give an exhaustive bibliography regarding augmentation quotients.

1. The Augmentation Terminals of Finite Abelian Groups

Passi [20] proved that if $G = H \oplus K$ where H, K are groups of coprime finite exponents, then $Q_n(G) \simeq Q_n(H) \oplus Q_n(K)$ for all $n \geq 1$. He also proved that if G is any Abelian group and Z is the infinite cyclic group, then $Q_n(G \oplus Z) \simeq \oplus \sum_{i=1}^{n} Q_i(G) \oplus Z$. An application of these results reduces the study of $Q_n(G)$ for finitely generated Abelian group G to the study of $Q_n(G)$ when G is a finite Abelian p-group, p a prime. The following result gives the periodic behaviour of $Q_n(G)$.

Theorem 1.1 ([2], [33]). *If G is a finite group then there exist integers n_0 and π such that $Q_n(G) \simeq Q_{n+\pi}(G)$ for all $n \geq n_0$. If, in addition, G is nilpotent of class c, then the least π with the above property divides the l.c.m. of $1, 2, \ldots, c$.*

Definition 1.2. When n_0 and π in the above theorem are the least possible, then the sequence $\{Q_{n_0}(G), Q_{n_0+1}(G), \ldots, Q_{n_0+\pi-1}(G)\}$ is called the stable behaviour of $Q_n(G)$. Also then π is called its period.

Clearly the period π of $Q_n(G)$ when G is Abelian is 1 so that there exists a positive integer n_0 such that $Q_n(G) \simeq Q_{n+1}(G)$ for all $n \geq n_0$. The eventually constant value $Q_{n_0}(G)$ is called the augmentation terminal of G and is denoted by $Q_\infty(G)$.

Singer [40] proved

Theorem 1.3. *Let G be direct sum of a_i cyclic groups of order p^i, $1 \le i \le m$. Then the order of $Q_\infty(G)$ is p^J where*

$$J = p^{t_1} + \cdots + p^{t_{m-1}} + \frac{p^{t_m+1} - 1}{p - 1} \quad with$$

$$t_i = a_1 + 2a_2 + \cdots + (i-1)a_{i-1} + i(a_1 + \cdots + a_m - 1), \quad 1 \le i \le m.$$

Proposition 1.4. *If N is the number of cyclic subgroups of the group G as in Theorem 1.3, then $N = J + 1$.*

Proof. The number of elements of order p^r in G is

$$p^{\left(\sum_{i=1}^{r} ia_i + r \sum_{i=r+1}^{m} a_i\right)} - p^{\left(\sum_{i=1}^{r-1} ia_i + (r-1) \sum_{i=r}^{m} a_i\right)}$$

$$= p^{t_r + r} - p^{t_{r-1} + r - 1}$$

and each cyclic subgroup of order p^r contains $p^r - p^{r-1}$ elements of order p^r. Therefore the number of cyclic subgroups of order p^r is

$$(p^{t_r + r} - p^{t_{r-1} + r - 1})/(p^r - p^{r-1}) = (p^{t_r + 1} - p^{t_{r-1}})/(p - 1).$$

Hence the number of non-identity cyclic subgroups of G is

$$\sum_{r=1}^{m} (p^{t_r+1} - p^{t_{r-1}})/(p-1) = \{\sum_{i=1}^{m-1} (p^{t_i+1} - p^{t_i}) + (p^{t_m+1} - 1)\}/(p-1) = J$$

so that $N = J + 1$, as the identity subgroup is another cyclic subgroup of G. □

We thus have

Theorem 1.5. *If G is a finite Abelian p-group, then $|Q_\infty(G)| = p^{N-1}$, where N is the number of cyclic subgroups of G.*

A simple proof of this result has been given by Hales and Passi [8].

Let G be a finite Abelian p-group. Define a finite Abelian p-group by generators and relators as follows : For each cyclic subgroup H of G choose a generator X_H and introduce a relation $pX_H = X_K$ if K is a subgroup of index p in H. Also introduce the relation $x_{\{e\}} = 0$. Clearly if G is a group of exponent p^m, then Q_G is also of exponent p^m. Moreover $|Q_G| = p^{N-1}$, where N is the number of cyclic subgroups of G. Encouraged by this and the value of the augmentation terminal of an elementary Abelian p-group obtained by Passi [20] Hales [5] conjectured and finally proved [6] that $Q_\infty(G) \simeq Q_G$ for every finite Abelian p-group. (Also cf. [4] for a particular case of this result.)

We observe

Proposition 1.6. *If $G = H \oplus K$ where H is a torsion group and K is a divisible group, then $Q_n(G) \simeq Q_n(H) \oplus Q_n(K)$.*

Proof. Let $a \in H, b \in K$ with order of a being m. Let $c \in K$ such that $b = c^m$. Then, modulo $I^3(G)$

$$(a - 1)(b - 1) \equiv m(a - 1)(c - 1) = 0, \quad \text{as} \quad m(a - 1) \in I^2(G).$$

Consequently every element of $Q_n(G)$ can be written as $u + v + I^{n+1}(G)$, where $u \in I^n(H), v \in I^n(K)$. The natural projections $G \longrightarrow H, G \longrightarrow K$ induce an epimorphism $Q_n(G) \longrightarrow Q_n(H) \oplus Q_n(K)$ which is then clearly a monomorphism and, hence, an isomorphism.   □

Let G be a divisible Abelian group with A as its torsion subgroup. Then $G = A \oplus B$ where B is a torsion free divisible Abelian group. In view of the above we have (also cf. [47])

$$Q_1(G) \simeq A \oplus B,$$
$$Q_n(G) \simeq Q_n(B) \simeq SP^n(B) \text{ for } n > 1.$$

Here for an Abelian group M, $SP^n(M)$ denotes the n-th symmetric power of M and equals $\underbrace{(M \otimes M \otimes \cdots \otimes M)}_{n\text{-terms}}/B$, where B denotes the subgroup of $M \otimes \cdots \otimes M$ generated by all elements of the form $a_1 \otimes \cdots \otimes a_n - a_{\sigma(1)} \otimes \cdots \otimes a_{\sigma(n)}$, $a_i \in M$ and σ any permutation of set $1, 2, \ldots, n$.

Let H be a normal subgroup of a group G. Write $\overline{G}$ for G/H. The natural projection $G \longrightarrow \overline{G}$ induces an epimorphism $\alpha : Q_2(G) \longrightarrow Q_2(\overline{G})$ with kernel $I^2(G) \cap (I^3(G) + ZGI(H))/I^3(G) = (I^3(G) + I(G)I(H) + I^2(G) \cap I(H))/I^3(G)$. It is fairly easy to see that $I^2(G) \cap I(H) = I(H \cap G') + I^2(H)$. Therefore Ker $\alpha = (I^3(G) + I(G)I(H) + I(H \cap G'))/I^3(G)$ and α induces an isomorphism $I^2(G)/(I^3(G) + I(G)I(H)) + I(H \cap G') \simeq Q_2(\overline{G})$. We then have

Proposition 1.7. *For any normal subgroup H of a group G,*

$$I^2(G)/(I^3(G) + I(G)I(H)) + I(H \cap G') \simeq Q_2(G/H).$$

Corollary 1.8 ([31]). *For any subgroup H of an Abelian group G,*

$$I^2(G)/(I^3(G) + I(G)I(H)) \simeq SP^2(G/H).$$

2. Stable Behaviour of $Q_n(G)$

Let

$$\zeta : G = H_1 \geq H_2 \geq \cdots \geq H_i \geq H_{i+i} \geq \cdots \tag{2.1}$$

be an N-series of a group G, i.e., a series of subgroups of G with the commutator subgroup $[H_i, H_j] \leq H_{i+j}$ for all $i, j \geq 1$. The N-series ζ introduces a weight function on the elements of G :

$$wt(x) = \begin{cases} n & \text{if } x \in H_n \backslash H_{n+1} \\ \infty & \text{otherwise.} \end{cases}$$

Let $\bigwedge_n$ be the Z-submodule of ZG spanned by all products $(g_1 - 1) \cdots (g_s - 1)$, $\sum_{i=1}^{s} wt(g_i) \geq n$, $g_i \in G$. Then

$$I(G) = \bigwedge_1 (G) \geq \bigwedge_2 (G) \geq \cdots \geq \bigwedge_i (G) \geq \cdots \tag{2.2}$$

is a series of ideals of ZG with the property that $\bigwedge_i \bigwedge_j \leq \bigwedge_{i+j}$ for all $i, j \geq 1$ and is called the canonical filtration of ZG w.r.t. N-series ζ.

Proposition 2.3. *If the series ζ is the lower central series of G, then $\bigwedge_n = I^n(G)$ for all $n \geq 1$.*

Proof. Let $H_i = G_i$, the i th term in the lower central series of G. Let $g_1, g_2, \ldots, g_s \in G$ with $\sum_{i=1}^{s} wt(g_i) \geq n$. Let $wt(g_i) = a_i$, $1 \leq i \leq s$. Then $g_i - 1 \in I^{a_i}(G)$ and so, $(g_1 - 1) \cdots (g_s - 1) \in I^{a_1 + a_2 + \cdots + a_s}(G) \leq I^n(G)$. Thus $\bigwedge_n \leq I^n(G)$. As $I^n(G)$ is spanned as a Z-module by elements of the form $(g_1 - 1) \cdots (g_n - 1)$ and $wt(g_i) \geq 1$ for every i, $I^n(G) \leq \bigwedge_n$. Hence $\bigwedge_n = I^n(G)$ for all $n \geq 1$.   □

Suppose that G is a finite p-group, p a prime and that the N-series (2.1) has the additional property that $H_i^p \leq H_{ip}$ for all $i \geq 1$. Let $t_i = $ rank of H_i/H_{i+1}, $i \geq 1$ and $n_0 = (p - 1)(t_1 + 2t_2 + \cdots + ct_c - c) + 1$ where c is the largest integer for which H_c is nontrivial. Then, Losey and Losey [19] proved that

2.4. (i) For all $n \geq n_0$, $Q_n(G)$ is an elementary Abelian p-group;
(ii) $Q_n(G) \simeq Q_{n+\pi}(G)$ for all $n \geq n_0$, where π is the period of $Q_n(G)$.

They also calculated the exact rank of $Q_n(G)$ for all $n \geq n_0$. Using this result they obtained the stable behaviour of $Q_n(G)$ for all groups of order p^3 [18] and some groups of order p^4. The stable behaviour of the remaining groups of order p^4 has been determined by Horibe and Tahara [11] and Tahara and Yamada ([44], [45]).

3. The Non-Abelian Group Case — General

Let G be any group. It is an elementary fact that $Q_1(G) = G/G_2$. The map $x \to x - 1 + I^3(G)$, $x \in G_2$, is a homomorphism : $G_2 \to Q_2(G)$ with kernel equal to the subgroup $\{x \in G_2 \mid x - 1 \in I^3(G)\} = G_3$ [17]. Therefore it induces a monomorphism $\alpha : G_2/G_3 \to Q_2(G)$ with Coker $\alpha \simeq I^2(G)/(I^3(G) + I(G_2)) = (I^2(G) + ZGI(G_2))/(I^3(G) + ZGI(G_2)) \simeq Q_2(G/G_2) \simeq SP^2(G/G_2)$ [21].

We thus have an exact sequence

$$1 \to G_2/G_3 \xrightarrow{\alpha} Q_2(G) \xrightarrow{\beta} SP^2(G/G_2) \to 0 \qquad (3.1)$$

where the homomorphism β is given by

$$(x - 1)(y - 1) + I^3(G) \to xG_2 \widehat{\otimes} yG_2, \quad x, y \in G.$$

For any Abelian group A we have an exact sequence

$$0 \to \langle a \otimes b - b \otimes a \mid a, b \in A \rangle \xrightarrow{i} A \otimes A \xrightarrow{\gamma} SP^2(A) \to 0 \qquad (3.2)$$

where i is the inclusion map and γ is the natural projection. That the sequence (3.1) splits was proved (i) by Sandling [33] when G is finite, (ii) by Losey [16] and by Bachmann and Grünenfelder [1] when G is finitely generated. Bachmann and Grünenfelder used homological methods for their proof. Again using cohomological methods Hales and Passi [7] proved that the sequence (3.1) splits when the sequence (3.2) with $A = G/G_2$ splits. They prove that the sequence (3.2) splits when $A = G/G_2$ (i) is direct sum of cyclic groups; or (ii) is divisible; or (iii) is completely decomposable torsion free; so that in each of the three cases the sequence (3.1) splits. However the sequence (3.2) does not always split. For example

3.3. Consider the group $Q \oplus Q$, Q the additive group of rationals. Write u for the element $(1,0)$ of this group and v for the element $(0,1)$. Let P be the set of odd primes and write $P = P_1 \cup P_2 \cup P_3$ where the union is disjoint and each P_i is infinite. Let $A \subseteq Q \oplus Q$ be the subgroup generated by the elements. $\{n^{-1}u \mid n$ is a square-free P_1-number$\}$, $\{n^{-1}v \mid n$ is a square free P_2-number$\}$ and $\{n^{-1}(u + v) \mid n$ is a square free P_3-number$\}$. For this group A, the subgroup $\langle x \otimes y - y \otimes x \mid x, y \in A \rangle$ is not a direct summand of $A \otimes A$ [7]. Using basic sequences and basic products Tahara [41] has obtained the structure of $Q_3(G)$ for a finite group G in terms of generators and relators.

3.4. Let $G_r(ZG) = \oplus \sum_{n \geq 0} Q_n(G)$ be the associated graded ring of ZG, $LG = \oplus \sum G_n/G_{n+1}$ the Lie ring associated with the lower central series $\{G_n\}$ and $U(LG)$ the universal enveloping ring of the Lie ring LG. Then there is a natural epimorphism

$$f_- : U(LG) \longrightarrow G_r(ZG)$$

induced by the map $G_n \to I^n(G)$, $a \mapsto a^{-1}$, $a \in G_n$. The homorphism (3.5) is an isomorphism in dimensions 1 and 2 (cf. [1]). The precise kernel of the epimorphsim (3.5) in dimension 3 has very recently been calculated in [10(a)] (as informed by the referee and also M. Hartl).

Theorem 3.6. *The kernel of the map* $f_3 : U_3(LG) \to Gr_3(ZG) (= Q_3(G))$ *is generated by the elements*

$$x_1^k \bar{x}_2 - \bar{x}_1 x_2^k + \binom{k}{2}(\bar{x}_1 \bar{x}_2^k - \bar{x}_1^k \bar{x}_2),$$

where $x_1, x_2 \in G, k \in Z$ *such that* $x_1^k, x_2^k \in G_2$ *and where* $\bar{x}_i = x_i G_2$.

The structure of the associated graded algebra $Gr(QG)$ where Q is the field of rational numbers is given by a result of D. Quillen [26(a)] which states that epimorphsim (3.5) becomes an isomorphsim on tensoring with Q. This result of Quillen and Poincaré–Birkhoff–Witt decomposition of the enveloping ring then yield the following result of Sandling and Tahara [35]

Theorem 3.7. *If* G *is a nilpotent group such that all the lower central quotients* G_i/G_{i+1} *are free Abelian for* $i < n$, *then* $Q_n(G) \simeq \oplus \sum \otimes_{i=1}^n SP^{a_i}(G_i/G_{i+1})$ *where the sum is taken over all non-negative integers* $a_1, a_2, \ldots, a_n$ *such that* $\sum_{i=1}^n ia_i = n$.

4. Polynomial Groups — A Generalization

The study of generalized polynomial groups $I^n(G)I(H)/I^{n+1}(G)I(H)$ where H is a (normal) subgroup of G is related to investigation of generalized Fox subgroups $G \cap (1 + I^n(G)I(H))$. There has not been enough progress on the study of these two problems. However, M. Curzio and C. K. Gupta [3] have very recently proved that the generalized Fox subgroup $G \cap (1 + I^2(G)I(H)) = K_G(H)[H, H, H][H \cap G', H \cap G']$, where $K_G(H)$ is a certain specifically defined subgroup contained in H'. We do not go into the relationship of the two problems but only report on the quotients. We have a simple observation:

Proposition 4.1 [27]. *If* H *is a normal subgroup of a group* G *then for all* $n \geq 1$

$$I(G)I^{n-1}(H)/I(G)I^n(H) \simeq Q_n(H) \oplus (I(G/H) \otimes Q_{n-1}(H)).$$

Pushing the argument leading to the above result a bit further, Karan and Vermani [29] prove

Theorem 4.2. *Let H be a subgroup of a group G. Then for all $n \geq 0$*

$$I(G)I^n(H)/(I^2(G)I^n(H) + I^{n+1}(H)) \simeq G/HG_2 \otimes Q_n(H).$$

Note the following interesting special cases of this result :

Corollary 4.3.
(a) $I(G)/(I^2(G) + I(H)) \simeq G/HG_2$;
(b) $I(G)I(H)/(I^2(G)I(H) + I^2(H)) \simeq G/HG_2 \otimes H/H_2$.

An application of Theorem 4.2 also gives

Corollary 4.4 [29]. *If G is the semidirect product $H \rtimes K$ of a normal subgroup H and a subgroup K, then for all $n \geq 0$, $I(G)I^n(K)/I^2(G)I^n(K) \simeq Q_{n+1}(K) \oplus ((G/KG_2) \otimes Q_n(K))$.*

This result (with positions of G and K interchanged) had also earlier been obtained by M. Khambadkone [14] when G is finite and both H and K are nilpotent.

Let G be any group, H any subgroup of G and $A = G/HG_2$. If the exact sequence

$$1 \to HG_2/G_2 \to G/G_2 \overset{p}{\to} A \to 1$$

where p is the natural projection, splits (this happens, for example, when $G = H \rtimes K$), then [29]

$$I(G)I(H)/I^2(G)I(H)$$
$$= ((G/HG_2) \otimes (H/H_2)) \oplus (I^2(G)I(H) + I^2(H))/I^2(G)I(H).$$
$$\tag{4.5}$$

The quotient $I(G)J/I^2(G)J$ where J is a left ideal in ZH has also recently been investigated by M. Hartl [10a].
   If $G = H \rtimes K$, then

$$I^2(H) \cap I^2(G)I(H) = I^3(H) + I([H, K])I(H)$$

and we have ([30], also cf [13] when H is finitely generated)

$$I(G)I(H)/I^2(G)I(H) = (K/K_2 \otimes H/H_2) \oplus I^2(H)/(I^3(H) + I([H, K])I(H)).$$
$$\tag{4.6}$$

This leads us in a natural way to consider the quotient groups $I^2(G)/(I^3(G) + I(G)I(H))$, where H a normal subgroup of G. It is an old result of Sandling [34] that

Proposition 4.7. *If B is a central subgroup of a group G, then $G \cap (1 + I^3(G) + I(G)I(H)) = G_3$.*

Some improvements of this result have been obtained by Passi [22], Passi and Sharma [25] and Karan and Vermani [28]. The problem has very recently been settled in [9] as:

Proposition 4.8. *If H is any subgroup of a group G, then $G \cap (1 + I^3(G) + I(G)I(H)) = G_3U$, where U is the subgroup $\langle [x^m, y] \mid x^m, y^m \in HG_2$ for some $m \geq 1, x, y \in G \rangle$ of G.*

The map $x \to x - 1 + I^3(G) + I(G)I(H)$, $x \in G_2$, H any normal subgroup of G induces a monomorphism

$$\alpha : G_2/G_3U \to I^2(G)/(I^3(G) + I(G)I(H))$$

with Coker $\alpha \simeq I^2(G)/(I^3(G) + I(G)I(H) + I(G_2)) \simeq SP^2(G/HG_2)$ (Proposition 1.7). We thus have an exact sequence.

$$1 \to G_2/G_3U \xrightarrow{\alpha} I^2(G)/(I^3(G) + I(G)I(H)) \xrightarrow{\beta} SP^2(G/HG_2) \to 0 \quad (4.9)$$

where the homomorphism β is given by

$$(x - 1)(y - 1) + I^3(G) + I(G)I(H) \to xHG_2 \widehat{\otimes} yHG_2, \quad x, y \in G.$$

Suppose that the normal subgroup H of G is such that $[H, G] \leq G_3U$ (this happens, for example, when G/H is a torsion group and $H = H^m$ for every $m \geq 1$ or when H is a central subgroup of G). Set $G/HG_2 = A$. The map

$$A \times A \to I^2(G)/(I^3(G) + I(G)I(H))$$

given by

$$(xHG_2, yHG_2) \to (x - 1)(y - 1) + I^3(G) + I(G)I(H), \quad x, y \in G,$$

is well defined and bilinear. It, therefore, induces a homomorphism

$$f : A \otimes A \to I^2(G)/(I^3(G) + I(G)I(H))$$

and $\beta f = \gamma$ (cf the exact sequence (3.2) with $A = G/HG_2$). We get a commutative diagram

$$
\begin{array}{ccccccc}
0 \to \langle a \otimes b - b \otimes a \mid a, b \in A \rangle & \xrightarrow{i} & A \otimes A & \xrightarrow{\gamma} & SP^2(A) \to 0 \\
\downarrow & & \downarrow f & & \| \\
0 \to G_2/G_3U & \xrightarrow{\alpha} & I^2(G)/(I^3(G) + I(G)I(H)) & \xrightarrow{\beta} & SP^2(A) \to 0
\end{array}
$$

from which it follows

Theorem 4.10. *If the normal subgroup H of G is such that $[H, G] \leq G_3U$, then the sequence (4.9) splits when the exact sequence (3.2) with $A = G/HG_2$ splits.*

Corollary 4.11 [48]. *For any integer $n \geq 1$,*

$$I^2(G)/(I^3(G) + I(G)I(G^n)) \simeq G_2/[G^n, G]G_3 \oplus SP^2(G/G_2G^n).$$

As a consequence of Theorem 4.10 we also get

Proposition 4.12. *If G/G_2 is divisible, then $I^2(G)/(I^3(G) + I(G)I(H)) = [H, G]/([H, G] \cap G_3)U \oplus G_2/[H, G]G_3 \oplus SP^2(G/HG_2)$.*

In this direction Khambadkone [13] had earlier obtained:

4.13. When $G = H \rtimes K$ and H is finitely generated such that H/H_2 has a positive uniqueness basis adapted to $[H, K]$, i.e., $\bar{x}_1, \bar{x}_2, \ldots, \bar{x}_m$ is a basis of H/H_2 so that $[H, K]$ is generated modulo H_2 by $x_1^{e_1}, x_2^{e_2}, \ldots, x_m^{e_m}$ where e_i are non negative integers such that $e_1 | e_2 | e_3 \ldots | e_m$, then $I^2(H)/(I^3(H) + I([H, K])I(H)) \simeq SP^2(H/[H, G]) \oplus (H_2/G \cap (1 + I^2(G)I(H))$.

From now on we assume that $G = H \rtimes K$.

Let $\{\bigwedge_i\}_{i=1}^\infty$ be the canonical filtration of $I(H)$ relative to the N-series $\mathcal{H} : H = H_{(1)} \geq [H, G] = H_{(2)} \geq \cdots \geq [H_{(m)}, G] = H_{(m+1)} \geq \cdots$ of H, where $H_{(m+1)} = [H_{(m)}, G]$ for $m \geq 1$. Define the subgroup $\mathcal{K}_i$, $i \geq 1$ of ZG by

$$\mathcal{K}_1 = \bigwedge_i I(K) \quad \text{and} \quad \mathcal{K}_m = \sum_{i=1}^m \bigwedge_i I^i(K) \quad \text{for } m \geq 2.$$

Khambadkone [15] proved that for all $m \geq 1$, $I^m(G)I(K)/I^{m+1}(G)I(K)$ is isomorphic to $Q_{m+1}(K) \oplus \mathcal{K}_m/\mathcal{K}_{m+1}$. Using the techniques and results of Sandling and Tahara [35] she proves

Theorem 4.14. *If G is finite and H is nilpotent, then $I^2(G)I(K)/I^3(G)I(K)$ is isomorphic to direct sum of groups $[K_3/K_4 \oplus (K/K_2 \otimes K_2/K_3) \oplus SP^3(K/K_2)]/R$ and $\{(H_{(2)}/H_{(3)} \oplus SP^2(H/H_{(2)})) \otimes K/K_2 \oplus (H/H_2 \otimes (K_2/K_3 \oplus SP^2(K/K_2))\}/R_3 (H, K)$ where $R, R_3(H, K)$ are certain well defined subgroups of the concerned groups.*

When G may not be finite, we have

Theorem 4.15 ([30], [32]). *If the exact sequence (3.2) with $A = H/[H, G]$ splits and either $H/[H, G]$ or K/K_2 is torsion free, then $I^2(G)I(K)/I^3(G)I(K) \simeq Q_3(K) \oplus (H/[H, G] \otimes (SP^2(K/K_2) \oplus K_2/K_3)) \oplus ((SP^2(H/[H, G]) \oplus [H, G]/[H, G, G]) \otimes K/K_2)$.*

Regarding the quotient $I^2(G)I(H)/I^3(G)I(H)$ Khambadkone [12] proves

Proposition 4.16. *If G is finite, then $I^2(G)I(H)/I^3(G)I(H) \simeq X \oplus Y$, where $X = (I^3(H) \oplus I([H, K])I(H))/(I^4(H) + I(H)I([H, K])I(H) + I([H, K, H])I(H) + I([H, K, K])I(H))$, and $Y = (I(K)I^2(H) + I^2(K)I(H))/(I(K)I^3(H) + I^2(K)I^2(H) + I^3(K)I(H) + I(K)I([H, K])I(H))$.*

If the subgroup H is nilpotent, then Y is isomorphic to $(K_2/K_3 \otimes H/H_2) \oplus (SP^2(K/K_2) \otimes H/H_2) \oplus (K/K_2 \otimes [H, K]/[H, G, G]) \oplus (K/K_2 \otimes SP^2(H/[H, G])) \simeq (Q_2(K) \otimes H/H_2) \oplus (K/K_2 \otimes [H, K]/[H, G, G]) \oplus (K/K_2 \otimes SP^2(H/[H, G]))$.

However, when the group G is not finite and H is not necessarily nilpotent, the subgroup Y in Proposition 4.16 is given by

Theorem 4.17 ([30], [32]). *If H/H_2 or K/K_2 is torsion free, then the direct summand Y of $I^2(G)I(H)/I^3(G)I(H)$ as in Proposition 4.16 is isomorphic to $(Q_2(K) \otimes H/H_2) \oplus (K/K_2 \otimes I^2(H)/(I^3(H) + I([H, G])I(H)))$.*

In case it is also assumed that $H/[H, G]$ is free Abelian, or H_2/H_3 is divisible, then the group $I^2(H)/(I^3(H) + I([H, G]))$ becomes isomorphic to $H_2/H_3U \oplus SP^2(H/[H, G])$, where U is the subgroup $\langle [x^m, y] \mid x^m, y^m \in [H, G]$ for some $m \geq 1$, $x, y \in H \rangle$ of H.

Partial information about the structures of the quotients $I^3(G)I(K)/I^4(G)I(K)$ and $I^3(G)I(H)/I^4(G)I(H)$ has been obtained in [31] and [47]:

Theorem 4.18. $I^3(G)I(K)/I^4(G)I(K) = Q_4(K) \oplus X$, *where the group X when both $H/[H, G]$ and K/K_2 are torsion free is isomorphic to direct sum of groups* $H/[H, G] \otimes Q_3(K)$, $((I([H, G]) + I^2(H))/(I([H, G, G]) + I([H, G])I(H) + I^3(H))) \otimes Q_2(K)$ and $(I([H, G, G]) + I([H, G])I(H) + I^3(H))/L \otimes K/K_2$, *where* $L = I([H, G, G, G]) + I([H, G, G])I(H) + I([H, G])I^2(H) + I^2([H, G]) + I^4(H)$.

Observe that when the subgroup H of G is Abelian, then

$$(I([H, G]) + I^2(H))/(I([H, G, G]) + I([H, G]I(H) + I^3(H)) \simeq$$

$$[H, G]/[H, G, G] \oplus SP^2(H/[H, G]).$$

Theorem 4.19. *If both H/H_2 and K/K_2 are torsion free, then $I^3(G)I(H)/I^4(G)I(H)$ is isomorphic to direct sum of the groups* $Q_3(K) \otimes H/H_2$, $Q_2(K) \otimes I^2(H)/(I^3(H) + I([H, G])I(H))$, $K/K_2 \otimes (I^3(H) + I([H, G])I(H))/(I^4(H) + I([H, G])I^2(H) + I([H, G, G])I(H))$ *and* $(I^4(H) + I([H, G])I^2(H) + I([H, G, G]))/(I^5(H) + I([H, G])I^3(H) + I^2([H, G])I(H) + I([H, G, G])I^2(H) + I([H, G, G, G])I(H))$.

References

[1]   F. Bachmann and L. Grünenfelder, Homological methods and the third dimension subgroup, Comment. Math. Helv. 47 (1972), 526–531.

[2]   F. Bachmann and L. Grünenfelder, The periodicity in the granded ring associated with an integral group ring, J. Pure Appl. Algebra 5 (1974), 253–264.

[3]   M. Curzio and C. K. Gupta, Second Fox subgroups of arbitrary groups, Bull. Canad. Math. Soc., to appear.

[4]   M. Goyal, The Augmentation Ideal of an Integral Group ring, Ph.D. Thesis, Kuruk-shetra University, Kurukshetra (India), 1981.

[5]   A. W. Hales, Augmentation terminals of finite Abelian groups, in: Abelian Group Theory (Proc. Honolulu, 1982–1983), Lecture Notes in Math. 1006, Springer-Verlag 1983, 720–733.

[6]   A. W. Hales, Stable augmentation quotients of Abelian groups, Pacific J. Math. 118 (1985), 401–410.

[7]   A. W. Hales and I. B. S. Passi, The second augmentation quotient of an integral group ring, Arch. Math. (Basel) 31 (1978), 259–265.

[8]   A. W. Hales and I. B. S. Passi, The augmentation quotients of finite Abelian p-groups, in: Contemp. Math. 93, Amer. Math. Soc., 1989, 167–171.

[9]   M. Hartl, On relative polynomial construction of degree 2 and applications, preprint.

[10]   M. Hartl, Polynomiality properties of group extensions with torsionfree Abelian kernel, preprint.

[10(a)] M. Hartl, On successive quotients of group ring filtrations induced by N-series, preprint.

[11]   K. Horibe and K. Tahara, The stable behaviour of the augmentation quotients of some groups of order p^4 I, Japan J. Math. 10 (1984), 137–157.

[12]   M. Khambadkone, Subgroup ideals in group rings I, J. Pure Appl. Algebra 30 (1983), 261–275.

[13]   M. Khambadkone, On the structure of augmentation ideals in group rings, J. Pure Appl. Algebra 35 (1985), 35–45.

[14]   M. Khambadkone, Augmentation quotients of semidirect products, Arch. Math. 45 (1985), 407–420.

[15]   M. Khambadkone, Subgroup ideals in group rings III, Comm. Algebra 14 (3) (1986), 411–421.

[16]   G. Losey, On the structure of $Q_2(G)$ for finitely generated groups, Canad J. Math. 25 (1973), 353–359.

[17]   G. Losey, N-series and filtrations of the augmentation ideal, Canad J. Math. 26 (1974), 962–977.

[18]   G. Losey and N. Losey, The stable behaviour of the augmentation quotients of the groups of order p^3, J. Algebra 60 (2) (1979), 337–351.

[19]   G. Losey and N. Losey, Augmentation quotients of some non-Abelian finite groups, Math. Proc. Camb. Philos. Soc. 85 (1979), 261–270.

[20]   I. B. S. Passi, Polynomial maps on groups, J. Algebra 9 (1968), 121–151.

[21]   I. B. S. Passi, Polynomial functors, Proc. Camb. Philos. Soc. 66 (1969), 505–512.

[22]   I. B. S. Passi, Polynomial maps on groups II, Math. Z. 35 (1974), 137–143.

[23]   I. B. S. Passi, The associated graded ring of a group ring, Bull. London Math. Soc. 10 (1978), 241–255.

[24]   I. B. S. Passi, Group Rings and Their Augmentation Ideals, Lecture Notes in Math. 715, Springer-Verlag, Berlin–Heidelberg 1979.

[25]   I. B. S. Passi and Sneh Sharma, The third dimension subgroup mod n, J. London Math. Soc. 9 (2) (1974), 176–182.

[26]   I. B. S. Passi and L. R. Vermani, The associated graded ring of an integral group ring, Math. Proc. Camb. Philos. Soc. 82 (1977), 25–33.

[26(a)] D. Quillen, On the associated graded ring of a group ring, J. Algebra 10 (1968), 411–418.

[27]   Ram Karan and L. R. Vermani, A note on augmentation quotients, Bull. London Math. Soc. 18 (1986), 5–6.

[28]   Ram Karan and L. R. Vermani, A note on polynomial maps, J. Pure Appl. Algebra 51 (1988), 169–173.

[29]   Ram Karan and L. R. Vermani, Augmentation quotients of integral group rings, J. Indian Math. Soc. 54 (1989), 107–120.

[30]   Ram Karan and L. R. Vermani, Augmentation quotients of integral group rings II, J. Pure Appl. Algebra 65 (1990), 253–262.

[31]   Ram Karan and L. R. Vermani, Augmentation quotients of integral group rings III, J. Indian Math. Soc. 58 (1992), 19–32.

[32]   Ram Karan and L. R. Vermani, Corrigendum: Augmentation quotients of integral group rings-II, J. Pure Appl. Algebra 77 (1992), 229–230.

[33]   R. Sandling, Modular augmentation ideals, Math. Proc. Camb. Philos. Soc. 71 (1972), 25–32.

[34]   R. Sandling, Dimension subgroups over arbitrary coefficient rings, J. Algebra 21 (1972), 250–265.

[35]   R. Sandling and K. Tahara, Augmentation quotients of group rings and symmetric powers, Math. Proc. Camb. Philos. Soc. 85 (1979), 247–252.

[36]   M. Singer, On the graded ring associated with an integral group ring, Comm. Algebra 3 (1975), 1037–1049.

[37]   M. Singer, Determination of the augmentation terminal for all finite Abelian grops of exponent 4, Comm. Algebra 4 (1976), 639–645.

[38]   M. Singer, On the augmentation terminal of a finite Abelian group, J. Algebra 41 (1976), 196–201.

[39]   M. Singer, Determination of the augmentation terminal for all Abelian groups of exponent 8, Comm. Algebra 5 (1977), 87–100.

[40]   M. Singer, Determination of the augmentation terminal for finite Abelian groups, Bull. Amer. Math. Soc. 83 (1987), 1321–1322.

[41]   K. Tahara, On the structure of $Q_3(G)$ and the fourth dimension subgroups, Japan. J. Math. 3 (1977), 381–394.

[42]   K. Tahara, The augmentation quotients of group rings and the fifth dimension subgroups, J.Algebra 71 (1) (1981), 141–173.

[43]   K. Tahara, augmentation quotients and dimension subgroups of semidirect products. Math. Proc. Camb. Phil. Soc. 91 (1982), 39–49.

[44]   K. Tahara and T. Yamada, The stable behaviour of the augmentation quotients of some groups of order p^4 II, Japan. J. Math. 10 (1984), 159–184.

[45]   K. Tahara and T. Yamada, The stable behaviour of the augmentation quotients of some groups of order p^4 III, Japan. J. Math. 11 (1985), 109–130.

[46]   L. R. Vermani, On polynomial groups, Proc. Camb. Philos. Soc. 68 (1970), 285–289.

[47]   L. R. Vermani and Ram Karan, Augmentation quotients of integral group rings, III-Corrigendum, J. Indian Math. Soc. 59 (1993), 261–262.

[48]   L. R. Vermani and Atul Razdan, Some remarks on augmentation ideals in group rings, to appear.

[49]   L. R. Vermani and Atul Razdan, Some intersection theorems and subgroups determined by certain ideals in integral group rings, Algebra Collo., to appear.

[50]   L. R. Vermani, Atul Razdan and Ram Karan, Some remarks on subgroups determined by certain ideals in integral group rings, Proc. Indian Acad. Sci. (Math. Sci.) 103 (1993), 249–256.

[51]   J. Wilson, Polynomial functors on Abelian Groups, unpublished.

On the Nielsen and Whitehead Methods in Combinatorial Group Theory and Topology

Heiner Zieschang

Abstract. This is a survey on concepts, results and applications related to the Whitehead algorithm, the Nielsen method and quadratic equations in free groups or free products of groups. The close relation between combinatorial group theory and topology is in the center of our considerations. Proofs are only sketched.

1991 Mathematics Subject Classification: 20-02, 20F10, 20F30, 20F34, 57M05, 57M25

1. Introduction

The basic concept of combinatorial group theory is the presentation of a group G by *generators* S and *defining relators* R, expressed by $G = \langle S \mid R \rangle$, together with some principal constructions as free groups, free products of groups, free products with amalgamation, HNN-groups, graphs of groups, commutator subgroups .... Closely related to it is the fundamental group of a cell-complex and its presentation obtained from the 1- and 2-dimensional skeleton together with the Seifert–van Kampen theorem and the covering property. Many group-theoretical statements have topological counterparts and often these admit much simpler, shorter and more intuitive proofs. Let us mention the subgroup theorems for free groups [Ni 21], [Sc 27], for free products [Ku 34], [BL 36], for free products with amalgamation [Ne 48], [KS 70], (which leads to the idea of a graph of groups [Se 80]), and the Reidemeister–Schreier method [Re 27], [Sc 27], [GZ 79].

Two other important methods of combinatorial group theory, the *Nielsen* and the *Whitehead method*, are not so simply and directly related to geometric considerations (although the original approach of Whitehead was a topological one). In the following, we discuss these theories and their applications, also mixing geometric-topological and group-theoretical considerations. The bibliography is quite large, but far from complete. Many more references can be found in the books [MKS 66], [LS 77], [ZVC 80], in particular in its Russian variant from 1989, [CZ 93], [GK 93], [Zi 88]. Let us mention here some other important methods and approaches related to combinatorial group theory and topology which will not be considered here (although nowadays great research in these fields is going on): small cancellation theory [Sc 73], word hyperbolic groups [Gr 88], [CDP 90], [GH 90], $\mathbb{R}$-trees [CM 87], fixed elements of automorphisms of free groups [BH 92], automatic groups [Ep 92], growth of groups [GK 93], ...

2. The Whitehead Method

In this section we will describe the Whitehead method and some of its applications to topology. The general problem is the following. Given two elements x, x' in a group G, decide whether there is an automorphism $\alpha \in \operatorname{Aut} G$ such that $\alpha(x) = x'$ and describe the stabilizer $\operatorname{Stab}(x) = \{\alpha \in \operatorname{Aut} G : \alpha(x) = x\}$ of x in $\operatorname{Aut} G$. Or more generally, given two finite systems $X = (x_1, \ldots, x_m)$, $X' = (x'_1, \ldots, x'_m)$ of elements of a group G, decide whether there is an $\alpha \in \operatorname{Aut} G$ such that $\alpha(x_i) = x'_i$ (or $\alpha(x_i) = w_i x'_i w_i^{-1}$, $w_i \in G$) for $i = 1, \ldots, m$ and describe the set of all such automorphisms.

Clearly, this can be solved for finitely generated abelian groups. For free groups J. H. C. Whitehead (1936) has founded an algorithm which we will describe next.

Let $F = F_n = \langle s_1, \ldots, s_n \mid \rangle$. In the following we use the free length of an element w of F with respect to this basis and denote it by $|w|$. Two elements of F are called **Nielsen-equivalent** if one is the image of the other under an automorphism of F. In a similar way the Nielsen equivalence of (ordered) subsets of F is defined.

Definition. The following automorphisms are called **Whitehead automorphisms**:

$$s_i \mapsto \begin{cases} s_1^{-\varepsilon} s_i & \text{for } 1 \le j \le n_1, \\ s_1^{-\varepsilon} s_i s_1^{\varepsilon} & \text{for } n_1 + 1 \le j \le n_2, \\ s_1^{-\varepsilon} s_i & \text{for } n_2 + 1 \le i \le n_3, \\ s_1^{-\varepsilon} s_i s_1^{\varepsilon} & \text{for } n_3 + 1 \le i \le n \\ & \text{where } 1 \le n_1 \le n_2 \le n_3 \le n, \quad \varepsilon = \pm 1, \end{cases}$$

or the automorphism obtained from one of these after a permutation of the subscripts; furthermore the automorphisms induced by permutations of the set $\{s_1, s_1^{-1}, \ldots, s_n, s_n^{-1}\}$.

Now we can formulate the theorem of Whitehead.

Theorem 2.1 (J. H. C. Whitehead [Wh 36, 36']). *Let F be a free group of finite rank.*

(a) *If $w \in F$ does not have minimal length in the set of elements Nielsen-equivalent to w then there is a Whitehead automorphism τ such that $|\tau(w)| < |w|$.*

(b) *If w_1, w_2 are Nielsen-equivalent and both of minimal length then there is a sequence of Whitehead automorphisms $\tau_1, \ldots, \tau_k$ such that $\tau_k \cdots \tau_1(w_1) = w_2$ and $|\tau_j \cdots \tau_1(w_1)| = |w_1|$ for $1 \le j \le k$.*

A direct consequence is the first statement of the following result; the second is due to J. McCool [Mc 75].

Theorem 2.2. *For $w \in F$ the stabilizer* Stab(w) *is finitely generated. Moreover, it admits a finite system of defining relators. Using the Whitehead method, a presentation of the automorphism group of a free group can also be found* [Mc 74].

The original proof of J. H. C. Whitehead [Wh 36, 36′] uses topological arguments. A combinatorial group theoretical proof is due to E.S. Rapaport [Ra 58]. A simpler one was given in the 1960's by G. Higgins and R. C. Lyndon and published finally in [HL 74]; it is based on quite simple combinatorial length calculations which lead to a *Peak Reduction Lemma*. Consider $w \in F$ and Whitehead automorphisms α, β such that

$$|\alpha^{-1}(w)| \leq |w| \geq |\beta(w)|$$

and there is at least one proper inequality. This situation is called a *peak*. Now there is a sequence $\gamma_1, \ldots, \gamma_r$ of Whitehead automorphisms such that

$$\beta\alpha^{-1} = \gamma_r \cdots \gamma_1 \quad \text{and} \quad |\gamma_k \cdots \gamma_1(w)| < |w| \text{ for } 1 \leq k < r.$$

The replacement $\beta\alpha^{-1}$ by $\gamma_r \ldots \gamma_1$ is called a *peak reduction*; thus the above statement is that there is a peak reduction. J. McCool noticed that a peak reduction needs at most four Whitehead automorphisms γ_j; a consequence is that the group of free automorphisms stabilizing an element is finitely presentable. For a geometric approach see also [Ho 79].

Examples 2.3.
(a) Let $F = F(s, t)$, consider the elements $u = s^3 t^5$ and $v = s^3 t^3 s^3 t^2$. Each is of minimal length in its Nielsen equivalence class; this is easily checked using 2.1 and the fact that one can consider conjugacy classes and, hence, that it suffices to study the effect of the automorphisms $s \mapsto s$, $t \mapsto ts^{\pm 1}$ and $s \mapsto st^{\pm 1}$, $t \mapsto t$ on the cyclic length. Each of these automorphisms increases the lengths of u and v, so these elements have minimal lengths in their Nielsen equivalence classes. Since the lengths are different they are not Nielsen equivalent (nor one to the inverse of the other).
(b) Consider $\prod_{j=1}^g [t_i, u_i] \in \langle t_1, u_1, \ldots, t_g u_g \mid \rangle$ or $v_1^2 \cdots v_g^2 \in \langle v_1, \ldots, v_g \mid \rangle$, where $[t, u] = tut^{-1}u^{-1}$. Both have minimal length in their equivalence classes. Roughly speaking, the canonical defining relator of the fundamental group of a closed surface has minimal length.

All generalizes to systems of elements or to conjugacy classes of elements or to systems of conjugacy classes in a free group with almost no changes in the formulation. Next we describe some applications of the Whitehead method to group theoretical and topological problems.

2.4. On handlebodies and Heegaard splittings. A *handlebody H of genus g* is obtained from a 3-dimensional ball B^3 by attaching g disjoint handles $D_j^2 \times I$:

$$H = B^3 \cup \bigcup_{j=1}^g (D_j^2 \times I), \ B^3 \cap (D_j^2 \times I) = \partial B^3 \cap (D_j^2 \times \partial I) = D_j^2 \times \{0, 1\} \text{ for } 1 \leq j \leq g,$$

where D_j^2 denotes a disk and I the unit segment. Moreover the $2g$ disks $D_j^2 \times \{0, 1\}$ on the 2-sphere ∂B^3 are pairwise disjoint and ∂H is an orientable surface. These conditions determine a handlebody up to homeomorphism. The regular neighbourhood of a bouquet of g circles in $\mathbb{R}^3$ is a handlebody of genus g.

In 3-dimensional topology handlebodies are often used, in particular, in *Heegaard decompositions (splittings)* of a 3-manifold M^3,

$$M^3 = H \cup H', \quad H \cap H' = \partial H = \partial H' \quad \text{where } H, H' \text{ are handlebodies,}$$

or *Heegaard diagrams* where M^3 is described by a system of disjoint simple closed curves on the boundary of a handlebody H; these curves are meridians of the second handlebody H', that is, they bound disks in H'. The following theorem offers a possibility to decide whether a system of elements in the free group $\pi_1 H$ can be represented by a system of disjoint simple closed curves on ∂H and whether two systems of curves are "homeomorphic" under a homeomorphism of the handlebody.

Theorem 2.5 [Zi 65]. *Let H_g be a handlebody of genus g. Fix a system μ of meridians of H_g and the resulting system Σ of free generators of $\pi_1 H_g \cong F_g$. If a system of elements of $\pi_1 H_g$ can be represented by a system of disjoint simple closed curves on ∂H_g then any equivalent system of homotopy classes of minimal length can also be represented by a system of disjoint simple closed curves on ∂H_g intersecting the meridians in μ in exactly the same form as the (freely reduced) words in the generators of Σ.*

Using the Whitehead method and geometrical arguments on systems of arcs on a 2-sphere with $2g$-holes, it is decidable in a finite number of steps whether a system of g (conjugacy classes of) elements of F_g can be realized by a system of pairwise disjoint simple closed curves on ∂H_g and, in addition, whether two such systems of curves are homeomorphic. 2.5 has several applications in 3-topology, for instance, W. Haken [Ha 79] used it to prove that a Heegaard splitting of a connected sum of 3-manifolds decomposes into Heegaard splittings of the summands. For Heegaard splittings see [Zi 88].

2.6. Recognizing one-relator-groups. Given two presentations

$$\langle s_1, \ldots, s_n \mid r(s) \rangle \quad \text{and} \quad \langle s_1, \ldots, s_m \mid q(s) \rangle$$

with one defining relation, decide whether they describe isomorphic groups.

Suppose first that one of the groups is free; by the Freiheitssatz of M. Dehn and W. Magnus [Ma 30], [ZVC 80, 2.11.8] the relation is a member of some system of free generators and this can be checked using the Whitehead method. So we may assume that both groups are not free. Then $m = n$. Magnus [MKS 66, p. 401] asked whether there is an automorphism α of the free group F_n such that $\alpha(r(s)) = q(s)^{\pm 1}$. It turns out that in general the question has a negative answer.

Example 2.7. Consider the group G of the torus knot $\mathfrak{t}(3, 5)$ and make the following Tietze transformations:

$$G = \langle s, t \mid s^3 t^5 \rangle = \langle s, t, z \mid t^2 z^{-1}, s^3 t^5 \rangle$$
$$= \langle s, t, z \mid t^2 z^{-1}, s^3 z^3 t^{-1} \rangle = \langle s, z \mid (s^3 z^3)^2 z^{-1} \rangle;$$

hence, by 2.3 (a), there are isomorphic one-relator-groups with non Nielsen-equivalent relators. In this case we will call the presentations non-Nielsen equivalent. This first example of a negative answer to Magnus' question is due to J. McCool and A. Pietrowski [MP 71] and H. Zieschang [Zi 70]. Later there were constructed many other examples of non-Nielsen equivalent one-relator presentations, even examples with infinitely many non-Nielsen-equivalent one-relator presentations, see [Br 76]. For torus knot groups, also using the Nielsen method, one obtains the full classification of all one-relator presentations.

Theorem 2.8.
(a) *Any pair of generators of a group $\langle s, t \mid s^a t^b \rangle$, $a, b \geq 2$ (for $\gcd(a, b) = 1$ a torus knot group) is Nielsen equivalent to exactly one of the following pairs:*

$$\{(s^\alpha, t^\beta) : 1 \leq 2\alpha \leq \beta a, \ 1 \leq 2\beta \leq \alpha b,$$
$$\gcd(a, \alpha) = \gcd(b, \beta) = \gcd(\alpha, \beta) = 1\}.$$

For all these pairs there are presentations with one or two defining relators. If one of the numbers α, β equals 1 then one relator suffices, [Zi 77].
(b) *The last condition is necessary, [Co 78].*
    Hence, there are only finitely many Nielsen equivalence classes of one-relator presentations for the torus knot groups, although there are infinitely many Nielsen equivalence classes of generating pairs.

The proof of 2.8 (a) is a rather unpleasant combinatorial one with a lot of case considerations; it would be nice to have more intelligible arguments, but this seems difficult since there remain only finitely many Nielsen-equivalence classes in the quotients obtained after introducing the relation $s^a = 1$, see [Zi 77, 4.14], or factoring out some higher commutator subgroup, see [Re 79]. This result is the algebraic background for the classification of Heegaard decompositions of genus 2 of the exterior of torus knots. A ***Heegaard splitting of genus g*** of the complement of a knot $\mathfrak{k}$ is given by a handlebody H of genus g and $g - 1$ attached 2-handles $D_j^2 \times I$:

$$\overline{S^3 - \mathcal{N}(\mathfrak{k})} = H \cup \bigcup_{j=1}^{g-1} (D_j^2 \times I), \quad H \cap (D_j^2 \times I) = \partial H \cap (D_j^2 \times I) = \partial D_j^2 \times \{0, 1\},$$

$(D_j^2 \times I) \cap (D_i^2 \times I) = \emptyset$ for $j \neq i$;

here $\mathcal{N}(\mathfrak{k})$ denotes the regular neighborhood of $\mathfrak{k}$. For torus knots the Heegaard splittings of the complement are classified up to homeomorphism and isotopy.

Theorem 2.9 [BRZ 86]. *The exterior of the torus knot* $\mathfrak{t}(p, q)$, $\gcd(p, q) = 1$, $p, q \geq 2$ *admits exactly three non-homeomorphic Heegaard splittings of genus 2, except if* $|p - q| = 1$ *when there is, up to homeomorphism, only one Heegaard splitting of genus 2 or if* $|p - q| \neq 1$, *but* $p \equiv \pm 1 \bmod q$ *or* $q \equiv \pm 1 \bmod p$ *when there are exactly two non-homeomorphic Heegaard splittings of genus 2.*

Any two Heegaard splittings of genus 2 of a torus knot exterior become homeomorphic after adding one trivial handle to both of them.

The torus knots form the only class of 1-tunnel knots (= knots with Heegaard splitting of genus 2) where Heegaard splittings are classified. For some 2-bridge knots and some pretzel knots there have been constructed up to four non-homeomorphic Heegaard splittings of genus 2 of the knot exterior, but it is not known whether these represent all types of Heegaard splittings (M. Kuhn [Ku 94], K. Morimoto and M. Sakuma [MS 91], J. Tipp [Ti 89]). M. Kuhn and J. Tipp found also many (but only finitely many) non Nielsen-equivalent presentations for the groups of these knots; it is neither clear that they found all possible one-relator presentations nor that the Heegaard splittings of genus 2 obtained are all those possible. For a given manifold, the number of non-homeomorphic Heegaard splittings of a fixed genus is finite, [Jo 90]. On the other hand there are one-relator-groups which admit only "one" defining relator, see the following theorem which has been proved for surface groups using the generalized Nielsen method from 3.4. It seems to be very difficult to decide whether a one-relator-group admits, up to free isomorphism, one or more one-relator presentations.

Theorem 2.10 [Zi 65]. *Let* $\langle x_1, \ldots, x_{2g} \mid R(x_j) \rangle$ *be a presentation of the fundamental group* $\langle t_1, u_1, \ldots, t_g, u_g \mid \prod_{j=1}^{g} [t_j, u_j] \rangle$ *of a closed orientable surface. Then there is an isomorphism between the free groups*

$$\varphi \colon \langle x_1, \ldots, x_{2g} \mid \rangle \to \langle t_1, u_1, \ldots, t_g, u_g \mid \rangle \quad \text{with } \varphi(R(x_j)) = \prod_{i=1}^{g} [t_i, u_i] \rangle.$$

A similar result is true for the fundamental groups of non-orientable closed surfaces. Hence, applying the Whitehead method, the fundamental group of a closed surface can be recognized among one-relator-groups.

This result has an interesting immediate application. If $h: S_g \rightarrow S_g$ is a homeomorphism then its **mapping class** is the set of all auto-homeomorphisms of S_g which are isotopic to h. The mapping classes form a group, nowadays often called the **homeotopy group**. Basic for the study of the mapping classes are the theorem of Nielsen that *any automorphism of $\pi_1 S_g$ is induced by an auto-homeomorphisms* and the theorem of Baer that *two homeomorphisms are isotopic if they induce the same homomorphism on $\pi_1 S_g$;* hence, *the homeotopy group of a closed surface is isomorphic to the outer automorphism group of its fundamental group.* See [ZVC 80, ch. 5]. In addition, *each automorphism of the fundamental group is induced by an automorphism of the free group which maps the defining relator to a conjugate of itself or its inverse.* (The automorphism of the free group is almost uniquely determined by the automorphism of $\pi_1 S_g$, [Zi 85].) So the results of J. H. C. Whitehead and J. McCool imply the following theorem.

Theorem 2.11. *The homeotopy group of a closed surface admits a finite presentation; of course, the same is true for the isomorphic group of outer automorphisms of the fundamental group. There is an algorithm to determine a presentation of the group considered.*

Another proof together with a more geometric algorithm was obtained by A. Hatcher and W. Thurston [HT 80]. The number of generators obtained when using the Whitehead method is quite large while arguments of Dehn and Lickorish reduce the number of generators to $2g$ or 4; recently B. Waynrub gave two generators for the mapping class group.

The Higgins–Lyndon proof of the Whitehead theorem can be extended to free products to obtain a Whitehead method also for this more general case. Let

$$G = \left(*_{j \in J} G_j\right) * F(S) = *_{i \in J \cup S} G_j$$

where no factor G_j, $j \in J$ is infinite cyclic and $F(S)$ denotes the free group with free generators S. Let $|x|$ denote the usual length of an element $x \in G$ corresponding to the free product decomposition. We consider the following types of automorphisms.

2.12. Special automorphisms of a free product.

(i)   Permutation automorphisms permute the factors G_j, $j \in J$ with fixed isomorphisms between possibly different factors and the identity if a factor is fixed or they permute $S \cup S^{-1}$.

(ii) An automorphism $\alpha: G \rightarrow G$ is called a **factor automorphism** if $\alpha(G_i) = G_i$, $\forall i \in J$ and $\alpha(s) = s$, $\forall s \in S$.

(iii) α is called a **Whitehead automorphism** if there is some fixed letter x which is either a non-trivial element of some G_j, $j \in J$ or is from $S \cup S^{-1}$ such that
(a)   for each factor G_j, $j \in J$ there is an $\eta_j \in \{0, 1\}$ such that $\alpha(y) = x^{-\eta_j} y x^{\eta_j}$ for all $y \in G_j$
(b)   for $s \in S$, $\alpha(s)$ is one of s, sx, $x^{-1}s$, $x^{-1}sx$.

Let α be a Whitehead automorphism with associated letter x as in (iii). Define

$$A = \{j \in J \;:\; \alpha(G_j) = x^{-1}G_j x\} \cup \{s \in S \cup S^{-1} \;:\; \alpha(s) = sx \text{ or } x^{-1}sx\} \cup A_0,$$

$$A_0 = \begin{cases} \{k\} & \text{if } x \in G_k, \\ \{x\} & \text{if } x \in S \cup S^{-1}; \end{cases}$$

then α is determined by A and x and, following [HL 74], we write $\alpha = (A, x)$.

(iv) β is a **multiple Whitehead automorphism** if there are Whitehead automorphisms
$\beta_p = (A_p, x_p)$, $p = 1, \ldots, r$ such that
   (a) $x_1, \ldots, x_r$ all lie in the same factor G_k, $k \in J$,
   (b) $\alpha = \beta_r \ldots \beta_1$,
   (c) if $p \neq q$ then $x_p \neq x_q$ and $A_p \cap A_q = \{k\}$.

Using similar calculations as in [HL 74] one obtains the following result.

Theorem 2.13 [CZ 84']. *Let G be as above a free product and let Ω be the set of all permutation, factor, Whitehead and multiple Whitehead automorphisms. Then for any cyclic words u and v and automorphism α of G such that $\alpha(u) = v$, there exist $\rho_1, \ldots, \rho_n \in \Omega$ such that $\alpha = \rho_n \ldots \rho_1$ and for some p, q, $0 \leq p \leq q \leq n$*
   (a) $|\rho_{i-1} \ldots \rho_1 u| > |\rho_i \ldots \rho_1 u|$, $\quad 1 \leq i \leq p$,
   (b) $|\rho_{j-1} \ldots \rho_1 u| = |\rho_j \ldots \rho_1 u|$, $\quad p+1 \leq j \leq q$,
   (c) $|\rho_{j-1} \ldots \rho_1 u| < |\rho_j \ldots \rho_1 u|$, $\quad q+1 \leq j \leq n$.

For two or three factors there is no need of multiple Whitehead automorphisms, but in general they cannot be avoided, see [CZ 84]. For the proof one can generalize the Peak-Reduction Lemma of Higgins–Lyndon to free products. In fact, the reduction can be done by an algorithm if the factors satisfy some decision properties. To a great extent the Whitehead–McCool theory can be generalized to free products, see [Co 87], [CZ 84', 84'', 87]. Till to now there are only few applications of this theory to topology, may-be, one can use the following example.

Let us finish this section with an example, quadratic words, which shows that there may be "more types" in free products than in free groups. Let us call the number of letters from factors G_j or generators $s \in S$ occurring in a (reduced) word w its length. An element w is called **minimal** if its length is minimal among the lengths of the images of w under automorphisms of $G = (*_{j \in J} G_j) * F(S)$; then the number of different factors G_j or generators $s \in S$ which occur in w is called the **rank** of w. The rank of an element is the minimal length of all words representing g or an image of G under an automorphism of $G = (*_{j \in J} G_j) * F(S)$. Using 2.13 one easily obtains the following result.

Proposition 2.14. *Again let $G = (*_{j \in J} G_j) * F(S)$ be a free product, no G_j infinite cyclic. An element $p_1 \cdots p_m \cdot f_1^2 \cdots f_n^2 \cdot \prod_{i=1}^{g} a_i b_i \bar{a}_i \bar{b}_i$ is minimal and of rank $m + n + 2g$ if*
   (i) *$p_1, \ldots, p_m$ are from distinct factors G_j, $j \in J$;*

(ii) $f_1, \ldots, f_n \in S$;

(iii) *either a_j and $\bar{a}_j$ are non-trivial elements of one factor $G_i(j)$ or $a_j \in S$ and*
$\bar{a}_j = a_j^{-1}$;

(iv) *a similar condition for the $b_j, \bar{b}_j$.*
   Here no subscript $k \in J$ and no letter $s \in S$ appears twice in (i)–(iv).

Examples 2.15. Let $G = A * B * C$ be a free product of pairwise non-isomorphic groups, none infinite cyclic. Then there is no automorphism of G reducing the length of the quadratic elements

$$w_1 = abc\bar{a}\bar{b}\bar{c} \text{ and } w_2 = ab\bar{a}c\bar{b}\bar{c}, \quad a, \bar{a} \in A, \ b, \bar{b} \in B, \ c, \bar{c} \in C$$

or mapping one to the other; hence they have rank 3. This would be different when A, B, C are infinite cyclic and $\bar{a} = a^{-1}$ etc.; then the word would not be minimal as is well known from the classification of orientable closed surfaces.

Similar in $G = A * B * C * D$, all factors not infinite cyclic and pairwise not isomorphic, the words

$$abcda^{-1}b^{-1}c^{-1}d^{-1} \quad \text{and} \quad aba^{-1}b^{-1}cdc^{-1}d^{-1}, \ a \in A, \ b \in B, \ c \in C, \ d \in D$$

are of minimal length among all their images under automorphisms of G and they are not equivalent, while for infinite cyclic groups A, B, C, D the words are equivalent.

If one tries to classify quadratic words in a general free product $G = \left(*_{j \in J} G_j \right) * F(S)$ by repeating the arguments to find the canonical presentation of fundamental groups of surfaces or Fuchsian groups one obtains words of the following form

$$p_1 \cdots p_m f_1^2 \cdots f_t^2 v \quad \text{or} \quad p_1 \cdots p_m \prod_{j=1}^{k} a_j b_j a_j^{-1} b_j^{-1} v$$

where
(a) $p_1, \ldots, p_m$ are from distinct factors $G_j, \ j \in J$,
(b) different letters are from different factors,
(c) a factor G_j or a free generator $s \in S$ appears in v twice or not at all and every couple is separated by another one; moreover there are no two couples corresponding to different $s \in S$ which separate one another.

Remember that for Fuchsian or NEC-groups the normal form contains only elements like the p_i, squares like the f_j^2 or commutators. In general, the v cannot be reduced to a system of squares or "commutators" and the number of non-equivalent types becomes arbitrarily large when, of course, the number of letters increases. A classification of all types is not yet known.

3. On the Nielsen Method

Given a finite system of elements $w_1, \ldots, w_m \in F_n = \langle s_1, \ldots, s_n \mid \rangle$ one can reduce the sum of the lengths to a minimum without changing the subgroup $\langle w_1, \ldots, w_m \rangle$ replacing the system by $w_1^*, \ldots, w_m^*$ obtained by a "Nielsen process"

$$w_i^* = w_i w_m^{\pm 1}, \quad w_j^* = w_j \text{ for } j \neq i \text{ and } m \text{ fixed.}$$

If one "considers the free group generated by" $w_1, \ldots, w_m$ then these correspond to special Whitehead automorphisms. A famous result of Nielsen is as follows.

Theorem 3.1 (J. Nielsen [Ni 21]). *Using a strict order $<$ obtained from the length and a lexicographical order one can monotonically reduce the given system to a minimal system; if the conditions about the order are correctly formulated the minimal system is uniquely determined.*

Proposition 3.2. *If the system $w_1, \ldots, w_m$ is minimal and does not contain the trivial element then it has the following Nielsen properties:*

$$|w_i^\varepsilon w_j^\eta| \geq |w_i|, \ |w_j| \text{ where } i \neq j \text{ or } i = j \text{ and } \varepsilon \neq \eta$$

$$|w_i^\varepsilon w_k w_j^\eta| > |w_i| + |w_j| - |w_k| \text{ where } i \neq k \neq j.$$

Here $1 \leq i, j, k \leq n$, $\varepsilon, \eta \in \{1, -1\}$.

Consequences. a) (Finitely generated) subgroups of free groups are free. (This was the first proof, by Nielsen, for this fundamental theorem.)

b) The rank of the subgroup generated by $w_1, \ldots, w_m$ is m.

c) If $x \in F_n$ is in the subgroup generated by $w_1, \ldots, w_m$ then there is a w_i^ε, $1 \leq i \leq n$, $\varepsilon = \pm 1$ such that $x w_j^\varepsilon < x$ and it is decidable whether $x \in \langle w_1, \ldots, w_m \rangle$ or not.

d) If $x = w_{i_1}^{\varepsilon_1} \cdots w_{i_\ell}^{\varepsilon_\ell}$ and neighbours are not formal inverses then

$$|x| \geq \begin{cases} |w_{i_j}| & \text{for } 1 \leq j \leq \ell, \\ \frac{1}{2}\left(|w_{i_1}| + |w_{i_\ell}|\right) + \ell - 2 & \text{if } n \geq 2. \end{cases}$$

In particular, from each factor $w_{i_j}^{\varepsilon_j}$ there remains at least one letter in the reduced word x.

e) The Nielsen transformations, applied to the generators $(s_1, \ldots, s_n)$ and considered as automorphisms, generate the group Aut F_n.

This cancellation method can be generalized to free products (R. C. Lyndon [Ly 63]) and one obtains similar results, clearly, a bit weaker, in particular the Grushko theorem [Gr 40]. It also can be applied to free products with amalgamation ([Zi 70], [Ro 77, 77'], [Ho 81], [CZ 87]) and to HNN-extensions [PR 78]; in theory, it could also be extended to graphs of groups. But in these cases there appear unpleasant difficulties. The main result for free products with amalgamation is as follows.

Theorem 3.4. *Let $G = *_{j \in J}(G_j; A)$ be a free product with amalgamated subgroup A. Use the usual length and normal form for elements and let X be a system of elements of G. Let X be minimal with respect to the length and some lexicographical order and with respect to Nielsen transformations. Then one has either the situation* (a) *similar to that for free groups or it occurs an obstruction or a collaps as described in* (b).

(a)   *In any freely reduced product of elements of X the following holds:*

   (i)   *the front half of the first and the back half of the last factor appear undisturbed in the normal form of the product;*

   (ii) *each factor of the product contains a letter which is not cancelled when the product is reduced to its normal form;*

   (iii) *the length of the normal form of the product exceeds or equals the maximum of the lengths of its factors.*

(b)   *One of the following collapses* (1)–(3) *or obstructions* (4) *occurs.*

   (1) *There exists $x = uzv^{-1}$ in X, where u is the front half and v^{-1} the back half of x, and $y_1, \ldots, y_m \in X \cap A$ such that $v_1^{-1} y_1 \cdots y_m v \in A$.*

   (2) *There exist $z_i \in X$, $i = 1, 2, \cdots \in X$, all of length at most 1 and not all of length 0, whose product is in A.*

   (3) *There exist $x_i \in X$, $1 \leq i \leq m$, all of odd length, such that*

$$x_1 = uz_1 v^{-1}, \quad x_i = vz_i v^{-1} \ (i = 2, \ldots, m-1), \quad x_m = vz_m^{-1} u^{-1}$$

      where u, v^{-1}, v, u^{-1} are the appropriate halves, $z_1, \ldots, z_m \in G_k$ for a fixed $k \in J$, and $z_1 \cdots z_m \in A$.

   (4) *There exist $x_i = uz_i u^{-1} \in X$, $1 \leq i \leq m$, where u, u^{-1} are the front and back halves of each x_i, and $x \in X$ whose front half is of the form usu' such that*

$$s^{-1} u^{-1} x_1 \cdots x_m us \in A \quad but \ u^{-1} x_1 \cdots x_m u \notin A.$$

Every exceptional case (b)(1) - (b)(4) appears in some examples of free products with amalgamated subgroup [Ro 77′]. If there are no obstructions and collapses as described in (b) (or in some weaker form) then one can describe the subgroups in a nice form (as in the case of free products).

3.5. Partition of system of elements. Define $X_k = \{x \in X \ : \ |x| = k\}$ and put

$$X = X^e \cup X^d, \quad X^e \cap X^d = \emptyset \quad \text{where } X^e = X_0 \cup X_1 \cup \bigcup_{\lambda \in \Lambda} X_\lambda^e$$

with

$$X_\lambda^e = \{x \in X \ : \ x = u_\lambda z u_\lambda^{-1}, \ |x| = 2|u_\lambda| + 1 > 1\} \subset X_\lambda$$

for a fixed potential half $u_\lambda \neq 1$, $u_\lambda \neq u_\nu$ if $\lambda \neq \nu$, $\lambda, \nu \in \Lambda$ and where no element of X^d lies in a conjugate of a factor.

Proposition 3.6. *Assume that X is a minimal system and that there are no obstructions or collapses (for X). Take the partition from above. Then*
(a) $\langle X \rangle = \langle X_0 \rangle * \langle X_1 \rangle *_{\lambda \in \Lambda} \langle X_\lambda^e \rangle * \langle X^d \rangle$;
(b) $\langle X_1 \rangle = *_{j \in J} \langle X_1 \cap G_j \rangle$;
(c) $\langle X^d \rangle$ *is a free group with basis X^d.*

Under the above assumptions the system X is more or less determined by the subgroup it generates. For a less general case the result is easier to formulate as follows.

Proposition 3.7. *Let $G = *_{j \in J}(G_j; A)$ and let X be a minimal system with $\langle X \rangle \triangleleft G$ and $\langle X \rangle \cap A = 1$. Then, using a partition of X as in 3.6,*
(a) $\langle X \rangle = \left(*_{j \in J} \langle X \cap G_j \rangle \right) * \left(*_{\lambda \in \Lambda} \langle X_\lambda^e \rangle \right) * \langle X^d \rangle$.
(b) *If Y is another minimal system and $\langle X \rangle = \langle Y \rangle$ then*
  (i) $\langle X_1 \rangle \cap G_j = \langle Y_1 \rangle \cap G_j$, $\forall j \in J$;
  (ii) $\langle X_\mu^e \rangle = \langle Y_\mu^e \rangle$, *for a suitable bijection of the index sets;*
  (iii) $\langle X^d \rangle = \langle Y^d \rangle$.

In particular, this is the case for free products [CZ 87']. For free groups, 3.7 implies that the minimal generating system is uniquely determined by the subgroup.

The Nielsen method for amalgamated free products has been developed to determine the rank of a Fuchsian group

$$G = \langle s_1, \ldots, s_m, t_1, u_1, \ldots, t_g, u_g \mid s_j^{a_j}, \ 1 \leq j \leq m, \ \prod_{j=1}^{m} s_j^{-1} \prod_{i=1}^{g} [t_i, u_i] \rangle,$$

a problem posed by Nielsen. The rank was expected to be $2g + m - 1$ if $m > 0$ and $2g$ for $m = 0$; this number is the minimal number of generators obtained from the shifts of a fundamental domain to a neighbour. However the following example of R. Burns, A. Karrass, A. Pietrowski and D. Solitar shows that this is not so.

Example 3.8. Consider $G = \langle s_1, s_2, s_3, s_4 \mid s_1^2, s_2^2, s_3^2, s_4^{2k+1}, s_1 s_2 s_3 s_4 \rangle$ and $x = s_1 s_2$, $y = s_3 s_4$. Then x, y generate G because

$$xy^{-1}x^{-1}y = s_1 s_2 \cdot s_3^{-1} s_1^{-1} \cdot s_2^{-1} s_1^{-1} \cdot s_1 s_3 = (s_1 s_2 s_3)^2 = s_4^{-2}$$

and, hence, $s_1 s_2 s_3 = s_4^{-1} = s_4^{2k} \in \langle x, y \rangle$ and it follows that $s_3 \in \langle x, y \rangle$ and thus $s_1, s_2 \in \langle x, y \rangle$.

It turns out that this is more or less the only exceptional case.

Theorem 3.9 [PRZ 75]. *Let*

$$G = \langle s_1, \ldots, s_m, t_1, u_1, \ldots, t_g, u_g \mid s_j^{a_j}, \ 1 \le j \le m, \ \prod_{j=1}^{m} s_j^{-1} \prod_{i=1}^{g} [t_i, u_i] \rangle;$$

assume $a_1 \le a_2 \le \cdots \le a_m$. *Then*

$$\text{rank } G = \begin{cases} 2g & \text{if } m = 0, \\ m - 2 & \text{if } g = 0 \text{ and } a_1 = \cdots = a_{m-1} = 2 \text{ and } a_m \text{ odd}, \\ 2g + m - 1 & \text{else}. \end{cases}$$

For NEC-groups the rank is not yet determined and this seems to be quite difficult if reflections occur since the algebraic rank may differ substantially from the geometric rank; there are examples where the algebraic rank is $2/3$ of the geometric rank, see [KZ 93]. For groups generated by three reflections M. Sakuma has determined those which are generated by two elements.

The exceptional case in 3.9 leeds to an unexpected answer to a question of Waldhausen, namely whether for a 3-manifold M^3 the rank of $\pi_1 M^3$ is equal to the minimal genus of Heegaard splittings of M^3. A proof of a positive answer must be extremely difficult since the case of rank 0 corresponds to the Poincaré conjecture.

3.10. On Seifert fibre spaces. Closely related to Fuchsian groups are Seifert fibre spaces and one can try to use the above example for these spaces. A Seifert fibre space M^3 admits a fibration into circles which is locally trivial except at a finite number of exceptional fibres. The basis is a surface. An exceptional fibre has as neighbourhood a solid torus where the fibration looks like the "parallels of a torus knot"; when running once along a normal fibre one passes the core of the solid torus a times and turns b times around it; here $\gcd(a, b) = 1$. Examples are 3-manifolds admitting an action of S^1. We restrict ourselves to the case where the total space M^3 and the base are orientable. From the Seifert–van Kampen theorem it follows that the fundamental group has the following presentation

$$\pi_1 M^3 = \langle s_1, \ldots, s_m, t_1, u_1, \ldots, t_q, u_g, h \mid [s_j, h], \ s_j^{a_j} h^{b_j} \ (1 \le j \le m),$$

$$[t_i, h], \ [u_i] \ (1 \le i \le g), \ \prod_{j=1}^{m} s_j \cdot \prod_{i=1}^{g} [t_i, u_i] \cdot h^e \rangle$$

where $a_j \ge 2$, $\gcd(a_j, b_j) = 1$. One can construct a Heegaard splitting of genus $m - 1 + 2g$ (if $m > 0$). In most cases this is the rank of $\pi_1 M^3$ since after cancelling the fibre generator h a Fuchsian group results of rank $m - 1 + 2g$. But – repeating the calculations for the case where $g = 0$, m even and all a_j equal 2 except one which is odd – the example 3.8 can be generalized and it can be shown that $m - 2$ elements suffice to generate the group (there suffice also $m - 2$ defining relators, see [BZ 84]). For $m = 4$ it has been proved in [BZ 84] that the minimal genus of a Heegaard splitting is 3 but the rank is 2. Recently Y. Moriah and J. Schultens announced a result which implies that rank and Heegaard genus differ also for $m \geq 6$.

4. Quadratic Equations in Groups

Definition 4.1. A *quadratic equation* in a group G has the form $W(x_1, \ldots, x_n) = h$ where $h \in G$ and each of the variables $x_1, \ldots, x_n$ appears exactly twice with exponents ± 1 in the word W. The equation is called *homogenous* if $h = 1$. A *solution* in G is a system of elements $g_1, \ldots, g_n \in G$ such that $W(g_1, \ldots, g_n) = h$. Of course, trivial subwords as $x_j x_j^{-1}$ have to be avoided in W. More precisely, W must have the form of the boundary of a disk obtained from a closed orientable or non-orientable surface by cutting along a system of curves. The best known words from surface theory are

$$\prod_{i=1}^{g} [x_{2i-1}, x_{2i}], \quad x_1 \cdots x_{2g} x_1^{-1} \cdots x_{2g}^{-1}, \quad \prod_{i=1}^{g} x_i^2,$$

that is, the canonical and symmetric form for an orientable closed surface S_g and the canonical form for a non-orientable closed surface N_g of genus g.

In order to deal with homogeneous quadratic equations one can develop a cancellation theory like the Nielsen method, see [Zi 64]. Clearly, if there is one solution of the equation then there are infinitely many others which are Nielsen-equivalent to the first one. The question arises whether there also are non-Nielsen-equivalent solutions and how many classes may occur. A more recent approach is to look to homomorphisms $\pi_1 S_g \to G$ or $\pi_1 N_g \to G$.

Definition 4.2. Let $\varphi, \psi : G \to H$ be homomorphisms between groups G and H. Then φ and ψ are called *equivalent* if there are automorphisms $\alpha \in \operatorname{Aut} G$, $\beta \in \operatorname{Aut} H$ such that the following diagram is commutative: The homomorphisms φ and ψ are called *strongly equivalent* if $\beta = \operatorname{id}_H$.

Classification has been achieved by several authors for the case where G is the fundamental group of a closed surface and H a free group F_r of rank r, see [CE 87], [GK 90], [GKZ 92], [Ly 59], [Ol 89], [Zi 64].

The main results are the following.

Proposition 4.3. *Given a homomorphism* $\varphi : \pi_1 S_g \to F_r$ *then there is an automorphism* $\alpha \in \operatorname{Aut} \pi_1 S_g$ *such that*

$$\varphi\alpha(t_1, u_1, \ldots, t_j, u_j, t_{j+1}, u_{j+1}, \ldots, t_g, u_g) = (x_1, 1, \ldots, x_j, 1, 1, 1, \ldots, 1, 1);$$

moreover, $\langle x_1, \ldots, x_j \rangle < F_r$ has rank j. If there is fixed some free basis of F_r the system $(x_1, \ldots, x_j)$ can be chosen Nielsen reduced and, thus, is uniquely determined by the subgroup $\langle x_1, \ldots, x_j \rangle$.

Corollary 4.4. *If there is an epimorphism* $\varphi \colon \pi_1 S_g \to F_r$ *then* $r \le g$. *Any two epimorphisms* $\pi_1 S_g \to F_r$ *are strongly equivalent.*

Two homomorphisms $\varphi, \psi \colon \pi_1 S_g \to F_r$ *are strongly equivalent if and only if* $\varphi(\pi_1 S_g) = \psi(\pi_1 S_g)$ *and they are equivalent if there is an automorphism* $\beta \in \mathrm{Aut}\, F_r$ *such that* $\beta \varphi(\pi_1 S_g) = \psi(\pi_1 S_g)$.

The first proof of these results apply the Nielsen method to quadratic equations in free groups, see [Zi 64]. For non-orientable surfaces there are more cases. Using the notation

$$[t, u]_\varepsilon = \begin{cases} [u, t] = tut^{-1}u^{-1} & \text{if } \varepsilon = 1, \\ tutu^{-1} & \text{if } \varepsilon = -1 \end{cases}$$

one obtains a result similar to Proposition 4.3.

Proposition 4.5. *Let* $\varphi \colon \pi_1 N_g \to F_r$ *be a homomorphism. Then there is a presentation*

$$\pi_1 N_g = \langle t_1, u_1, \ldots, t_\gamma, u_\gamma, t_e \mid \prod_{i=1}^{\gamma} [t_i, u_i]_{\varepsilon(i)} \cdot t_e^2 \rangle,$$

where $\gamma = [\frac{g}{2}]$, $\varepsilon(i) = \pm 1$, *and* t_e *appears only if* g *is odd, such that*

$$\varphi(t_1, u_1, \ldots, t_j, u_j, t_{j+1}, u_{j+1}, \ldots, t_\gamma, u_\gamma, t_e) = (1, x_1, \ldots, 1, x_j, 1, 1, \ldots, 1, 1, 1),$$

and $x_1, \ldots, x_j$ *span a subgroup of* F_r *of rank* j. *(If* g *is even then the last 1 in the right-handed expression above does not occur.) Relative to some fixed basis the system* $(x_1, \ldots, x_j)$ *has the Nielsen property and, thus, is uniquely determined by the subgroup* $\langle x_1, \ldots, x_j \rangle < F_r$.

Here $[r]$ denotes the largest integer $n \le r$. However, with respect to equivalence or strong equivalence there are more cases of epimorphisms than for the orientable case.

Theorem 4.6. *Let* q *denote the number of classes of epimorphisms* $\pi_1 N_g \to F_r$ *with respect to strong equivalence and* p *the number of classes of such epimorphisms with respect to equivalence. Then* $r \le [\frac{g}{2}]$ *and*
(a) $q = p = 1$ *if* $g \equiv 1 \bmod 2$,
(b) $q = 2^r, p = 2$ *if* $g \equiv 0 \bmod 2$, $2r < g$,
(c) $q = 2^r - 1, p = 1$ *if* $g = 2r$.

This theorem is from [GK 90]; in [GKZ 92] there is given a shorter geometric proof.— The approaches for the results above can be generalized to the case where the image is also a surface group (this has not been studied) or where the image is a free product of groups or even a free product with amalgamated subgroups, but there are not yet known such strong results as Theorem 4.6. Let me quote only some preliminary conclusions from common research with C. Hayat-Legrand.

Proposition 4.7. *Let S denote an orientable or non-orientable closed surface. An epimorphism $\varphi\colon \pi_1 S \to A * B$ is strongly equivalent to an epimorphism $\bar{\varphi}\colon \pi_1 S \to A*B$ such that $S = S'\cup S''$, $S'\cap S'' = \gamma$ is a simple closed curve, and $\bar{\varphi} i_\#{}'(\pi_1 S') < A$, $\bar{\varphi} i_\#{}''(\pi_1 S'') < B$ for the inclusions $i'\colon S' \hookrightarrow S$, $i''\colon S'' \hookrightarrow S$.*

This can be proved by combinatorial group theoretical arguments using the Nielsen cancellation method. Following arguments from J. Stallings [St 62] there is also a more pleasant topological proof: Realize A and B by 2-complexes C_A and C_B, take a basepoint $*$ and connect C_A and C_B with $*$ by segments to obtain a space C. Then φ can be realized by a continuous mapping $f\colon S \to C$. By homotopic deformations f first is made transversal to $*$ such that the preimage of $*$ consists of a system of pairwise disjoint simple closed curves and then the number of components can be reduced to one since φ is surjective.– Clearly, 4.7 can be generalized to free products with more than two factors. The genus/rank restrictions from above are consequences of 4.7. More generally:

Proposition 4.8. *Let $G = *_{i=1}^{r} G_i$, $G_i \neq 1$, be a free product and $\varphi\colon \pi_1 S_g \to G$ an epimorphism. Then $g \geq r$. If in addition $G_i \not\cong \mathbb{Z}_2$ for $1 \leq i \leq r$ then the existence of an epimorphism $\varphi\colon \pi_1 N_g \to G$ implies $\left[\frac{g}{2}\right] \geq r$.*

It would be nice to find a relationship between the number of (strong) equivalence classes of epimorphisms of fundamental groups of surfaces to $A * B$ and those to the factors. Perhaps one can use these methods for the case when $A * B$ is the connected sum of 3-manifolds. There is also possible some generalization to a free product with cyclic amalgamated subgroup; this is of interest because of results on surfaces of T. Delzant [De 93] and A. Rezhnikov [Re 93].

References

[BH]   Bestvina, M., Handel, M., Train tracks and automorphisms of free groups, Ann. of Math. 135 (1992), 1–51.

[Br 76]   Brunner, A. M., A group with an infinite number of Nielsen inequivalent one relator presentations, J. Algebra 42 (1976), 81–84.

[BRZ 88] Boileau, M., Rost, M., Zieschang, H., On Heegaard decompositions of torus knot exteriors and related Seifert fibre spaces, Math. Ann. 279 (1988), 553–581.

[BZ 84]   Boileau, M., Zieschang, H., Heegaard genus of closed orientable Seifert 3-manifolds, Invent. Math. 76 (1984), 455–568.

[CDP 90]   Coornaert, M., Delzant, T., Papadopoulos, A., Géométrie et Théorie des groupes, Lecture Notes in Math. 1441, Springer-Verlag, Berlin–Heidelberg–New York 1990.

[CE 87]   Comerford, I. P., Edmunds, C. C., Solutions of equations in free groups, in: Proc. Conf. Group Theory Singapore 1987 (K. N. Cheng and Y. K. Leong, eds.), Walter de Gruyter, Berlin 1989, 347–355.

[CM]   M. Culler, J. W. Magnus, Group actions on $\mathbb{R}$-trees, Proc. Lond. Math. Soc. 55 (1987), 571–604.

[Co 78]   Collins, D., Presentations of the amalgamated free product of two infinite cycles, Math. Ann. 237 (1978), 233–241.

[Co 87]   Collins, D., Peak reduction and automorphisms of free groups and free products, in: Combinatorial Group Theory and Topology (S. M. Gersten, J. R. Stallings, eds.), Annals Math. Studies 111, Princeton Univ. Press, Princeton N.J. 1987, 107–120.

[CZ 84]   Collins, D. J., Zieschang, H., On the Whitehead method in free products, Contemp. Math. 33 (1984), 141–158.

[CZ 84′]   Collins, D. J., Zieschang, H., Rescuing the Whitehead method for free products, I: Peak reduction, Math. Z. 185 (1984), 487–504.

[CZ 84″]   Collins, D. J., Zieschang, H., Rescuing the Whitehead method for free products, II: The algorithm, Math. Z. 186 (1984), 335–361.

[CZ 87]   Collins, D. J., Zieschang, H., A presentation for the stabilizer of an element in a free product, J. Algebra 106 (1987), 53–77.

[CZ 87′]   Collins, D. J., Zieschang, H., On the Nielsen method in free products with amalgamated subgroups, Math. Z. 197 (1987), 97–118.

[CZ 93]   Collins, D. J., Zieschang, H., Combinatorial group theory and fundamental groups, in: Encyclopaedia Math. Sci. 58, Algebra VII (A. N. Parshin, I. R. Shafarevich, eds.), Springer-Verlag, Berlin–Heidelberg–New York 1993, 1–166; Russ. Var.: Itogi Nauki i Tekhniki, Sovrem. Probl. Mat. 58, Algebra–7, VINITI, Moskva 1990, 5–190.

[De 93]   Delzant, T., L'image d'un groupe dans un groupe hyperbolique, preprint IRMA, Strasbourg 1993.

[Ep]   Epstein, David B. A., Word Processing in Groups, Jones and Bartlett Publ., Boston 1992.

[Fu 75]   Funcke, K., Gegenbeispiele zu einer Vermutung von Magnus, Math. Z. 141 (1975), 205–217.

[Ge 87]   Gersten, S. M. (ed.), Essays in Group Theory, Math. Sci. Res. Inst. Publ. 8, Springer-Verlag, New York–Berlin–Heidelberg 1987.

[GH 90]   Ghys, E., de la Harpe, P., Sur les groupes hyperbolique d'après Mikhael Gromov, Progr. Math. 83, Birkhäuser, Boston–Basel–Berlin, 1990.

[GK 90]   Grigorchuk, R. I., Kurchanov, P. F., Classification of epimorphisms of fundamental groups of surface onto free groups, Mat. Zametki 48 (1990), 26–35 (in Russian); English transl.: Math. Notes Acad. Sci. USSR 48 (1990), 736–742.

[GK 93]   Grigorchuk, R. I., Kurchanov, P. F., Some questions on group theory related to geometry, in: Encyclopaedia of Math. Sci. 58, Algebra VII (A. N. Parshin, I. R. Shafarevich, eds.), Springer-Verlag, Berlin–Heidelberg–New York 1993, 167–232; Russ. Var.: Itogi Nauki i Techniki, Sovrem. Probl. Mat. 58, Algebra–7, 191–261, VINITI, Moskva 1990.

[GKZ 92]   Grigorchuk, R. I., Kurchanov, P. F., Zieschang, H., Equivalence of homomorphisms of surface groups to free groups and some properties of 3-dimensional handlebodies, Contemp. Math. 131 (1992), 521–530.

[Gr 88]   Gromov, M., Hyperbolic Groups, in: Essays in Group Theory, (S.M. Gersten, ed.), Math. Sci. Res. Inst. Publ. 8, Springer-Verlag, New York–Berlin–Heidelberg 1987, 75–263.

[Gr 40]   Grushko, I. A., Über die Basen eines freien Produktes von Gruppen, Mat. Sb. 8 (1940), 169–182.

[GZ 79]   Gramberg, E., Zieschang, H., Order reduced Reidemeister–Schreier subgroup presentations and applications, Math. Z. 168 (1979), 53–70.

[Ha 79]   Haken, W., Various aspects of the 3-dimensional Poincaré problem, in: Topology of manifolds, Markham, Chicago 1979, 140–152.

[HL 74]   Higgins, G., Lyndon, R. C., Equivalence of elements under automorphisms of a free group, J. London Math. Soc. 24 (1974), 247–254.

[Ho 79]   Hoare, A. H. M., Coinitial graphs and Whitehead automorphisms, Candad. J. Math. 31 (1979), 112–123.

[Ho 81]   Hoare, A. H. M., Nielsen method in groups with a length function, Math. Scand. 48 (1981), 153–164.

[HT 80]   Hatcher, A., Thurston, W., A presentation of the mapping class group of a closed orientable surface, Topology 19 (1980), 221–237.

[Jo 90]   Johannson, K., Heegaard surfaces in Haken 3-manifolds, Bull. Amer. Math. Soc. 23 (1990), 91–98.

[KS 70]   Karrass, A., Solitar, D., The subgroups of a free product of two groups with an amalgamated subgroup, Trans. Amer. Math. Soc. 150 (1970), 227–250.

[Ku 94]   Kuhn, M., Tunnels of 2-bridge links, J. Knot Theory Ramifications, to appear.

[Ku 34]   Kurosch, A. G., Die Untergruppen der freien Produkte von beliebigen Gruppen, Math. Ann. 109 (1934), 647–660.

[Ly 59]   Lyndon, R. C., The equation $a^2 b^2 = c^2$ in free groups, Michigan Math. J. 6 (1959), 89–94.

[KZ 93]   Kaufmann, R., Zieschang, H., On the rank of NEC groups, London Math. Soc. Lecture Note Ser. 173, Cambridge Univ. Press, Cambridge 1993, 137–147.

[LS 77]   Lyndon, R. C., Schupp, P. E., Combinatorial Group Theory, Ergeb. Math. Grenzgeb. 89, Springer-Verlag, Berlin–Heidelberg–New York 1977.

[Ly 65]   Lyndon, R. C., Grushko's theorem, Proc. Amer. Math. Soc. 16 (1965), 822–826.

[Ma 30]   Magnus, W., Über diskontinuierliche Gruppen mit einer definierenden Relation, J. Reine Angew. Math. 163 (1930), 141–165.

[Mc 74]   McCool, J., A presentation for the automorphism group of a free group of finite rank, J. London Math. Soc. 8 (1974), 259–266.

[Mc 75]   McCool, J., On Nielsen's presentation of the automorphism group of a free group, J. London Math. Soc. 10 (1975), 265–270.

[Mc 75']   McCool, J., Some finitely presented subgroups of the automorphism group of a free group, J. Algebra 18 (1975), 205–213.

[MKS 66]   Magnus, W., Karrass, A., Solitar, D., Combinatorial Group Theory: Presentations of Groups in Terms of Generators and Relations, Interscience Publishers, John Wiley & Sons, Inc., New York–London–Sidney 1966.

[MP 71]   McCool, J., Pietrowski, A., On free products with amalgamation of two infinite cyclic groups, J. Algebra 18 (1971), 377–383.

[MS 91]   Morimoto, K., Sakuma, M., On unknotting tunnels for knots, Math. Ann. 289 (1991), 143–167.

[Ne 48]   Neumann, H., Generalized free products with amalgamated subgroups, Amer. J. Math. 70 (1948), 590–625.

[Ni 18]   Nielsen, J., Die Isomorphismen der allgemeinen, unendlichen Gruppe mit zwei Erzeugenden, Math. Ann. 78 (1918), 385–397.

[Ni 19]   Nielsen, J., Über die Isomorphismen unendlicher Gruppen ohne Relation, Math. Ann. 79 (1919), 269–272.

[Ni 21]   Nielsen, J., Om Regning med ikke kommutative Faktoren og dens Anvendelse i Gruppeteorien, Mat. Tidsskr. B (1921), 77–94.

[Ni 24]   Nielsen, J., Die Isomorphismengruppen der freien Gruppen, Math. Ann. 91 (1924), 169–209.

[Ol 89]   Ol'shanski, A. Yu., Diagrams of homomorphisms of surface groups. Sib. Mat. Zh. 30 (6) (1989), 150–171; English transl.: Siber. Math. J. 30 (6) (1989), 961–979.

[PR 78]   Peczynski, N., Reiwer, W., On cancellations in HNN-groups, Math. Z. 158 (1978), 79–86.

[PRZ 75]   Peczynski, N., Rosenberger, G., Zieschang, H., Über Erzeugende ebener diskontinuierlicher Gruppen, Invent. Math. 29 (1975), 161–180.

[Ra 58]   Rapaport, E. S., On free groups and their automorphisms, Acta Math. 99 (1958), 139–163.

[Re 79]   Reiwer, W., Die Nielsenäquivalenzklassen der metabelsch gemachten Torusknotengruppen, Archiv Math. 33 (1979), 310–324.

[Re 93]   Rezhnikov, A., Quadratic equations in groups from the global geometry viewpoint, preprint 1993.

[Ro 77]   Rosenberger, G., Anwendungen der Nielsenschen Kürzungsmethode in Gruppen mit einer definierenden Relation, Monatshefte Math. 84 (1977), 55–68.

[Ro 77′]   Rosenberger, G., Bemerkungen zu einer Arbeit von H. Zieschang, Archiv. Math. 24 (1977), 623–627.

[Ro 78]   Rosenberger, G., Alternierende Produkte in freien Gruppen, Pacific J. Math. 78 (1978), 243–250.

[Ro 78′]   Rosenberger, G., Über alternierende Wörter in freien Produkten mit Amalgam, Archiv Math. 31 (1978), 417–422.

[Ro 80]   Rosenberger, G., Gleichungen in freien Produkten mit Amalgam, Math. Z. 173 (1980), 1–12.

[Ro 83]   Rosenberger, G., Remarks on a paper by R. C. Lyndon, <<Quadratic equations in free products with amalgamation>>, Archiv Math. 40 (1983), 200–207.

[Ro 84]   Rosenberger, G., Über Darstellungen von Elementen und Untergruppen in freien Produkten, in: Groups—Korea, Kyoungju 1983, Lecture Notes in Math. 1098, Springer-Verlag, Berlin–Heidelberg–New York 1984, 142–160.

[Se 80]   Serre, J.-P., Trees, Springer-Verlag, Berlin–Heidelberg–New York 1980.

[St 62]   Stallings, J., On fibering certain 3-manifolds, in: Topology of 3-manifolds (M. K. Fort, ed.), Prentice-Hall, Englewood Cliffs 1962, 95–100.

[Ti 89]   Tipp, J., Heegaard-Zerlegungen der Außenräume von 2-Brückenknoten und Nielsenäquivalenzklassen der Gruppendarstellungen, Diss. Bochum 1989.

[Wh 36]   Whitehead, J. H. C., On certain sets of elements in a free group, Proc. London Math. Soc. 41 (1936), 48–56.

[Wh 36′]   Whitehead, J. H. C., On equivalent sets of elements in a free group, Ann. of Math. 37 (1936), 782–800.

[Zi 64]   Zieschang, H., Alternierende Produkte in freien Gruppen I, II, Abh. Math. Sem. Univ. Hamburg 27 (1964), 13–31; 28 (1965), 219–233.

[Zi 65]   Zieschang, H., On simple systems of paths on complete pretzels, Mat. Sb. 66 (1965), 230–239; English transl.: Amer. Math. Soc. Transl. (2) 92 (1970), 127–137.

[Zi 70]   Zieschang, H., Über die Nielsensche Kürzungsmethode in freien Produkten mit Amalgam, Invent. Math. 10 (1970), 4–37.

[Zi 77]   Zieschang, H., Generators of the free product with amalgamation of two infinite cyclic groups, Math. Ann. 227 (1977), 195–221.

[Zi 84]   Zieschang, H., On subgroups of a free product of cyclic groups, Trudy Mat. Inst. Steklov 154 (1983), 284–295; English transl.: Proc. Steklov Inst. Math. 4 (1984), 305–316.

[Zi 85]   Zieschang, H., A note on the mapping class groups of surfaces and planar discontinuous groups, in: Low Dimensional Topology, London Math. Soc. Lecture Note Ser. 95, Cambridge Univ. Press, Cambridge 1985, 206–213.

[Zi 88]   Zieschang, H., On Heegaard diagram of 3-manifolds, Soc. Math. France, Astérisque 163-164 (1988), 247-280.

[ZVC 80]   Zieschang, H., Vogt, E., Coldeway, H.-D., Surfaces and Planar Discontinuous Groups, Lecture Notes in Math. 835, Springer-Verlag, Berlin–Heidelberg–New York 1980; enlarged edition published in Russian by Nauka 1988.

List of Participants

Bernhard Amberg (Department of Mathematics, University of Mainz, 55122 Mainz, Germany)

Jin Hee Bae (Department of Mathematics, Pusan National University, Pusan 609-735, Korea)

Young Gheel Baik (Department of Applied Mathematics, Pusan National Susan University, Korea)

Won Jeong Baek (Department of Mathematics, Pusan National University, Pusan 609-735, Korea)

Sun Sik Byon (Department of Mathematics, Pusan National University, Pusan 609-735, Korea)

Ji Su Byun (Department of Mathematics, Department of Mathematics, Pusan National University, Pusan 609-735, Korea)

Jese Ma. Balmaceda (Department of Mathematics, University of Phillippines, Phillippines)

William A. Bogley (Department of Mathematics, Oregon State University, Corvallis, OR 97331-4605, USA)

Paul R. Brown (Department of Mathematics, University of Berkeley, USA)

Colin M. Campbell (Department of Mathematics, University of St. Andrews, Fife, St. Andrews, Scotland)

Hong Rae Cho (Department of Mathematics, Pusan National University, Pusan 609-735, Korea)

Jung Rae Cho (Department of Mathematics, Pusan National University, Pusan 609-735, Korea)

Ji Young Choi (Department of Mathematics, Pusan National University, Pusan 609-735, Korea)

Seul Hee Choi (Department of Mathematics, Kyong-Gi University, Seoul, Korea)

Martin Dunwoody (Faculty of Mathematical Studies, University of Southampton, Southampton SO9 5NH, England)

Benjamin Fine (Department of Mathematics, Fairfield University, Fairfield, Connecticut 06430, USA)

Koji Fujiwara (MSRI, UC Berkeley, CA 94720, USA)

Dietmar Garbe (Department of Mathematics, University of Bielefeld, Bielefeld, Germany)

N.D. Gilbert (Department of Mathematical Sciences, University of Durham, Durham DH1 3LE, U.K.)

Bill Grosso (Department of Mathematics, University of Berkeley, USA)

Narain D. Gupta (Department of Mathematics and Astronomy, University of Manitoba, Winnipeg, Manitoba, R3T 2N2, Canada)

Marcel Hagelberg (Laboratoire de Topologie et Géometrie, U.R.A. CNRS No.1408, Université Paul Sabatier MIG, 118 route de Narbonne, F-31062 Toulouse Cedex, France)

Hayat-Legrand (Laboratoire de Topologie et Géometrie, URA 1408, UFR MIG, UPS, 118 route de Narbonne, 31 062 Toulouse Cedex, France)

Heinz Helling (Department of Mathematics, University of Bielefeld, 33615 Bielefeld, Germany)

Sang Jung Ha (Department of Mathematics, Pusan National University, Pusan 609-735, Korea)

Woo Chorl Hong (Department of Mathematics, Pusan National University, Pusan 609-735, Korea)

Jim Howie (Department of Mathematics, Heriot-Watt University, Edinburgh EH14 4AS, Scotland)

Chan Huh (Department of Mathematics, Pusan National University, Pusan 609-735, Korea)

Won Huh (Department of Mathematics, Pusan National University, Pusan 609-735, Korea)

Noboru Ito (Department of Mathematics, Meijo University, Nagoya, Tenpaku, Japan)

Young Ho Im (Department of Mathematics, Pusan National University, Pusan 609-735, Korea)

Hak Jong Jeon (Department of Mathematics, Pusan National University, Pusan 609-735, Korea)

Byeong Woo Jeong (Department of Mathematics, Pusan National University, Pusan 609-735, Korea)

Kwang Woo Jeong (Department of Mathematics, Pusan National University, Pusan 609-735, Korea)

David L. Johnson (Department of Mathematics, University of Nottingham, Nottingham, England)

Jin Sook Kang (Department of Mathematics, Pusan National University, Pusan 609-735, Korea)

Young Hee Kang (Department of Mathematics, Pusan National University, Pusan 609-735, Korea)

Naoki Kawamoto (Department of Mathematics, Maritime Safety Academy, Kure 737, Japan)

Ann Chi Kim (Department of Mathematics, Pusan National University, Pusan 609-735, Korea)

Chol On Kim (Department of Mathematics, Pusan National University, Pusan 609-735, Korea)

Dae Hyun Kim (Department of Mathematics, Pusan National University, Pusan 609-735, Korea)

Hyun Jung Kim (Department of Mathematics, Pusan National University, Pusan 609-735, Korea)

Jai Heui Kim (Department of Mathematics, Pusan National University, Pusan 609-735, Korea)

Ji Yae Kim (Department of Mathematics, Pusan National University, Pusan 609-735, Korea)

Jung Hee Kim (Department of Mathematics, Pusan National University, Pusan 609-735, Korea)

Pan Soo Kim (Department of Mathematics, Pusan National University of Education, Korea)

Goansu Kim (Department of Mathematics, Kangnung National University, Kangnung, Korea)

Soo Hwan Kim (Department of Mathematics, Pusan National University, Pusan 609-735, Korea)

Sung Kun Kim (Department of Mathematics, Pusan National University, Pusan 609-735, Korea)

Yeong Ho Kim (Department of Mathematics, Pusan National University, Pusan 609-735, Korea)

Laci Kovacs (Mathematics Department, Faculty of Science, Australian National University, Canberra ACT 0200, Australia)

Jin Kee Lee (Department of Mathematics, Pusan National University, Pusan 609-735, Korea)

Jung Yup Lee (Department of Mathematics, Pusan National University, Pusan 609-735, Korea)

Sang Min Lee (Department of Mathematics, Pusan National University, Pusan 609-735, Korea)

Yong Hun Lee (Department of Mathematics, Pusan National University, Pusan 609-735, Korea)

Hae Suk Lyu (Department of Mathematics, Pusan National University, Pusan 609-735, Korea)

Toru Maeda (Department of Mathematics, Kansai University, 3-3-35 Yamate-cho, suita, 564, Japan)

Osamu Maruo (Department of Mathematics, Faculty of Education, Hiroshima University, Higashi-Hiroshima, Japan)

Alexander Makhnev (Inst. Math. and Mechanics, UB RAS, 620219 Ekaterinburg, Kovalevskoy 18, Russia)

Jens L. Mennicke (Department of Mathematics, University of Bielefeld, 33615 Bielefeld, Germany)

Qaiser Mushtag (Department of Mathematics, Universiti Brunei Darussalam, Bandar Seri Begawan, Negara Brunei Darussalam)

B. H. Neumann (Mathematics Department, Faculty of Science, Australian National University, Canberra ACT 0200, Australia)

Walter D. Neumann (Department of Mathematics, University of Melbourne, Parkville, Victoria 3052, Australia)

Markku Niemenmaa (Department of Mathematics, University of Oulu, 90570 Oulu, Finland)

Daniela Nickolova (Bulgarian Academy of Sciences, Institute of Mathematics with Computing Center str. Acad. G. Bonchev, b1.8 1113 Sofia, Bugaria)

Luz Nochefranca (Department of Mathematics, University of Philippines, Philippines)

A. Yu. Ol'shanskii (Department of Mathematics, Moscow State University, Moscow, Russia)

In Hyok Park (Department of Mathematics, Pusan National University, Pusan 609-735, Korea)

Jong Yeoul Park (Department of Mathematics, Pusan National University, Pusan 609-735, Korea)

Young Sik Park (Department of Mathematics, Pusan National University, Pusan 609-735, Korea)

Jung Pil Park (Department of Mathematics, Pusan National University, Pusan 609-735, Korea)

Sun Hyang Park (Department of Mathematics, Pusan National University, Pusan 609-735, Korea)

Sun Ja Park (Department of Mathematics, Pusan National University, Pusan 609-735, Korea)

Cheryl E. Praeger (University of Western Australia, Nedlands, W.A., 6009, Australia)

Stephen J. Pride (Department of Mathematics, University of Glasgow, Glasgow, G12 8QW, Scotland)

Akbar H. Rhemtulla (Department of Mathematical Sciences, University of Alberta, Edmonton, Alberta, T6G 2G1, Canada)

J. Rohlfs (Mathematisch-Geographische Fakultät, Katholische Universität Eichstätt, Ostenstrasse 28, 85027 Eichstätt, Germany)

Gehard Rosenberger (Fachbereich Mathematik, Universität Dortmund, 44221 Dortmund, Germany)

Kyoji Saito (Research Institute for Mathematical Sciences, University of Kyoto, 606 Kyoto, Japan)

Takanori Sakamoto (Department of Mathematics, Fukuoka University of Education, 729 Akama, Munakata City, Fukuoka Pref., 811-41 Japan)

Tae Yeong Seo (Department of Mathematics, Pusan National University, Pusan 609-735, Korea)

Gyeong-Sig Seo (Department of Mathematics, Chonpook National University, Korea)

Gi-Yeon Sin (Department of Mathematics, Pusan National University, Pusan 609-735, Korea)

Hyo-Seob Sim (Department of Mathematics, Pusan National University of Engineering, Korea)

Kwang Ho Sohn (Department of Mathematics, Pusan National University, Pusan 609-735, Korea)

Hyeon Jong Song (Department of Applied Mathematics, Pusan National Susan University, Korea)

Tai Sung Song (Department of Mathematics Education, Pusan National University, Pusan 609-735, Korea)

K. P. Shum (Department of Mathematics, Chinese University of Hong Kong, Hong Kong)

John Stallings (Department of Mathematics, University of Berkeley, USA)

Yaroslav P. Sysak (Mathematical Institute, Ukrainain Academy of Science, Ul. Repina 3, 252601 Kiev, Ukraine)

F. C. Y. Tang (Department of Pure Mathematics, University of Waterloo, Waterloo, Ontario, N2L 3G1, Canada)

Ken-Ichi Tahara (Department of Mathematics, Aichi University of Education, Igaya, Kariya 448, Japan)

Richard Thomas (Department of Mathematics and Computer Science, University of Leicester, University Road, Leicester LE1 7RH, England)

Dao-Rong Ton (Department of Mathematics and Physics, Hohia University, Jingsu Province, Nanjing, China)

L. R. Vermani (Department of Mathematics, Kurukshetra University, Kurukshetra - 132 119 (Haryana), India)

A. Vesnin (Institute of Mathematics, Novosibirsk 630090, Russia)

Ki Mun Woo (Department of Mathematics, Pusan National University, Pusan 609-735, Korea)

Craig Wotherspoon (School of Mathematical Computational Sciences, University of St Andrews, St Andrews KY16 9SS, Scotland)

Hiroyashi Yamaki (Institute of Mathematics, University of Tsukuba, Ibaraki 305, Japan)

Kyeong Mi Yu (Department of Mathematics, Pusan National University, Pusan 609-735, Korea)

Heiner Zieschang (Fakultät für Mathematik, Ruhr-Universität Bochum, 44780 Bochum, Germany)